Applied Condition Monitoring

Volume 7

About this Series

The book series Applied Condition Monitoring publishes the latest research and developments in the field of condition monitoring, with a special focus on industrial applications. It covers both theoretical and experimental approaches, as well as a range of monitoring conditioning techniques and new trends and challenges in the field. Topics of interest include, but are not limited to: vibration measurement and analysis; infrared thermography; oil analysis and tribology; acoustic emissions and ultrasonics; and motor current analysis. Books published in the series deal with root cause analysis, failure and degradation scenarios, proactive and predictive techniques, and many other aspects related to condition monitoring. Applications concern different industrial sectors: automotive engineering, power engineering, civil engineering, geoengineering, bioengineering, etc. The series publishes monographs, edited books, and selected conference proceedings, as well as textbooks for advanced students.

More information about this series at http://www.springer.com/series/13418

Cristina Verde · Lizeth Torres
Editors

Modeling and Monitoring of Pipelines and Networks

Advanced Tools for Automatic Monitoring and Supervision of Pipelines

Springer

Editors
Cristina Verde
Instituto de Ingeniería-UNAM
Ciudad Universitaria
Coyoacán
Mexico

Lizeth Torres
Instituto de Ingeniería-UNAM
Ciudad Universitaria
Coyoacán
Mexico

ISSN 2363-698X ISSN 2363-6998 (electronic)
Applied Condition Monitoring
ISBN 978-3-319-55943-8 ISBN 978-3-319-55944-5 (eBook)
DOI 10.1007/978-3-319-55944-5

Library of Congress Control Number: 2017935556

Printed on acid-free paper

This Springer imprint is published by Springer Nature
The registered company is Springer International Publishing AG
The registered company address is: Gewerbestrasse 11, 6330 Cham, Switzerland

Preface

A pipeline network, which is by far the most efficient way to transport fluids despite the expense of appropriate maintenance, is a complex system with many types of components and consumers. Therefore, pipeline networks have reached high levels of importance for transportation all over the world.

It is crucial for industries and societies that networks operate properly by considering the growing need for efficient interconnecting fluid systems. This task is not easy, because one must concurrently ensure a safe fluid supply and the fulfillment of the varying demands of consumers. This task becomes more complicated with the emergence of leaks blockage and the defects in sensors, actuators, which can generate the deterioration and malfunction of the whole network.

Network breakdowns induce high economic losses and environmental damage, which make the design and implementation of monitoring systems essential for the opportune detection and localization of faults to safeguard the network.

The monitoring of networks is a concern normally tackled by interdisciplinary groups comprising scientists, researchers, and industrial engineers from diverse knowledge fields. These include fluid mechanics, instrumentation, automatic control, signal processing, computing, and civil engineering. The main problems that these interdisciplinary groups find in the design of automatic supervision systems for pipeline networks are the following.

- The imprecise knowledge about the network parameters that significantly changes from their design values because of the manufacturing execution, the installation process and the aging deterioration.
- The reduced number of available sensors with high sensitivity and wide bandwidth that allows the monitoring in real time of every network component.
- The high number of fault scenarios which could occur in a real network.
- The deterioration of the components during the life cycle of the installations.
- The occurrence of unpredictable natural phenomena such as earthquakes, tornadoes, and hurricanes which can damage the whole system.

- The changing structure of the network caused by either scheduled or unplanned events.
- The varying demand of fluid during time intervals.

For the safety of pipeline networks, specific software tools have been developed over the past three decades which are complementary to the traditional supervisory control and data acquisition systems (SCADA). Commonly, such tools are comprised of fault detection and location (FDI) algorithms, based on mathematical models of the fluid or routines for processing signals, and they take into account a limited number of available variables from the pipeline.

Note that some faults to be detected require active detection, i.e., by demanding that supervision systems act upon the pipeline network in periodical intervals or at critical times by using test signals to generate, for instance, transient responses of the fluid for detecting abnormal events. Thus, there are a considerable number of research groups around the world with different backgrounds who are dedicating efforts to propose feasible automatic monitoring and supervision systems for networks.

Diverse abnormal conditions have been studied, specific sensors have been developed, and some maneuvers have been proposed to look for abnormal conditions. The most common fault scenarios are leaks, blockages, sequential faults over time, and illegal extractions.

This monograph presents some tools for improving the automatic monitoring and supervision of networks, and it could be suitable for a wide variety of users who are interested in the new technology for the monitoring of pipelines.

The main goals of the monograph are the following:

- To introduce fault detection techniques in pipelines to a new audience;
- To present new algorithms to manage the monitoring of pipeline networks for more general fault scenarios;
- To present techniques for designing auxiliary signals to improve the identification of parameters and faults in pipelines;
- To present fault detection techniques in pipelines that have been validated in real applications;
- To integrate results developed by the civil engineers and the safe process communities.

The monograph is a compiled summary of research topics and applications studies done by international experts in the field of automatic monitoring for pipelines' networks. The content has been organized in 12 chapters related to leak detection and location, modeling of networks, auxiliary signals for parameter identification, design of monitoring systems, adequate placement of sensors in water demand networks, and related issues. The authors are civil engineers and diagnosis researchers from institutions of different countries such as France, Spain, Portugal, Venezuela, Mexico, Poland, Canada, Cuba, and Colombia.

The monograph was designed to provide key knowledge, useful techniques and tools to mathematicians, computer scientists, and engineers concerned with the issues of design, as well as the adjustment and operation of robust monitoring systems for pipeline networks through software. It is hoped that the content will be helpful reference text for newcomers to the field of computational pipeline monitoring (CPM) by using advanced automatic technologies as well as those with some knowledge of the subject.

This monograph was possible through the effort of the authors to present the topics in a clear and uniform manner. Our purpose during the preparation was that readers could follow the content even if they have different backgrounds. Furthermore, chapters can be read independently despite being arranged in a natural sequence. First, the issue of ad hoc models for fault detection in a pipeline and the problems' formulation are discussed. As the second group of contributions, specific leaks' diagnosis systems are introduced for a single pipeline, and finally some solutions to the complex problem of leak location in pipeline networks with multiple leaks and uncertain demand are suggested. This facilitates the understanding according to the interest of the reader.

The editors are indebted to many people for the support of this project. In particular, the financial support received from the Universidad Nacional Autónoma de México with the research grant DGAPA-IT100716 *Identificación automática de fallas en redes de transporte vía modelos dinámicos*. We like to thank Mr. Marcos Quiñones for the integration of manuscripts and his effort during the final period during the production of the monograph. We acknowledge as well the editing and production staff at Springer for their excellent work. In particular, the support of Ms. Leontina di Cecco. Finally, we especially would like to thank Ms. Mary-Ann Hall for proofreading the text.

Mexico City, Mexico
December 2016

Cristina Verde
Lizeth Torres

Contents

Chapter 1
Introduction

Cristina Verde

Abstract This chapter serves as a background of monitoring systems for pipeline networks by applying software-based fault diagnosis tools and briefly describes each contribution of this monograph. The term computational pipeline monitoring (CPM) refers to algorithmic monitoring tools that are used to enhance the abilities of the pipeline network's operators in recognizing anomalies which may be indicative of products' loss. The presented advanced methods and issues associated with models and features of pipeline networks are limited to scenarios of leaks and blockages in a single pipeline and pipeline networks. The framework of the procedures considers the physical variables associated with the flow process as measurements. In particular, pressure, flow, and temperature sensors are assumed to be located only at specific points of the pipeline networks. Thus, this chapter introduces the reader to the main theme of the monograph which is the analysis and design of advanced online automatic monitoring systems for pipeline networks by considering leaks and blocks as abnormal events. To simplify the understanding of the specific topics, the general fault detection and isolation (FDI) background is roughly presented in this chapter by citing tutorial books related to the FDI issues.

1.1 Introduction

To maintain a high level of safety performance, reliability, availability, maintenance, and longevity in transport processes, it is important that automatic systems promptly and continuously detect errors and abnormal events. Moreover, the sources and severity of a malfunction must be diagnosed. Prompt diagnosis of the pipelines' malfunctions are strategically important because of the economic and environmental demands required for distributing companies of products transported around the world. In particular, leaks and blockage can lead to shortages or environmental catastrophes.

C. Verde (✉)
Instituto de Ingeniería UNAM, CP 04510 Mexico City, Mexico
e-mail: verde@unam.mx

C. Verde and L. Torres (eds.), *Modeling and Monitoring of Pipelines and Networks*, Applied Condition Monitoring 7,
DOI 10.1007/978-3-319-55944-5_1

Leaks are one of the major concerns in chemical and oil industries. Water leaks are usually associated with aging infrastructure and improper maintenance. Diverse pipeline monitoring methods and leak detection techniques have been developed over the last 40 years. These developments are primarily driven by environmental and financial factors. Zhang (1996) reviewed various pipeline leaks' detection methodologies by dividing the study into biological, hardware, and software tools. In principle, the leak detection task in pipelines can be fulfilled through periodic maintenance, which is a very expensive and difficult task for subterranean or sea pipelines; continuous monitoring by sensors and software, which considers all the possible causes and interrelations of the observed physical effects, is also possible. The amount of abnormal conditions which can occur in a real network installation represents, however, a challenge for monitoring systems' designers.

Today, there is a variety of available commercial leak detection techniques ranging from simple physical inspection and acoustic methods to infrared thermography and optic sensors located along the entire pipeline which are complementary to the traditional supervisory control and data acquisition system (SCADA) (Knudsen 2013).

In terms of cost, a survey conducted by the Canadian National Research Council reports that an average of 82 million dollars is spent every year to repair broken water mains, which is as estimated 2,500 dollars in repair costs for each pipe failure. On the other hand, the U.S. Environmental Protection Agency (EPA 2009), after conducting a national survey among the American water systems, found that one will need to invest about 200.8 billion dollars in the transmission and distribution of systems over the next 20 (2006–2026). The major expense is dedicated to repairing the water transportation infrastructure. According to the EPA, the single largest category of need is the replacement of existing water distribution pipes (about 77 billion). Furthermore, in terms of volume loss, Colombo and Karney (2002) reported the following typical ranges of unaccounted water, with leakage as the dominant component: 9–30% in Europe and 43 and 56% s for Malaysia and Bangladesh, respectively.

Since the American Petroleum Institute (1995) published norms and regulatory laws under the umbrella of computational pipeline monitoring (CPM), there has been a surge of interest in research and applications of feasible methods by software for fault detection in pipeline networks. The complexity and performance of the CPM tools vary significantly according to the considered abnormal scenarios, and they are divided into the following: flow/pressure change detection and mass/volume balance, dynamic model-based system, and pressure point analysis methods. Colombo et al. (2009) presented a complete literature review of transient-based leak detection methods by software for water pipelines. It is usual now in technical meetings related to process safety and integrity of the control community to discuss pipeline networks diagnosis (8th IFAC-Safeprocess 2012; 9th IFAC-Safeproocess 2015; 3rd SysTol 2016). Therefore, there is an interest in the topic from a practical point of view as well as the feasibility to apply advanced software-based diagnosis methods by considering the available potential instrumentation, communications' technologies, and systematic tools developed in the field of automatic control. Unfortunately, verification of these techniques, while not entirely absent, is shown to be even now generally lacking in real installation and complex faults scenarios. Moreover, many theoretical problems and multiple faults' scenarios remain without a solution.

As an example, an important scenario is the case of multiple leaks, in which the methods reported by Billman and Isermann (1987), Shields and Daley (2001), Geiger et al. (2000) and Dinis et al. (1999) have a common property of a false alarm. If they are applied in the case of multiple leaks with similar conditions at the ends of the pipeline and with no branched junction in between, all of them deliver the same wrong position. The reasons for this fact are the following (Verde and Visairo 2004):

- The impossibility of isolating two or more leaks with steady-state variables because of a nonunique solution for the mass equation since the wave speed of the fluid has no influence on the steady-state condition.
- The parametrization of a virtual leak at a position which generates the same steady-state flows at the ends of the line, as an infinite set of multiple leaks' cases (Verde et al. 2014).

As a consequence, the leaks' positions can only be located online during the transient response caused by the simultaneous events. This means multiple abnormal scenarios with similar effects in the measurements in steady state are still a challenge, even when applying new technologies.

To asses the diagnosis methods, the following key attributes must be considered: leak sensitivity, false alarm rate, location estimation capability, robustness with respect to operational change, maintenance requirement, and cost. The common problem for most software methods is the high false alarm rate and the location estimation: the generation of leak alarm when the pipeline is under normal operation or the inability to determine the event position. Both facts are undesirable because of the following:

- Confidence of the operators in the automatic monitoring system is reduced.
- Extra work is generated for the operators.
- Real abnormal scenarios may be disregarded.
- The cost to complementing the diagnosis with network visual inspection and hardware.

Thus, from a technical point of view innovative supervision systems for a pipeline must improve both hardware and software aspects. Thus, more accuracy and rapid sensors together with new algorithms assuming coupled fault conditions with a low false alarm rate must be developed. This task is of great interest to governments, owners, and users.

The content of the monograph is focused on a range of combined processes of fault detection and location in pipelines, based on the use of analytical and signal's models of the fluid and experimental data of physical installations. Moreover, important issues associated with the type of models which can generate a false alarm and dependent on the physical parameters and properties of the transported fluid are discussed. These issues are usually neglected in practical applications. The monograph title *Modeling and Monitoring of Pipelines and Networks* has been chosen to reflect the wide scope of the material.

We hope this monograph helps to design robust monitoring systems for pipeline networks and can be useful for interdisciplinary groups which include fluid mechanics, instrumentation, automatic control, signal processing, computing, materials and civil engineering fields. The chapters have been written with the following intentions:

- Introduce fault detection techniques in pipelines to a new audience.
- Bring relevant conditions to approximate fluid behavior in pipelines with a finite-dimensional discretized model.
- Present new algorithms to manage the monitoring of pipeline networks for more general fault scenarios.
- Present techniques for designing auxiliary signals for improving the identifiability of parameters and faults in pipelines.
- Present fault detection techniques in pipelines that have been validated in real applications.
- Integrate results developed by the civil engineers and the SAFEPROCESS communities.

Before describing the book's structure, some terminology used in the SAFEPROCESS community is presented in the next section.

1.2 Background of Fault Detection and Pipelines' Diagnosis

The automatic control community has developed a general theory for the fault detection and identification (FDI) of dynamic systems which allows the generation of a fault symptom from a given set of measured signals by software and the identification of the fault parameters from a specific symptom pattern. The reader can consult for specific details of the FDI theory the tutorial books (Isermann 2006; Blanke et al. 2006).

The fundamental concept for fault detection is the diagnostic principle which is based on a comparison between a nominal behavior model and data obtained from the real physical system. Therefore, under normal conditions the comparison produces zero error called the residual. On the contrary, a deviation of the abnormal condition in the real process produces a residual deviated from zero called the fault symptom, and an alarm is generated. Figure 1.1 shows the generic principle of fault detection which can be applied independently from the type of model. Diverse types of models can be handled for the diagnostic principle implementation. Methodologies have been developed specifically for qualitative, quantitative and process historical models. Assumptions and very detailed descriptions of the procedures used for these models are reported in the publications of Venkatasubramanian et al. (2003a, b, c) respectively.

According to Patton et al. (1989), the FDI schemes are basically signal processing techniques which employ state estimation, parameters' identification, adaptive filtering, variable threshold logic, statistical decision theory, and various combinatorial and logical operations. The schemes are designed under the assumption that either

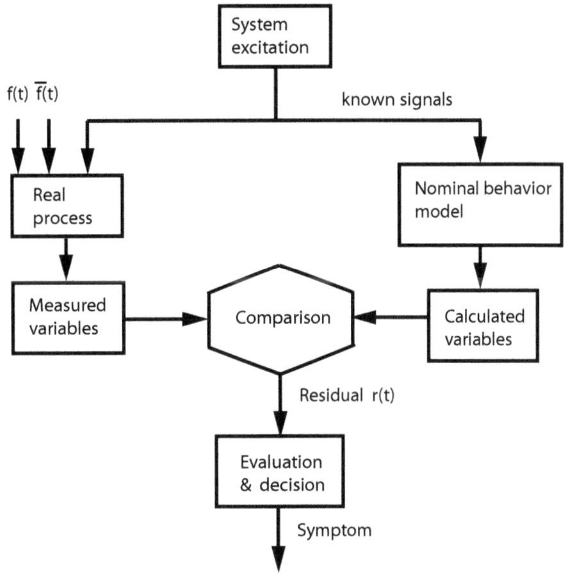

Fig. 1.1 General description of the diagnostic principle

the dynamic of the system being monitored is known with reasonable precision or that it is possible to determine certain features and parameters of the process from historical data. Thus, the key to achieving a robust fault's detection and identification is to provide a generic nominal model with a reasonable degree of precision and to have a set of measurements which deviate from the nominal values when the faults are present. This means, the measurements must be sensitive to the faults and robust against disturbances. The more simple or primitive the model and the less insensitive the sensors and data are to the faults, the poorer is the diagnosis capability.

Looking at the case of leaks' detection in a pipeline network, one can see that currently nonlinear dynamic models describing the fluid process are suggested, since the friction factor and the fluid dynamic depend on the operation conditions and external disturbances. Therefore, classic monitoring techniques based on variables' limits with a threshold are inadequate in identifying the two parameters associated with each leak, position, and outflow. The leak outflow can be easily estimated by the residual defined as the difference between inflow and outflow, but not the leak position.

According to Wylie and Streeter (1978), each branch junction or leak in a pipeline increases both the number of boundary conditions in the analytical model and the set of partial differential equations (PDEs). This fact modifies the model structure and the number of state variables. Moreover, since the flow for each pipeline section between leaks is different, even in steady state, the friction function changes as well in each section. Thus, each leak introduces three unknown parameters in the fluid model: position, loss flow, and friction factor. This is an atypical fault scenario, and few tools developed by the FDI community can be applied. This fact is one of the reasons why specific methods have been proposed for the monitoring of a pipeline.

Most of the algorithms for leak location in a single pipeline are based on the flow process model described by the mass and moment equations reported by Chaudhry (2014). Two frameworks have been used to deal with the problem and apply in some sense the diagnostic principle of Fig. 1.1. The transient leaks' procedure is one of the frameworks, and it was suggested by Brunone (1999). The idea is to approximate the PDE model as a linear transfer matrix of infinite dimension, and by introducing maneuvers, the transient response of the piezometric head is analyzed in the frequency domain to identify the events. The other framework consists to approximating the PDEs by computational techniques to obtain a nonlinear finite-dimension nonlinear state-space description, and via the diagnosis principle the leak position is estimated (Billman and Isermann 1987). Both frameworks are developed for a pipeline instrumented only with pressure and flow sensors at the ends, and they can be complementary. In the case of the analysis of the transient response, the main practical problems are the generation of disturbances which produce the pressure wave and the bandwidths and sensitivity required for sensors and actuators in real installations.

Visairo and Verde (2003) reported a fact regarding the location of multiple leaks in a single line by assuming a nonlinear state-space model of finite dimension with only measured variables at the ends. By assuming three possible leaks with known positions, there is a lack of isolability for the leak located in between the other two. As a consequence, the detection task of two possible leaks—one located upstream and the other downstream of the middle one—has no solution if the pressure and outflow at the middle leak are unknown.

Regarding the leak scenarios, in addition to the case of sequential faults reported by Delgado-Aguiñaga et al. (2016) and the scheme presented in Chap. 7, a suggestion has been made for the case of a leak in a branched pipeline (Verde et al. 2016) which is validated in a water pilot pipeline of 160 [m]. These results considered a finite-dimension state-space model.

By using the infinite-dimension model, Lee et al. (2005) proposed capturing the patterns with leaks of the frequency response diagram with a valve maneuvre in the line. The relevance of this contribution is the derivation of the analytical model which describes the pattern of the resonance peaks with leaks. In the contributions of Ferrante and Brunone (2003a, b), methods are presented based on the transient response of a pressure signal by using the Fourier transform and wavelet, respectively. Moreover, since appropriate unsteady-state tests give rise to a small discontinuous overpressure; based on this fact, Brunone and Ferrante (2001) reported a method with the capacity to detect two leaks, one after the other. All these methods have practical limitations, however, if dominant nonlinear effects are present in the fluid; the sensors have a limited bandwidth and sensitivity, and the pressure signal is noisy. Thus, the maneuver must be designed to improve the feasibility of the methods in real pipelines.

Recently, Da Silva et al. (2005) published a method based on a fuzzy classification of the transient response with multiple leaks; however, the amount of historical data with leaks in the training could be a limitation in a real case since the historical data with abnormal events are not always present.

As a consequence, despite the presence of diverse monitoring procedures for a single pipeline, there is a lack of satisfactory diagnosis algorithms for the general multiple leaks' scenario, and the transient response features of the pressure in the fluid is the key to obtaining robust leaks' detectors.

In spite of the great interest in leaks' detection and the location of complex networks where the demand in each node is uncertain, few satisfactory results have been reported in this complex scenario. The main reasons are the lack of precise analytical models for real pipeline networks and the low number of pressure and flow sensors installed in a network. Diverse research groups from different fields are trying to understand and develop technologies for the monitoring of pipeline networks.

In general, the identification of an analytical model for a real network is a complex task because of the large amount of parameters, constraints, and uncertainties involved. If an analytical model of the network is, however, available, the diagnostic principle of Fig. 1.1 can be applied based on fault signatures of the network (Pudar et al. 1992). A fault signature is a matrix that describes the typical measured pressure and flow variations at some points of the networks produced by a single fault, e.g., a leakage of certain magnitude at one network node with a specific demand outflow pattern.

It's crucial in any FDI problem the number and the sensors' location required for achieving satisfactory fault diagnosis results. Pérez et al. (2011) cope with this problem by using the fault signature approach, and the sensors' placement was formulated as an optimization problem by using separability measures for the fault signatures. The effect of network uncertainties can be reduced under this approach by using a time horizon analysis (Casillas et al. 2015) or Monte Carlo simulations (David Steffelbauer and Fuchs-Hanusch 2016).

The fault signature approach for leak location in real large-scale networks is not feasible, however, because of the amount of required signatures for describing the possible leaks' position, magnitude, and network conditions; e.g., demand patterns and operation conditions. In addition, model recalibration is often required because of the network updating and seasonality issues, e.g., network topology changes and demand behavior variability.

One can say that even when a network analytical model could be obtained, the design of monitoring tools by software remains a challenge for pipeline networks. On the other hand, the standard SCADA systems allow the development of data-driven-based fault diagnosis methods since real-time data are available from the instrumentation (Venkatasubramanian et al. 2003c). This knowledge offers an alternative for overcoming the limitations of the fault signature approach. Two methods are described in the monograph which consider this kind of model. Moreover, the relevant sensors' placement problem especially for demands' networks can also be formulated in the framework of artificial intelligence without analytical models.

Data-driven models based on artificial neural networks, principal component analysis, support vector machines, and clustering methods have been proposed by Gertler et al. (2010), Mounce et al. (2011), Arsene et al. (2012), Palau et al. (2012),

Ye and Fenner (2014), Wu et al. (2016). The application of these works is limited to at least one of the following conditions: many flows' sensors are installed or the pressure head is measured in most network nodes. Uncertainties are only associated to the measurements, or the demand variability is neglected throughout the day. Hence, contributions which guarantee the fault detection while avoiding such assumptions are further required. Moreover, a disadvantage of the data-driven model, is that the historical database is not usually available for describing all the demand network scenarios and then only nominal models are built having the consequence that faults can only be detected but not identified.

Conclusion, as indicated in the case of transient response diagnosis methods, in demand networks all the methods have practical limitations in the real networks. Thus, practical aspects must be included in the procedures to improve the robustness of the pipeline net monitoring.

1.3 Monograph Description

This monograph presents recent works from the perspective of leaks' detection in pipelines with nonintrusive devices, and the presentations are organized in five theme groups:

- Leak monitoring by transient response of infinite-dimension linear flow model.
- Modeling issue for faults' scenarios in pipelines and maneuvers' design which maximizes the abnormal effects in a pipeline.
- Leaks' detection algorithm for single and multiple leaks using observer tools.
- Fault detection for demand networks by using historical data.
- Sensors' placement for fault location in demand networks.

Chapter 2 could be considered an extension of this introductory chapter, since it treats in detail the techniques of transient leak detection by considering both active and passive systems. The authors examine material of the state of the art in this theme, providing a summary of what has been shown and discussed recently, along with opinions and diverse reflections of which methods could have great promise for deployment and commercialization.

Chapter 3 is dedicated to studying the important numerical issues when an approximated analytical model of the fluid dynamics is applied for fault detection. The model parameters' effects are deeply analyzed from a numerical point of view for a single pipeline. The main contributions of this chapter are the stability analysis of the numerical model and the derived necessary condition of the discretized model by introducing the low-order central difference approximation for time and spatial variables.

In Chap. 4, the authors present diverse one-dimensional models of a pipeline with leaks and blockages and the analysis of the models' behavior is validated by simulations.

In Chap. 5 the problem of fault diagnosis in a single pipeline is addressed by using a model-based fault diagnosis framework and algorithms of observers' theory developed in the control community. The diagnosis of a fault in the pipeline is reformulated by a spatial discretization of the partial difference equations and by considering the parameters associated with a leak or a blockage as unknown states of an augmented dynamic continuous system. Thus, the augmented model allows the application of dynamic observers' tools to estimate the parameters of a leak or blockage. An advantage of these observers is the theoretical base, since it is possible to design generic observers for leak and blockage under the local observability condition.

Chapter 6 copes with the problem of designing off-line optimal maneuvers' sequences in pipeline actuators for the parameters' identification. Specifically friction, wave speed, and longitude are identified by considering a Liénard-type model for the fluid behavior description. The key to the approach is the condition for achieving regularly persistent input (the maneuvers' sequence) in the pipeline such that the minimal deviation of the nominal fluid boundary conditions during the sequence time is achieved.

Chapter 7 deals with the issues of leaks' recursive location when the events occur sequentially. Specifically, the chapter discusses three problems associated with the monitoring of a single pipeline. The first analysis is related to the sensibility analysis of the friction factor, since any model for pipeline monitoring introduces an error if a constant friction along the entire pipeline is assumed. By taking into account this fact, the extension of the equivalent relation in steady state between one leak and multiple leak models is derived. Finally, a recursive scheme is proposed which includes interrelated dynamic models describing the fluid by preserving their structure for any number of leaks, and only the parameters are adapted each time a new leak occurs. The validation of the algorithm with a set of leaks simulated by Pipeline Studio (2013) shows successful results.

Chapter 8 covers the important problem of leaks in gas pipelines. The framework of the procedure is the interconnection of two linearized quadripole models with a leak in between. The quadripole models were calibrated with operational data supplied by REN Gasodutos, a Portuguese gas company. A case study was built, and leakages of diverse magnitude were simulated and relative small errors, around 3% were obtained.

Chapters 9 to 11 mainly describe some methods for detecting and locating leaks in water demand networks by using historical data in normal and abnormal conditions. Thus, Chap. 9 presents a feature extraction method for leak's detection in water distribution networks with demand patterns which allow the application of data driven methods as principal component analysis (PCA). Chapter 10 discusses the problem of leak localization in water distribution networks using pressure models and classifiers tools taken from the artificial intelligence field. Finally, Chap. 11 introduces a method for finding the optimal sensor placement for classifier-based leak localization in drinking water networks.

Finally in Chap. 12, the authors demonstrate the power and feasibility of advanced computational pipeline monitoring tools when they are applied off-line for identifying the position of a leak in a real pressurized pipeline network. In particular, an instrumented section of approximately 365 [m] long and 0.9 [m] of inner diameter of the Guadalajara City aqueduct was monitored during diverse time intervals to detect the position of a leak by using a discrete extended Kalman filter. An interesting point is that the leak position was determined on the basis of four sets of data which generated diverse leak locations. The final decision was obtained by considering the dominant tendency of the estimations.

References

3rd SysTol (Ed.). (2016). In *2016 3rd Conference on Control and Fault-Tolerant Systems (SysTol)*.

8th IFAC-Safeprocess (Ed.). (2012). In *IFAC symposium on fault detection, supervision and safety for technical processes (Safeprocess)*.

9th IFAC-Safeproocess (Ed.). (2015). In *IFAC symposium on fault detection, supervision and safety for technical processes (Safeprocess)*.

American Petroleum Institute (1995). API-1130 computational pipeline monitoring.

Arsene, C. T. C., Gabrys, B., & Al-dabass, D. (2012). Decision support system for water distribution systems based on neural networks and graphs theory for leakage detection. *Expert Systems with Applications*, *39*(18), 13214–13224.

Billman, L., & Isermann, R. (1987). Leak detection methods for pipelines. *Automatica*, *23*(3), 381–385.

Blanke, M., Kinnaert, M., Lunze, J., & Staroswiecki, M. (2006). *Diagnosis and fault tolerant control* (2nd ed.). Berlin: Springer.

Brunone, B., & Ferrante, M. (2001). Detecting leaks in pressurised pipes by means of transients. *Journal of Hydraulic Research*, *39*(5), 539–547.

Brunone, B. (1999). Transient test-based technique for leak detection in outfall pipes. *Journal of Water Resources Planning and Management ASCE*, *125–5*, 302–306.

Casillas, M. V., Garza-Castañ6n, L. E., & Puig, V. (2015). Sensor placement for leak location in water distribution networks using the leak signature space. *9th IFAC symposium on fault detection, supervision and safety of technical processes* (pp. 214–219). Paris, France: IFAC.

Chaudhry, H. M. (2014). *Applied hydraulic transients*. New York: Springer.

Colombo, A. F., & Karney, B. W. (2002). Energy and costs of leaky pipes: Toward comprehensive picture. *Journal of Water Resources Planning and Management (ASCE)*, *128*, 441–450.

Colombo, A. F., Lee, P., & Karney, B. W. (2009). A selective literature review of transient-based leak detection methods. *Journal of Hydro-Environment Research*, *2*, 212–227.

Da Silva, H. V., Morooka, C. K., Guilherme, I. R., da Fonseca, T. C., & Mendes, J. R. P. (2005). Leak detection in petroleum pipelines using fuzzy system. *Journal of Petroleum Science and Engineering*, *49*, 223–238.

David Steffelbauer, B., & Fuchs-Hanusch, D. (2016). Efficient Sensor Placement for Leak Localization Considering Uncertainties. *Water Resources Management*, *30*(14), 5517–5533.

Delgado-Aguiñaga, J., Besançon, G., Begovich, O., & Carvajal, J. E. (2016). Multi-leak diagnosis in pipelines based on extended kalman filter. *Control Engineering Practice*, *49*, 139–148.

Dinis, J. M., Wojtanowicz, A. K., & Scott, S. L. (1999). Leak detection in liquid subsea flowlines with no recorded feed rate. *Journal of Energy Resources Technology by ASME*, *121*, 161–166.

EPA (2009). Drinking water infrastructure needs survey and assessment. Technical Report, U.S. Environmental Protection Agency (EPA).

Ferrante, M., & Brunone, B. (2003a). Pipe system diagnosis and leak detection by usteady-state tests. 1. Harmonic Analysis. *Advances in Water Resources*, *26*(1), 95–105.

Ferrante, M., & Brunone, B. (2003b). Pipe system diagnosis and leak detection by unsteady-state tests. 2. *Wavelet analysis Advances in Water Resources*, *26*, 107–116.

Geiger, G., Gregoritza, W., & Matko, D. (2000). Leak detection and localisation in pipes and pipelines. In *European symposium on computer aided process engineering-10* (Vol. 2, pp. 781–786).

Gertler, J., Romera, J., Puig, V., & Quevedo, J. (2010). Leak detection and isolation in water distribution networks using principal component analysis and structured residuals. In *Conference on Control and Fault Tolerant Systems* (pp. 1–6). Nice, France: IEEE.

Isermann, R. (2006). *Fault diagnosis system*. Heidelberg: Springer.

Knudsen, O. Ø. (2013). Pipelines carry out their own health checks. Technical Report, SINTEF. http://www.sintef.no/en/latest-news/pipelines-carry-out-their-own-health-checks/.

Lee, P., Vitkovsky, J., Lambert, M., Simpson, A., & Liggett, J. (2005). Leak location using the pattern of the frequency response diagram in pipelines: A numerical study. *Journal of Sound and Vibration*, *284*, 1051–1075.

Mounce, S. R., Mounce, R. B., & Boxall, J. B. (2011). Novelty detection for time series data analysis in water distribution systems using support vector machines. *Journal of Hydroinformatics*, *13*(4), 672–686.

Palau, C., Arregui, F., & Carlos, M. (2012). Burst detection in water networks using principal component analysis. *Journal of Water Resources Planning and Management*, *138*(1), 47–54.

Patton, R. J., Frank, P. M., & Clark, R. N. (1989). *Fault diagnosis in dynamic systems: Theory and applications*. New York: Prentice Hall.

Pérez, R., Puig, V., Pascual, J., Quevedo, J., Landeros, E., & Peralta, A. (2011). Methodology for leakage isolation using pressure sensitivity analysis in water distribution networks. *Control Engineering Practice*, *19*, 1157–1167. doi:10.1016/j.conengprac.2011.06.004.

Pipeline Studio (2013). Software. In *Energy solutions international*. http://www.energy-solutions.com/.

Pudar, B. R. S., Member, A., & Liggett, J. A. (1992). Leaks in pipe networks. *Journal of Hydraulic Engineering*, *118*(7), 1031–1046.

Shields, D., & Daley, S. (2001). Design of nonlinear observers for detecting faults in hydraulic subsea pipelines. *Control Engineering Practice*, *9*, 297–311.

Venkatasubramanian, V., Rengaswamyd, R., Yin, R., & Kavuri, S. (2003a). A review of process fault detection and diagnosis: Part i: Quantitative model based methods. *Computers and Chemical Engineering*, *27*, 293–311.

Venkatasubramanian, V., Rengaswamyd, R., Yin, R., & Kavuri, S. (2003b). A review of process fault detection and diagnosis: Part ii: Qualitative model and search strategies. *Computers and Chemical Engineering*, *27*, 313–326.

Venkatasubramanian, V., Rengaswamyd, R., Yin, R., & Kavuri, S. (2003c). A review of process fault detection and diagnosis: Part iii: Process history-based methods. *Computers and Chemical Engineering*, *27*, 326–346.

Verde, C., Molina, L., & Torres, L. (2014). Parametrized transient model of a pipeline for multiple leaks location. *Journal of Loss Prevention in the Process Industries*, *29*, 177–185.

Verde, C., Torres, L., & González, O. (2016). Decentralized scheme for leaks' location in a branched pipeline. *Journal of Loss Prevention in the Process Industries*, *43*, 18–28.

Verde, C., & Visairo, N. (2004). Identificability of multi-leaks in a pipeline. In *Proceedings of the American Control Conference 2004*. ISBN-0-7803-8336-2.

Visairo, N., & Verde, C. (2003). Leak isolation conditions in a pipeline via a geometric approach. In *3rd IFAC-SAFEPROCESS symposium* (pp. 1023–1028).

Wu, Y., Liu, S., Wu, X., Liu, Y., & Guan, Y. (2016). Burst detection in district metering areas using a data driven clustering algorithm. *Water Research*, *100*, 28–37.

Wylie, E. B., & Streeter, V. L. (1978). *Fluid Transients*. McGraw-Hill International Book Co.

Ye, G., & Fenner, R. A. (2014). Weighted Least Squares with Expectation Maximization Algorithm for Burst Detection in U. K. Water Distribution Systems. *Journal of Water Resources Planning and Management*, *140*(4), 417–424. doi:10.1061/(ASCE)WR.1943-5452.0000344.

Zhang, J. (1996). Designing a cost effective and reliable pipeline leak detection system. In *Pipeline Reliability Conference, Houston, USA, November 19-22*.

Chapter 2
An Overview of Transient Fault Detection Techniques

Xinge Xu and Bryan Karney

Abstract This chapter overviews the theory and strategies of transient fault detection, considering both active and passive systems, and contrasting the more common frequent approaches with time-domain methodologies. The chapter contends that real complex systems may have mimics, where one characteristic can locally impersonate another. The chapter seeks to examine the "state of play" in these areas, providing a factual summary of what has been shown and demonstrated to date, along with a more speculative set of reflections about challenges and about which methods appear to the authors to have the greatest promise for deployment and commercialization.

2.1 Introduction

Transient events occur whenever flow or pressures' conditions change in a pressurized conduit. These unsteady flow conditions are created by local adjustments to the systems' operation. Through a combination of pipeline and system characteristics, the first created waves propagate from their origin to other parts of the system, undergoing reflections and refractions along the way and possibly inducing secondary changes in system status where they are received.

What makes these waves important is that they are the mechanism of change: the way a system changes its status from what it is currently doing to a sequence of complex intermediate states on the way to what it will do next. The waves bring "news" of the change and send the signals that are required to achieve the next equilibrium state. What makes these waves dangerous is that large pressure variations are sometimes produced. What makes the transient events particularly interesting is the fact that they both convey and send signals. Thus, by listening to the waves, one can potentially learn a great deal about system states and how they are coupled to physical attributes like the presence of a leak, blockage or the state of repair of a pipe. What makes the waves difficult, though, is really the flip side of the same things that

X. Xu · B. Karney (✉)
University of Toronto, 35 St. George street, Toronto, ON, Canada
e-mail: karney@ecf.utoronto.ca

C. Verde and L. Torres (eds.), *Modeling and Monitoring of Pipelines and Networks*, Applied Condition Monitoring 7,
DOI 10.1007/978-3-319-55944-5_2

make them powerful; they have some sensitivity to many systems' characteristics. Thus, the problem of determining what physical characteristic is linked to what signal response can be quite challenging in realistic cases.

2.1.1 Flow Characteristics

Transient Flow. Transient flow is often defined as the intermediate-state flow describing the transition between two steady states (Chaudhry 1979). Any change or disturbance, whether it is planned or accidental, can initiate transient conditions. Common boundary conditions that may introduce transients in pipeline systems include sudden changes in pump or valve settings, starting or stopping of pumps, and changes in the level of reservoir (Wylie and Streeter 1978).

Typically, the term transient flow describes an unsteady fluid flow phenomenon in pipeline systems. If the fluid is water, the phenomenon is known as water hammer effect. The instantaneous pressure rise in a pipe system initially having a steady flow is sometimes caused by the sudden closure of the valve. In this case, the pressure change is directly proportional to the velocity change. The basic water hammer equation has been derived to express the pressure head produced by the surge in the pipes, as shown in (2.1).

$$\Delta H = -\frac{a}{g}\Delta V \tag{2.1}$$

where ΔH = the head produced by surge (m), a = wave velocity (m/s), g = gravitational acceleration (m/s^2), and ΔV = the sudden change in velocity (m/s).

Transient flow is often represented in a single-pipe system with an upstream of a constant level reservoir and a valve or another constant level reservoir with a valve at the downstream end. The former system configuration is referred to as the reservoir-pipeline-valve (RPV) system while the latter is known as the reservoir-pipeline-reservoir (RPR) system. The RPV system has received intense attention by researchers because of its simplicity. The layouts of the two types of system configuration applied for fault detection are shown in Figs. 2.1 and 2.2. According to (2.1), the pressure surge wave propagates along the pipe after the instantaneous closure of the valve. At least one pressure transducer is located close to the valve to detect the transient signal from the reservoir and from the fault.

Steady Oscillatory Flow. The successful application of many related fault detection methods requires periodic (e.g., sinusoidal) valve operations to establish steady oscillatory flow to the system. Steady oscillatory flow, or periodic flow, occurs if the flow conditions are repeated at every fixed time interval (Chaudhry 1979). The minimum time interval of a repetitive condition is called as the oscillating period.

Considering a simple RPV system as illustrated in Fig. 2.1, the downstream valve is open and closed sinusoidally at a prescribed frequency. In a pipe system with friction, the flow oscillates at the same frequency with the valve, while the

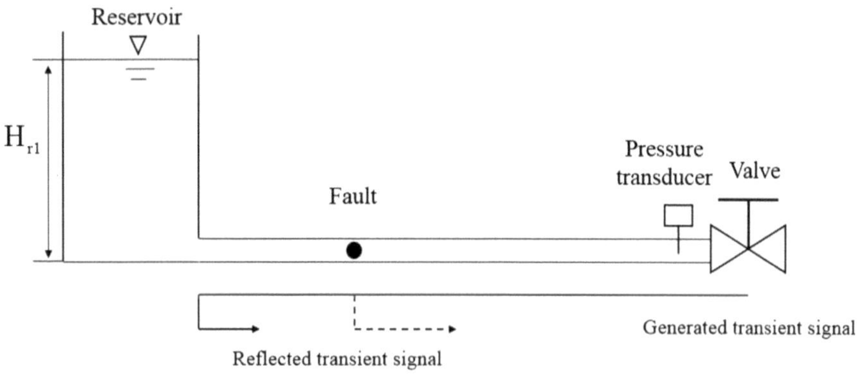

Fig. 2.1 RPV system layout with a fault

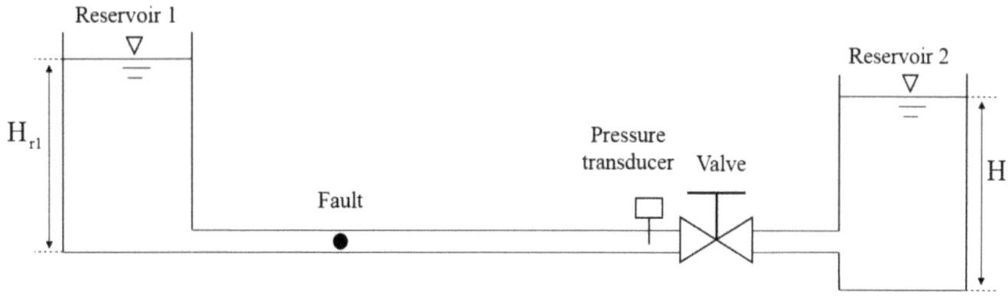

Fig. 2.2 RPR system layout with a fault

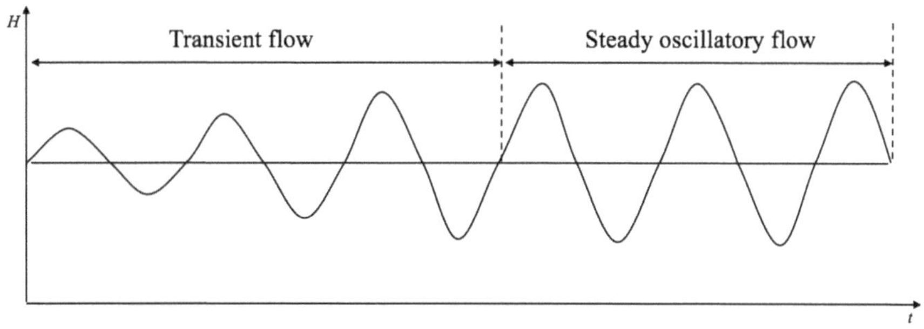

Fig. 2.3 The development of steady oscillatory flow at the downstream valve

amplitude increases until the energy provided by the oscillating valve is dissipated. The steady oscillatory state is established as shown in Fig. 2.3. The periodic excitation of the valve generates a complex transient flow until conditions settle into a constant amplitude response at the excitation frequency.

2.1.2 Governing Equations

Hydraulic transient models are developed based on the conservation rules of mass and momentum that govern transient flows. Since the flow velocity and pressure head of transient flow in closed conduits are functions of both time and distance, the governing equations are simplified one-dimensional partial differential equations that describe the unsteady flow in pressurized pipes (Chaudhry 1979; Wylie and Streeter 1978):

$$\frac{\partial H}{\partial t} + \frac{a^2}{gA}\frac{\partial Q}{\partial x} = 0 \tag{2.2}$$

$$\frac{1}{gA}\frac{\partial Q}{\partial t} + \frac{\partial H}{\partial x} + h_f = 0 \tag{2.3}$$

where H = instantaneous piezometric head at the centerline of the pipe, t = time, A = pipe cross-sectional area, Q = instantaneous flow, and x = distance along the pipe axis.Furthermore, h_f is the head loss per unit length and is usually expressed as (2.4)

$$h_f = \frac{fQ|Q|}{2gDA^2} \tag{2.4}$$

where f = Darcy–Weisbach friction factor and D = pipe diameter.

Equations (2.2) and (2.3) are commonly referred to as the continuity and momentum equations, which are developed in the assumed system of elastic pipes and weakly compressible fluids. The flow velocity also must be significantly lower than the wave speed if the advective terms are to be ignored.

The governing equations can be solved to perform the transient simulations either in the time or in the frequency domain. By assuming that the flow and pressure vary sinusoidally from a steady mean value, (2.2) and (2.3) can be solved in the frequency domain. The instantaneous head and flow of steady oscillatory flow are functions of time and are separated into two components: mean flow (Q_0)/ pressure head (H_0), and flow (q^*)/ pressure head (h^*) deviations from the mean. Figure 2.4 depicts how the flow at a certain section varies with time and repeats in accordance with the period.

$$Q = Q_0 + q^* \tag{2.5}$$

$$H = H_0 + h^* \tag{2.6}$$

In practice the deviation from the mean state is usually assumed to vary sinusoidally with time (Chaudhry 1979), which can be described as follows:

$$q^* = \mathrm{Re}\left(q\,(x)\,e^{j\omega t}\right) \tag{2.7}$$

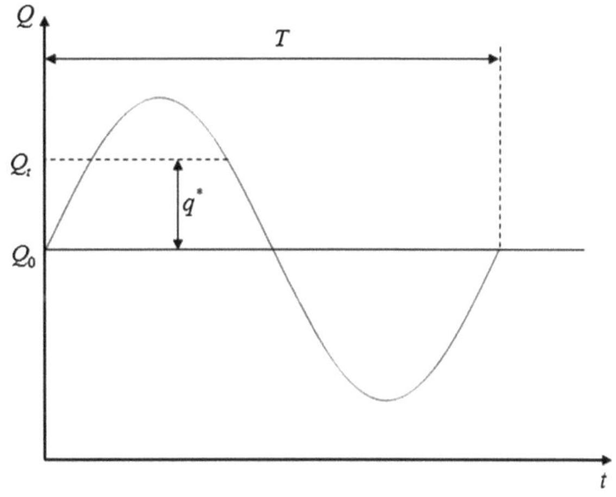

Fig. 2.4 Instantaneous flow varies sinusoidally with time

$$h^* = \text{Re}\left(h(x)\, e^{j\omega t}\right) \tag{2.8}$$

where $j = \sqrt{-1}$, q and h are complex variables as the function of x, Re is the real part of the complex variable, and ω is the frequency of oscillations, which is usually expressed in rad/s or $\omega = 2\pi/T$ if T is the period in seconds.

Substituting the average and oscillating components of head and flow into (2.2) and (2.3), the linearized equations are derived as (2.9) and (2.10) for the frequency analysis.

$$\frac{\partial q^*}{\partial x} + \frac{gA}{a^2}\frac{\partial h^*}{\partial t} = 0 \tag{2.9}$$

$$\frac{\partial h^*}{\partial x} + \frac{1}{gA}\frac{\partial q^*}{\partial t} + Rq^* = 0 \tag{2.10}$$

where R = unit friction term. For laminar flow,

$$R = \frac{32\upsilon}{gAD^2} \tag{2.11}$$

where υ = fluid viscosity. For turbulent flow,

$$R = \frac{fQ_0}{gDA^2} \tag{2.12}$$

Typically, the numerical method utilized in the time domain is the method of characteristics (MOC), by which the partial differential equations are transformed into a system of ordinary differential equations. The MOC is a well-developed method which describes nonlinear equations and various boundary conditions of simple to complex pipe systems. The commonly used schemes for the analysis in frequency domain are the transfer matrix method (TM) and the impedance method (IM). The theorem of the above methods has been explored in detail by several standard references (Chaudhry 1979; Wylie and Streeter 1978).

2.1.3 Detection Principle

Transient-based methods for the purpose of pipe fault detection all employ the same principle, which is to extract information about potential pipe or system faults by analyzing the measured trace of fluid transient behavior.

The internal and external characteristics of the pipeline can affect the transient response by altering the flow and pressure in the system. Considering pipeline features such as constrictions, expansions, ends, branches, valves, junctions and bends, leaks, blockages and deteriorations, there are many faults that commonly exist in pipe systems. The occurrence of a leak is a hydraulic phenomenon but one associated with various troubles. An amount of pressurized fluid released from the leaks, providing transient protection for the system and modifying the character of transient pressure wave. The pressure fluctuations can be identified collectively in the time or frequency domain to determine the location and size of the leak.

As the transient signal propagates throughout the pipe network, theoretically the information concerning the integrity and features of the pipe system can be detected hydraulically by using the transient signal as a kind of probe. The reflected transient signal and its damping pattern are the critical properties to be accessed for achieving fault detection. Their magnitude is generally proportional to the level of deterioration at the fault. After measuring the transient response at accessible locations along the pipeline, potential fault information is extracted. Typically, pipe system behaviors are studied in order to infer the system state (e.g., flow or pressure) under the assumption that all system characteristics are known. By contrast, all fault detection techniques can be regarded as solving an inverse problem, in which the system state is analyzed to determine the unknown system parameters such as faults and pipeline features.

Pipe fault detection techniques based on fluid transients have gained popularity over the last decades. Intensive numerical simulations, laboratory verifications and a few field tests have so far been conducted by researchers, generally based on the conviction that transient-based methods are superior to the other techniques. One reason for this belief is that they are nonintrusive and cost-effective. A nonintrusive technique used for evaluating the internal surface condition of pipelines is necessary for their planning and rehabilitation. When the fluid flow is in a transient state, the information about a long pipe can be obtained in a very short period of time at a great distance from the measurement point because the transient wave quickly travels through the fluid filled pipe. Another advantage is, compared to the performance of a pipe system under steady state, the system under transient state provides a vast amount of data, which is rewarding for the purpose of fault detection because the problem can be solved more accurately. Moreover, the results from fault detection using the transient are less susceptible to pipe friction factors. Theoretically at least, the fault detection and system calibration can be conducted simultaneously without knowing the precise friction value. Of course, these benefits are not the whole story, however.

2.1.4 Major Considerations and Categorizations

A number of transient-based techniques for pipeline fault detection have been developed in the last two decades (Wang et al. 2001; Colombo et al. 2009; Puust et al. 2010). These methods share some characteristics with each other but each has its own special advantages and disadvantages.

Objectives. While developing transient fault detection techniques, it is vital to have a clear picture of the objectives that should be met. The possibility of balancing the expectations of all these objectives must be continually considered.

- reliability—the technique is expected to show its effectiveness over time, and the performance of monitors should be stable in long-term identification of faults.
- accuracy—it is evident that the errors in detecting and identifying false positives and negatives can lead to unnecessary pipe maintenance costs. Although developing a perfectly accurate fault detection method is difficult, a high degree of detection accuracy is obviously desirable.
- cost—After completing the fault diagnostics, the economic aspect associated with the application of the proposed method is evaluated considering the costs of deployment, maintenance, and the management of diagnostic errors. One of the goals in developing a fault detection technique is to reduce the operational cost while improving the performance.
- sensitivity—For the cost saving, developing an efficient method which can detect small anomalies from a long distance is important. The sensitivity of a fault detection method is optimized by increasing its sensitivity with respect to the target faults and by enhancing its ability to eliminate noise from other disturbances.
- acceptance—Fault detection techniques are an application-oriented approach, which means the acceptance of the method must be established ideally both in the lab and in realistic field installations to prove its performance in laboratory or practical applications. The techniques should have simple and clear procedures, and be nondisruptive to the system.

Based on these objectives, various transient fault detection techniques are presented, with the following major concerns addressed in different ways.

Transient generation. The transient signal that is analyzed in the test can rely upon natural sources of fluid excitation, generated by an induced event (e.g., pump failure or pipe leak/burst) (Misiunas et al. 2005) or artificial events, such as injecting a prescribed transient signal into the pipe flow typically through valve operations. The subsequent behavior of the transient event is analyzed to acquire the system properties. The fault detection performed in passive systems obtain the transient signals directly from the system and exploit special features to detect and locate leaks and other faults. The systems that take advantage of information from transient events generated artificially are referred to as active systems, where the injected signals are customized and must be distinct from background noise.

Fault-induced effect. The transient signal contains two types of fault-induced effect. The reflection and damping effect, which can be utilized alone or can consider two types of information at the same time.

Domain type. After the time history of the pressure transient signal (i.e., the piezometric head) is measured by the transducers, the analysis of the data can be completed in the time domain or/and frequency domain. The time-domain methods analyze data straightforwardly in the time domain, while the frequency-domain methods require mathematical conversion.

Analysis approach. Signal processing and hydraulic transient model simulation are analysis approaches that are being considered. Signal processing is a method to extract information from measured data and compare it with the data sets from a fault-free benchmark on the basis of properties of leak-induced effect on a flow or pressure signal. Hydraulic transient models simulate data and reproduce pressure traces in the time or frequency domain. Based on the degree of coincidence between the data from accurately modeled systems with faults and measured data, information related to faults is determined. Two detection procedures with different configurations can be developed based on the propagation analysis of the pressure signal. A signal processing approach locates the generator and the transducer of the transient at the end of a pipeline, and the inverse method using a hydraulic model generates the transient and measures the pressure response at multiple locations along the pipe.

In summary, according to the manner in which transient signals are utilized, the methods described in existing literature can be divided into four types: the transient reflection method (TRM), the transient damping method (TDM), the system response method (SRM), and the inverse transient method (ITM). These methods are briefly introduced in the following sections. Their key features and categorization are systematically summarized in Fig. 2.5.

2.2 Current Transient Fault Detection Techniques

2.2.1 Transient Reflection Method (TRM)

Generally, a transient pressure wave is partially reflected, partially transmitted and partially absorbed wherever the system shows discontinuity (Burn et al. 1999). Faults and pipeline features cause additional reflections of transient pressure waves as shown in Fig. 2.1 and may create multiple wave paths in the pipe system. Figure 2.6 shows the hypothetical pressure trace at a transducer. The transient wave is detected by

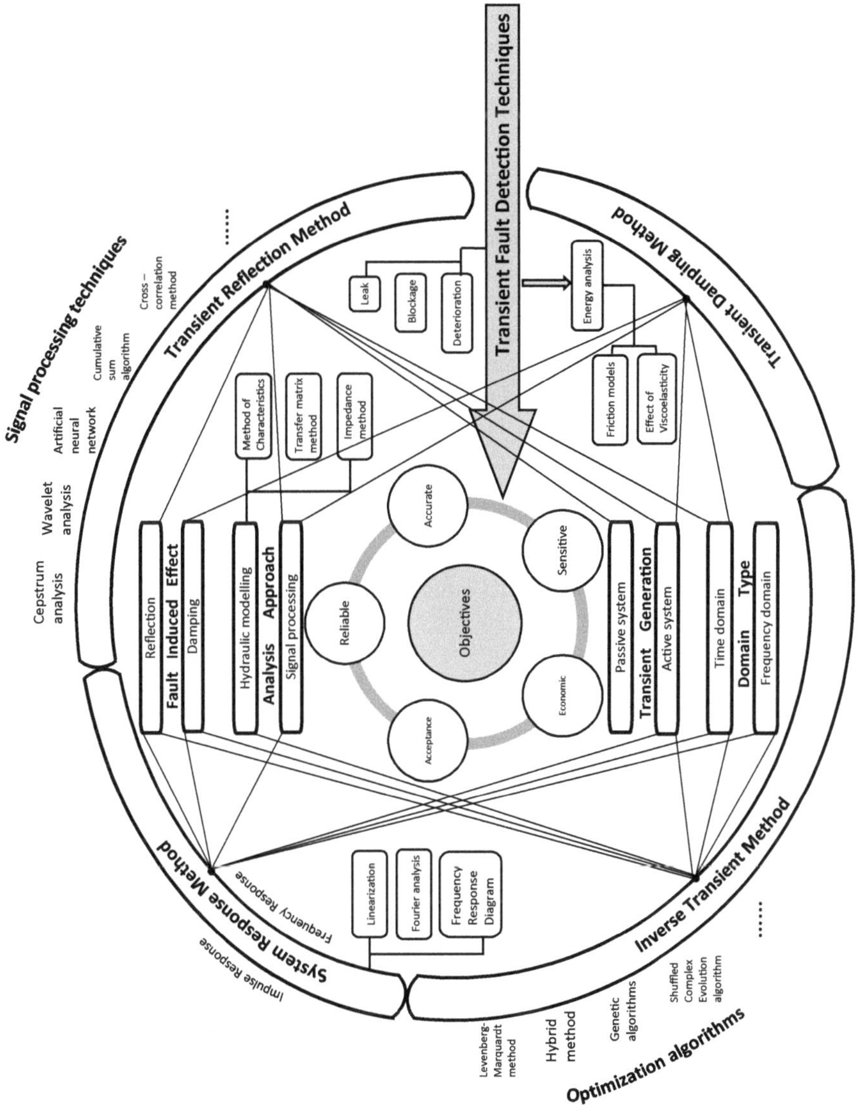

Fig. 2.5 Summary of current transient fault detection techniques

the transducer at t_1, and the passage of the reflected signal, normally the first signal reflected, is recorded at t_2. In addition, the distance between measurement point and fault location can be given by multiplying the half wave traveling time T by the wave speed.

The transient reflection method (TRM), often referred to as the time-domain reflectometry (TDR) technique, relies on differentiation of a reflected signal by identifying the discrepancies between measured results with the fault-free benchmark results. The benchmark results can be obtained from a fault-free laboratory system, or from an accurate numerical model of the pipe system. For a typical system with unknown characteristics, however, it is hard to determine the leak-free benchmark: only the changes from the benchmark results can be detected.

Jönsson and Larson (1992) were one of the first to use the generated transient to detect leaks by measuring the arrival time of a reflected wave. Brunone (1999) introduced the theory and verified it by an experiment in a single polyethylene pipe, while the background noise disturbed the identification of signals. The experimental validation was improved by Brunone and Ferrante (2001) to identify the leak location more accurately. Beck et al. (2005) used the cross-correlation method to reduce the problem of disturbance and detected more pipeline features in a T-junction network. Meniconi et al. (2011a) applied the TRM to detect the location of illegal side branch in a laboratory complex pipe system. The experiment performed well, but uncertainty about factors like friction and unaccounted for reflections in real systems may complicate practical application. Recent studies have improved the application of TRM by utilizing methods and algorithms such as cepstrum analysis (Taghvaei et al. 2006), wavelet analysis (Ferrante and Brunone 2003; Ferrante et al. 2007), cumulative sum method (Lee et al. 2007), and artificial neural network (ANN) (Stoianov et al. 2001).

Artificial intelligence methods have also attracted attention. For example, an artificial neural network (ANN) constructs relations between input and output data without any explicit mathematical model and has been effectively used to solve many classification problems including transient fault diagnosis. An ANN technique was applied for leak detection in a liquefied gas pipeline with 74,668 m in length and 0.203 m in diameter (Belsito et al. 1998). The ANN training uses data generated by a transient model using different leak and flow conditions. The method could detect and locate leaks down to 1% of the total flow rate in about 100 s according to the numerical results. When using measured data, however, the average error of the detected leak locations increases to about 7000 m for a leak of 1% of the total flow and 1645 m for a leak of 10% of the total flow (Belsito et al. 1998). Similarly to the problems that exist in the inverse methods, without the accurate simulation of the transients using a transient model, the information concerning leaks could be lost in the training process. A novel leak detection method was developed by integrating the wavelet transforms with artificial neural networks for signal identification and leak feature extraction (Stoianov et al. 2001). A supervised Kohonen network is applied to classify wavelet coefficients of the pressure response for transient events in a pipeline rig. Their potential has only been demonstrated using laboratory rigs, however.

The TRM has a basic concept and simple procedure, and previous works have also shown that the analysis of the reflected transient signal is effective in identifying pipeline features and faults (Beck et al. 2005; Lee et al. 2007; Ferrante et al. 2009; Gong et al. 2012b); nevertheless, its verification so far has rarely been extended to field tests. The accuracy of the method would be much lower without the assumption of a single-phase flow and rigid pipes. In addition, complicated geometries like loops in real complex pipe networks create complex reflected patterns that are less distinguishable. Therefore, it is difficult to obtain satisfactory results from pipe networks. The TRM does not require a precise mathematical model; however, it requires fault-free benchmark data sets from the controlled pipe system to extract and classify signal features. Thus, its scheme is not applicable to existing real pipe systems because the faults in the system would particularly perturb the benchmark data sets. It is also difficult to detect the integrity of the system if the induced transient signal in a fault-free system is not perfectly regular and reproducible.

2.2.2 Transient Damping Method (TDM)

The transient signal that propagates in a pipeline decays because of the presence of friction, faults, and pipeline features. The faults usually modify the damping pattern, so they can be detected by comparing the induced damping pattern with the fault-free benchmark in the same pipe system. Figure 2.6b uses the presence of leak as an example.

The transient damping method is a fault detection approach designed solely on the damping effect of the transient signal. It was first proposed by Wang et al. (2002). The approach introduced by Wang et al. performed a Fourier series analysis to solve the linearized differential equations. The approach detected the leak by utilizing the first two Fourier components in the transient signal based on the fact that the damping

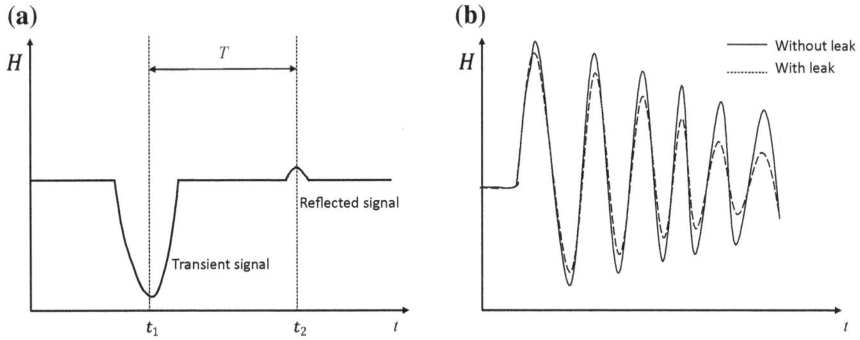

Fig. 2.6 Pressure trace modified by leak-induced reflection **a** and damping **b** effect

ratio of the two components is the same for friction factor, while the ratios are different for the leak factor. Wang et al. (2005) investigated another partial blockage detection method using the damping of fluid transient. The proposed method locates the blockage by the different damping ratios and uses the magnitude of the damping rate to determine the blockage size.

TDM is a simple and creative method. It has been applied in numerical simulation and lab experiments, which shows its ability to successfully locate and quantify the faults with small errors (Wang et al. 2005, 2002). Its application is restricted to single-pipe systems for damping is not only caused by friction and faults, so other physical features in complex systems would make the identification of damping difficult or even impossible. TDM assumes a linear system with steady friction. A study by Nixon et al. verified this assumption through a transient model and indicated the importance of unsteady friction in the applicability of the TDM (Nixon et al. 2006). The accuracy of the method is not influenced only if the unsteady friction effect is represented correctly in a simple system (Nixon et al. 2006; Nixon and Ghidaoui 2007). The TDM would be suitable in the field only if the difficulty of modeling unsteady friction is overcome. Aside from the characteristics of leaks, the leak-induced damping rate also relies on pipe pressure, the shape of the transient signal and the location of the generated transient, whereas the blockage-induced damping rate only depends on the position and size of the blockage. Because the damping functions induced by leaks and blockages have different modes (Wang et al. 2005), the accuracy of detecting leaks and blockages differs depending on the measurement location, making it hard to detect all faults at the same time.

2.2.3 System Response Method (SRM)

The main principle of the system response method is to utilize all the information (i.e., reflection and damping effects) contained in the transient signal of the system response to identify and locate the faults by comparing the results from a pipeline with and without faults.

The pipe system is considered as a transformation that can produce an output of the measured pressure response by using each given input of the transient injected to system. A complicated input signal is divided into a series of weighted unit impulses, and the overall system response is obtained by a process known as convolution, which adds the contributions from the entire input signal. The function relating the input and output signals indicates all information pertaining to the behavior and features of the pipe system.

The impulse response function converts the output signal in the trace into sharp impulses with well-defined spikes (Fig. 2.7). In the time domain, the system output can be calculated for any input by a convolutional integral:

$$y(t) = \int_{-\infty}^{+\infty} x\left(t'\right) I\left(t - t'\right) dt' \tag{2.13}$$

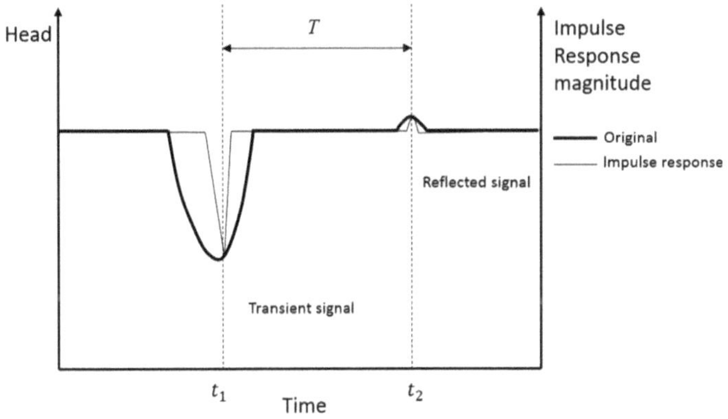

Fig. 2.7 Comparison between original reflected transient signal and impulse response function

where $I(t)$ is the impulse response function (IRF) of the system, and t' is the previous time step.

The application of the impulse response method (IRM) in fault detection was first presented by Liou (1998), who extracted the impulse response of the system by using the cross-correlation method and applied the technique in a real-time pipeline leak detection. The study is limited by its assumption of a linear system and can only easily be applied in single pipeline. Vitkovsky et al. (2003) extracted impulse response from a pipe system with a leak and blockage and compared it to the response from the fault-free pipe system. The numerical study shows that the use of IRF increased the accuracy of fault location. The genetic algorithm was integrated into the impulse response method by Kim (2005) in a study in which the location and size of a leak were calibrated. The study also considered the impact of unsteady friction by assuming different friction under the conditions of laminar and turbulent flow. Lee et al. (2007) validated the impulse response method by an experiment and demonstrated the impact of signal bandwidth and background noise on the extracted IRF. Gong et al. (2012a) utilized IRF to detect the distributed deterioration by determining the variation of pipe wall thickness. This novel distributed deterioration detection method was validated in the lab on a single pipeline.

Other sources of disturbance in real pipe systems, such as pumps, valves and turbines, can also modify the pressure trace by making it smooth, so identifying the fault-induced information becomes more difficult. By employing IRF, the conversion of output from pressure trace to sharp impulse enables an easier and more accurate estimation of signal arrival time. The IRF also has the advantage that it is independent of the input signal shape and is specific for each pipe system (Lee et al. 2007).

After applying a Fourier transform, the system response function in (2.13) is transferred to the frequency domain, known as frequency response function (FRF).

$$F(\omega) = \frac{Y(\omega)}{X(\omega)} \tag{2.14}$$

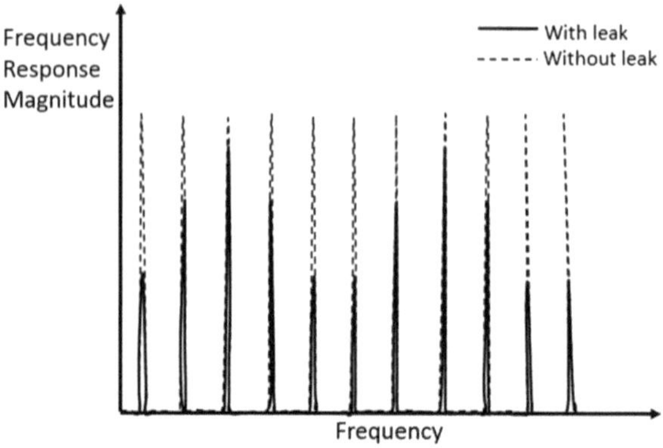

Fig. 2.8 Comparison between the frequency response of the intact and leak pipe

where ω is frequency, $F(\omega)$ is the Fourier transform of the function, $X(\omega)$ is the system input, and $Y(\omega)$ is the system output.

Mpesha et al. (2001) first proposed the frequency response method for pipe leak detection. The presence of a leak is revealed in the frequency response diagram by a relatively lower amplitude peak than an intact pipe, which is shown in Fig. 2.8. The numerical tests which used several different construction forms of pipe network verified the applicability of this method in detecting and locating small leaks in pipe systems. A new approach was put forward by Covas et al. (2005) by using the standing wave difference method for leak detection. The method is based on the steady oscillatory state generated by a sinusoidal valve operation, the principles of spectrum analysis, and the analysis of system frequency response. Although the method is effective and promising, it has some practical difficulties and has not been validated by experiment or field test. Lee et al. (2005) proposed an analytical solution in locating single as well as multiple leaks by explaining the nature of resonance peaks in a frequency response diagram. Duan et al. (2011b) presented the method in series' pipelines of different diameters and indicated the effect that the junction has on leak-induced information. Meniconi et al. (2011b) investigated the interaction between transient waves and in-line-valves within a viscoelastic pipeline and used wavelet functions to locate the blockage. Recently, Duan et al. (2014) observed the blockage-induced changes to the resonant frequencies and developed another method for blockage detection. The resonant frequency shifts caused by blockages in a pipe system is studied by conducting a transient wave perturbation analysis. The accuracy of the method is validated in the numerical application and experimental tests.

Unlike the IRM, the performance of FRM is closely related to the shape of the input transient. FRM analyzes transient response in the frequency domain, and it is equivalent to the impulse response method in the time domain in describing the system response from a transient excitation.

2.2.4 Inverse Transient Method (ITM)

The inverse transient method is a well-known and powerful approach for the fault detection and parameter calibration in pipe systems. The method involves the following procedures. First, the transient state is initiated by opening or closing a controlled valve over a short time period. Afterwards, transient flow characteristics (mostly the pressure fluctuations) are measured at multiple selected points in the system. The measurement points are usually located close to the transient generation point. The transient condition of the pipe system is then numerically modeled as a function of pipeline features, fault parameters and friction factors. The successful application of the method hinges on the modeling of pipe system behavior under transient state. After that, defining a nonlinear programming problem with an objective function to minimize the absolute difference between the actual measured data and simulated data from transient model must be achieved. An optimization tool is finally employed to solve the problem. The location and magnitude of the faults are determined by the minimization of the deviation between the actual measured data and simulated data by the transient model as shown in (2.15).

The objective function for optimization can be expressed as

$$\min:\ \mathrm{OF} = \sum_{i=1}^{N} \left| H_i^m - H_i^p \right| \tag{2.15}$$

where OF is the objective function, i is the time step points, N is the number of data points, H_i^m is the measured pressure head, and H_i^p is the predicted pressure head.

Various algorithms have been used to solve the objective optimization function of (2.15). Liggett and Chen (1994) introduced an inverse transient method for pipe network leak detection and calibration for the first time. They utilized the Levenberg–Marquardt (LM) method to solve the inverse problem. A simple small pipe network consisting of 11 pipes and seven nodes was theoretically developed to examine the model generated data. Their work shows that LM applied in ITM has good performance. Nash and Karney (1999) applied ITM in a series-connected pipeline and conducted a sensitivity analysis of the calibration results. Vítkovský et al. (2000) applied an improved unsteady friction model to increase the accuracy of transient simulation. The study improved the solution of the inverse problem by using a genetic algorithm (GA) instead of LM in previous research. The obtained numerical results of leak detection in a small pipe network was satisfactory. Kapelan et al. (2003) developed a novel optimization method in which two previously used methods, LM and GA, are combined into a hybrid genetic algorithm. The more reliable results were obtained after analyzing a benchmark pipe network. Stephens et al. (2004) used ITM to detect air pockets and blockage for the first time and successfully verified the ITM approach in the field. Vítkovský et al. (2007) presented the further improvement of ITM in exploring data and model errors by reducing the minimization complexity. The analysis was proved by experimental observations of detecting both single and multiple leaks in a laboratory pipeline. Covas and Ramos (2010) assessed the effectiveness of the technique by running ITM with the data collected from the laboratory and the field. Soares et al. (2011) verified ITM by an experiment in a complex system using PVC pipes with viscoelastic behavior. The experiment located leaks with a satisfactory accuracy, which was between 4 to 15% of the total length. More recently, Kim (2014) detected leaks and blockages considering unsteady friction in a branched RPR system. The impedance method for ITM was developed for the numerical verification. The results indicated that the impact of unsteady friction causes difficulty in locating faults, but the calibration of leak parameters was accurate.

ITM utilizes both the fault-induced reflection and the damping effect in producing the transient trace. Given enough measurements, the ITM can be used for pipe system fault detection as well as parameter calibration. The mathematical-related problems for ITM have been well exploited since the technique was introduced. Compared to other methods, the application of ITM has generality, since it can be applied without the restriction of topography and system configuration. Therefore, it is promising to extend the application of ITM to complex pipe networks in field. More efforts should be directed to experimental and field validation.

Traditionally, the measured data is analyzed by MOC and optimization algorithms such as GA, while the transient models are developed based on steady friction and the calculated damping rate (Wylie and Streeter 1978). The unsteady friction effect and boundary conditions that are not accurately considered in transient models can result in undesirable deviations between modeled and measured data. The key challenge in the satisfactory application of ITM lies in the accuracy of the transient model.

2.3 Critical Remarks

2.3.1 Evaluation of Techniques

Transient fault analysis usually needs to use the information generated by transient events, which is usually induced by several cycle-driven devices, such as modulation of valves. The difference between time-domain analysis (TDA) and frequency-domain analysis (FDA) is the way of analyzing the pressure monitoring data. The TDA method analyzes the pressure monitoring data directly in time domain, which does not involve any mathematical conversion, so it is intuitive and precise. In contrast to TDA, FDA must convert the pressure signal from time domain to frequency domain by using methods such as the fast Fourier transform and wavelet transform.

In most of the early studies, the transient flow is simulated and analyzed in the time domain by using the MOC, which must discretize the system in both time and space. In order to capture all the propagating waves and increase the model's accuracy, the system must be discretized into short reaches, resulting in tiny computational time steps, which with earlier computers rendered the MOC method computationally uneconomical. Modern computer technologies now allow us to do the calculations without these concerns, but the TDA is still practically limited to simple systems and boundary conditions. The FDA has become more popular in recent years. Pressure monitoring data does not only show the change over time, but also contains the information such as frequency and phase. The FDA method does not need discretization and is computationally very fast. It has been shown that transient fault diagnostics in the frequency domain allow the fault to be isolated from complex system phenomena such as unsteady friction and pipe wall viscoelasticity. Moreover, it provides an increased tolerance to random noises (Lee et al. 2006; Duan et al. 2011a). The fault parameters are more clearly evident in the frequency response diagram of the system and, therefore, the predicted fault parameters would be more reliable.

Note that TDA and FDA are applied under different demands and circumstances such that neither of them has a full advantage over the other. The FDA has some limitations associated with linearization of the governing equation as well as the detection of multiple faults directly from the system frequency response diagram. The TDA is a powerful analysis approach which is superior in describing the time history of transient response in pipe systems, and is effectively applied when the time history of the transient needs to be determined.

Advances in the development of the four principle techniques are summarized in Table 2.1. It is observed that all the techniques are feasible and accurate in numerical studies or in the laboratory, but are lacking any kind of comprehensive validation in the field. Although development is still in process, it is obvious that the application of methods like TDM is restricted, since TDM can only be applicable in simple pipe system, even if the fault detection accuracy is high. TRM is simple to use and apply, so its theory has been verified in various experimental systems of many studies. In complex systems, however, it does not have high accuracy and efficiency compared to the SRM. ITA is the most generally applicable method in dealing with complex pipe networks, but the general system condition must be quite well known.

Each one of these techniques requires specific and suitable system characteristics to trade off in complexity, precision and costs, so various techniques exist in this field. The overview of diverse transient-based fault detection techniques given in Sect. 2.2 also reveals that there is no perfect method that always meets all requirements, for each technique has its merits and drawbacks under different scenarios.

2.3.2 *Obstacles in Application*

The limitation of transient-based fault detection techniques can be concluded from the previous development and application of each method. Furthermore, the current techniques face the difficulty of identifying and characterizing a transient response from systems with complex configurations. The obstacles in solving the problem include any or all of the following issues.

System Assumptions. Most transient models assume ideal system conditions such as constant system demand, known pipe materials, and steady friction in the application

Table 2.1 Significant researches for principle techniques, Num* = numerical

Technics	References	Method/algorithm applied	System configuration	Pipe feature/fault	Key attribute		
					Num*	Lab	Field
TRM	Belsito et al. (1998)	ANN training	a single pipe	leak	√		√
	Brunone (1999)	MOC, inverse calculation	a single outfall pipe	leak	√	√	
	Stoianov et al. (2001)	wavelet analysis, ANN	six loop system	leak	√		√
	Beck et al. (2005)	cross-correlation method	T-junction pipe network	bend, joint, leak	√	√	
	Misiunas et al. (2005)	cumulative sum algorithm	a dead-end branch pipe	leak	√	√	√
	Lee et al. (2007)	cumulative sum algorithm	simple RPR system	leak	√	√	
	Ferrante et al. (2009)	wavelet analysis; Lagrangian model	Y-shape pipe network	dead end, leak, Y-junction	√	√	
	Gong et al. (2012b)	MOC	simple RPV system	distributed deterioration	√	√	
	Tuck et al. (2013)	MOC, visual comparision	simple RPR system	blockage	√	√	
TDM	Wang et al. (2002)	MOC; Fourier series analysis	simple RPR system	leak	√	√	
	Wang et al. (2005)	Fourier series analysis	simple RPR system	blockage	√	√	
SRM	Liou (1998)	impulse response; cross-correlation method	simple RPV system	leak	√		
	Mpesha et al. (2001)	frequency response, transfer matrix method	single, series, parallel, branched RPV system	leak	√		
	Vitkovsky et al. (2003)	impulse response, transfer matrix method	simple RPR system	leak, blockage	√		
	Lee et al. (2005)	frequency response, transfer matrix method	simple RPR system	multiple leak	√		
	Kim (2005)	impulse response, genetic algorithm	simple RPR system	leak	√		
	Covas et al. (2005)	frequency response, standing wave difference method	simple RPV system	leak	√		
	Lee et al. (2007)	impulse response CUSUM algorithm	simple RPR system	leak	√	√	

(continued)

Table 2.1 (continued)

Technics	References	Method/algorithm applied	System configuration	Pipe feature/fault	Key attribute		
					Num*	Lab	Field
	Sattar et al. (2008)	frequency response, MOC, transfer matrix method	simple RPV system	blockage	√	√	
	Duan et al. (2011b)	frequency response, transfer matrix method	complex pipe series	multiple leak	√		
	Gong et al. (2012a)	impulse response, sensitivity analysis	a single pipeline	distributed deterioration	√	√	
	Duan et al. (2014)	frequency response, analytical derivation	simple RPV system	extended blockages	√	√	
ITM	Liggett and Chen (1994)	Levenberg–Marquardt method	small pipe network	leak	√		
	Nash and Karney (1999)	Levenberg–Marquardt method	series RPR system	N/A	√		
	Vítkovský et al. (2000)	genetic algorithm	small pipe network	leak	√		
	Lee (2002)	SCE algorithm	simple RPR system	leak	√		
	Kapelan et al. (2003)	hybrid genetic algorithm	a looped pipe network	leak	√		
	Stephens et al. (2004)	Shuffled Complex Evolution global search algorithm	field network	air pocket, blockage	√		√
	Vítkovský et al. (2007)	Model Error Compensation Approach, Model Parsimony Approach	simple RPR system	single and multiple leak	√	√	
	Shamloo and Haghighi (2009)	Sequential Quadratic Programming method, backward MOC	simple RPV system	leak	√		
	Covas and Ramos (2010)	Levenberg–Maquardt method, genetic algorithms	complex pipe system	leak		√	√
	Soares et al. (2011)	Levenberg–Maquardt method, genetic algorithms	complex PVC pipe system	leak	√	√	
	Kim (2014)	impedance method genetic algorithm	branched RPR system	leak, blockage	√		

of fault detection techniques. Specifically, the assumption of steady friction and elastic behavior of pipe material in the experimental application (Wylie and Streeter 1978) results in deviation in phase and damping between measured and calculated results. By applying this, the results are far from satisfactory since the model does not correctly consider the unsteady friction and viscoelasticity effects. Steady or quasi-steady friction loss is a reasonable assumption for slow transients, while it is less valid for rapid transient events in fault detection techniques. Duan et al. (2010) also demonstrated the dominant effect of viscoelasticity with respect to unsteady friction in plastic pipes. The energy loss resulting from friction and fluid viscosity is uncertain, which is critical for the modeling of transient flow in real pressurized pipe system. Therefore, there have been a number of studies exploring the field of unsteady friction in the last two decades. Previous studies have pointed out, however, that unsteady friction leads to a more rapid attenuation of a transient signal (Chaudhry 1979). The improved unsteady friction transient model has not been developed to ascertain the accuracy and effectiveness of fault detection.

Transient Excitation. Transient signals in pipe system are excited by adjusting system elements such as in-line valves, side-discharge valves, or pumps. The first pressure transient oscillation is commonly analyzed for fault detection, so the periodic operation of the valve to generate steady oscillatory flow is required. These methods show that fault detection techniques based on odd harmonics of frequency are superior among the FRF-based methods (Lee et al. 2005; Duan et al. 2011b; Gong et al. 2013). The accuracy of the method increases with the number of resonant peaks in frequency response, so the input signals that have a wide bandwidth of frequency containing a number of harmonics are preferable. The requirement is not only for the frequency-domain method, but also the need for the application of time-domain fault detection techniques. Producing signals with resonant frequency is a risky decision, however, for it is likely to damage the pipe system. The practical difficulty of a rapid periodic maneuver operation at high frequency is also obvious. A desirable signal-to-noise-ratio (SNR) is difficult to achieve because of the limitation of existing transient generators. Modeling valve operations, especially steady oscillatory operation, can also be challenging in the field tests.

Signal Attenuation. All the applications of transient-based fault detection methods, regardless of whether they are used in the time or frequency domain, rely on the reflection information in transient trace (Lee et al. 2014). The energy contained in the transient signal attenuates along the propagation of the transient wave, so it is likely that in real-life application, the transient signal is incapable of traveling to the fault location and carrying fault-induced reflection information to the measurement point. Moreover, since the magnitude of reflection is proportional to the level of pipe's internal damage at a fault location, the small areas subjected to damage, such as cracks and rust patch, might not cause significant reflection, as it can be noticed on transient reflection. The problem of effectively extracting a useful weak transient signal created by a small leak or the decay along the pathway is still to be solved.

Hydraulic Impersonation. The complicated geometries, including junctions, branches and loops, in complex pipe systems create complex wave paths, which make the faults-related information difficult to extract. Previous work has improved the detection techniques and extended the experiments to identify pipeline features as well as the faults. Applying the methods in field tests, however, is still difficult. Reflections from multiple sources in a single pipe can distort the pattern of response trace. A normal pipe feature can introduce a pressure trace which is a good mimic of the response of a fault. In addition, system noises arise from a vast range of disturbances and events (e.g., pump failures, a change in demands) in a complex system and overshadow the useful transient pressure signals. Duan et al. (2014) indicated the highly nonlinear and nonstationary conditions of transient signals in complex systems for detecting multiple faults. The above-mentioned vexing phenomenon of hydraulic impersonation exists in all real pipe systems. Currently used data analysis techniques have the limitation of accurately distinguishing true positive from false positive results in fault diagnosis applications.

Other Factors. According to the previous real-life application of transient-based fault detection techniques in the literature, other limitations of current methods can also attributed to the following practical complications.

- The selection and relative position of the transient measurement and generation points. Features and faults close to the transducer are prone to be discernible;
- The automatic monitoring and detection of transient signals;
- Error in measuring actual wave speed; and
- The difficulty in determining the exact arrival time of transient signal because of the intrinsic error of time lag.

In brief, the developed techniques perform much better in ideal numerical simulations, since the hydraulic condition, valve operation, experimental uncertainties, and model inaccuracy can easily hinder the validation of transient-based fault detection techniques in realistic cases.

2.4 Promising Research Directions

On the basis of the discussions in previous sections, for better operating of pipe systems in industries, the following research directions are promising in achieving deployment and commercialization of transient-based fault detection techniques in the future.

Application in Complex System with Multiple Faults. All transient-based leak detection techniques face difficulties in locating faults in systems with a high level of complexity. The validations are expected to extend from a single-pipe system to a multiple-series pipe system, even to the pipe networks. Some consideration must be given on how the current methods can handle the practical complexity. The methods are mostly applied only in laboratory conditions in which the pre-known faults and features exist in grounded pipe. Fault detection applied to pipe systems in complex underground situations with unknown multiple types of faults is inapplicable in previous studies, and must be explored in the future. A comprehensive understanding of the system behavior for developing new effective transient-based detection techniques is critical.

Combination of ITM and FRM. The merits of the frequency response method in identifying the transient system characteristics are highlighted in recent research. Transient hydraulic simulation in the frequency domain does not require numerical computations' procedures and presents the presence of a fault evidently in the frequency response diagram. Previous studies mostly utilized the frequency response method for fault detection without the optimization. Therefore, the inverse transient frequency method (ITFM), which is the combination of ITM and FRM, would make the fault detection more precise and reliable compared to using ITM in the time domain. Formulating the system frequency response with the inverse problem has became a frontier in the field.

Higher Harmonics. Many experimental tests verify that obtaining more resonant peaks from the frequency response diagram can improve the fault detection performance. Hence, the rapid maneuvers resulting in the generation of high harmonics are preferable for application in the field. The rapid valve operation involved in the generation of high frequency, however, may cause the destructive water hammer in the pipe system. Furthermore, more distortion in high frequency components of the transient signal would be present in the field. Research which focuses on high harmonics is also lacking, since the mechanical device that can generate measurable high frequencies is currently unavailable. Further research in this field is needed to validate the applicability and to improve the performance of frequency-domain methods in practice.

2.5 Conclusions

This chapter introduces and evaluates a number of currently developed transient-based techniques of fault diagnosis in pipe systems. The fault detection principle and the method for analyzing the transient trace are presented. There is no unified framework for the identification and classification of these techniques. To tackle this problem, a systematic diagram is created to summarize the knowledge in the research area. The possible solutions with respect to the criterion of performance are different, since the selection of an optimal technique can only be performed considering the detection objectives, i.e., the need for reliable, economic, and robust methods applicable to complex and dynamic systems.

Selected literature on the development and verification of the described principal transient fault detection techniques are provided. The development of diagnostic techniques are based on numerical transient modeling and the theories of estimation and identification of fault information by the signal processing. The four discussed methods are mostly widely used in the active systems. Particular attention has been paid to the way of analyzing measured data, including the analysis in time domain and frequency domain.

The main drawback to all the proposed fault detection techniques is that they are remain difficult to apply in real and complex systems. This is caused by the fact that for each method one must give ideal system assumptions prior to its deployment, which reduces the accuracy and generality of the method. In addition, the problem of signal attenuation and technical difficulty in transient excitation are always challenging in real applications.

The possibility of joining the frequency response method with inverse analysis has vital advantages, which will be a subject for further investigations. The process of developing pipe fault detection techniques is a complex optimization problem. Computational requirements and errors grow together with system complexity. In this case, the fault detection is usually connected with the phenomenon of hydraulic impersonation. Further research on the techniques presented in the chapter is expected to address such issues, and certainly much research work still remains be done for minimizing the problem of both false positive and false negative field results.

References

Beck, S., Curren, M., Sims, N., & Stanway, R. (2005). Pipeline network features and leak detection by cross-correlation analysis of reflected waves. *Journal of Hydraulic Engineering*, *131*(8), 715–723.

Belsito, S., Lombardi, P., Andreussi, P., & Banerjee, S. (1998). Leak detection in liquefied gas pipelines by artificial neural networks. *American Institute of Chemical Engineers AIChE Journal*, *44*(12), 2675.

Brunone, B. (1999). Transient test-based technique for leak detection in outfall pipes. *Journal of Water Resources Planning and Management*, *125*(5), 302–306.

Brunone, B., & Ferrante, M. (2001). Detecting leaks in pressurised pipes by means of transients. *Journal of Hydraulic Research*, *39*(5), 539–547.

Burn, S., DeSilva, D., Eiswirth, M., Hunaidi, O., Speers, A., & Thornton, J. (1999). *Pipe leakage-future challenges and solutions*. Australia: Pipes Wagga Wagga.

Chaudhry, M. H. (1979). *Applied hydraulic transients*. Technical report: Springer.

Colombo, A. F., Lee, P., & Karney, B. W. (2009). A selective literature review of transient-based leak detection methods. *Journal of Hydro-environment Research*, *2*(4), 212–227.

Covas, D., & Ramos, H. (2010). Case studies of leak detection and location in water pipe systems by inverse transient analysis. *Journal of Water Resources Planning and Management*, *136*(2), 248–257.

Covas, D., Ramos, H., & De Almeida, A. B. (2005). Standing wave difference method for leak detection in pipeline systems. *Journal of Hydraulic Engineering*, *131*(12), 1106–1116.

Duan, H., Lee, P., Ghidaoui, M. S., & Tuck, J. (2014). Transient wave-blockage interaction and extended blockage detection in elastic water pipelines. *Journal of Fluids and Structures*, *46*, 2–16.

Duan, H. F., Ghidaoui, M., Lee, P. J., & Tung, Y. K. (2010). Unsteady friction and visco-elasticity in pipe fluid transients. *Journal of Hydraulic Research*, *48*(3), 354–362.

Duan, H. F., Lee, P. J., Ghidaoui, M. S., & Tung, Y. K. (2011a). Extended blockage detection in pipelines by using the system frequency response analysis. *Journal of Water Resources Planning and Management*, *138*(1), 55–62.

Duan, H. F., Lee, P. J., Ghidaoui, M. S., & Tung, Y. K. (2011b). Leak detection in complex series pipelines by using the system frequency response method. *Journal of Hydraulic Research, 49*(2), 213–221.

Ferrante, M., & Brunone, B. (2003). Pipe system diagnosis and leak detection by unsteady-state tests. 2. wavelet analysis. *Advances in Water Resources, 26*(1), 107–116.

Ferrante, M., Brunone, B., & Meniconi, S. (2007). Wavelets for the analysis of transient pressure signals for leak detection. *Journal of Hydraulic Engineering, 133*(11), 1274–1282.

Ferrante, M., Brunone, B., & Meniconi, S. (2009). Leak detection in branched pipe systems coupling wavelet analysis and a lagrangian model. *Journal of Water Supply Research and Technology-AQUA, 58*(2), 95–106.

Gong, J., Lambert, M. F., Simpson, A. R., Zecchin, A. C., et al. (2012a). Distributed deterioration detection in single pipes using the impulse response function. In *WDSA 2012: 14th Water Distribution Systems Analysis Conference, 24–27 September 2012 in Adelaide, South Australia* (p. 702). Engineers Australia.

Gong, J., Simpson, A. R., Lambert, M. F., Zecchin, A. C., Kim, Y. I., & Tijsseling, A. S. (2012b). Detection of distributed deterioration in single pipes using transient reflections. *Journal of Pipeline Systems Engineering and Practice, 4*(1), 32–40.

Gong, J., Zecchin, A. C., Simpson, A. R., & Lambert, M. F. (2013). Frequency response diagram for pipeline leak detection: Comparing the odd and even harmonics. *Journal of Water Resources Planning and Management, 140*(1), 65–74.

Jönsson, L. & Larson, M. (1992). Leak detection through hydraulic transient analysis. In *Pipeline Systems* (pp. 273–286). Springer.

Kapelan, Z. S., Savic, D. A., & Walters, G. A. (2003). A hybrid inverse transient model for leakage detection and roughness calibration in pipe networks. *Journal of Hydraulic Research, 41*(5), 481–492.

Kim, S. (2014). Inverse transient analysis for a branched pipeline system with leakage and blockage using impedance method. *Procedia Engineering, 89*, 1350–1357.

Kim, S. H. (2005). Extensive development of leak detection algorithm by impulse response method. *Journal of Hydraulic Engineering, 131*(3), 201–208.

Lee, P. J., et al. (2002). Leak detection in pipelines using an inverse resonance method. *Proc., 2002 Conf. on Water Resources Planning and Management*. Reston, Va: ASCE.

Lee, P. J., Duan, H. F., Tuck, J., & Ghidaoui, M. (2014). Numerical and experimental study on the effect of signal bandwidth on pipe assessment using fluid transients. *Journal of Hydraulic Engineering, 141*(2), 04014074.

Lee, P. J., Lambert, M. F., Simpson, A. R., Vítkovskỳ, J. P., & Liggett, J. (2006). Experimental verification of the frequency response method for pipeline leak detection. *Journal of Hydraulic Research, 44*(5), 693–707.

Lee, P. J., Vítkovskỳ, J. P., Lambert, M. F., Simpson, A. R., & Liggett, J. A. (2005). Leak location using the pattern of the frequency response diagram in pipelines: A numerical study. *Journal of Sound and Vibration, 284*(3), 1051–1073.

Lee, P., Lambert, M., Simpson, A., Vítkovsky, J., & Misiunas, D. (2007). Leak location in single pipelines using transient reflections. *Australian Journal of Water Resources, 11*(1), 53–65.

Liggett, J. A., & Chen, L. C. (1994). Inverse transient analysis in pipe networks. *Journal of Hydraulic Engineering, 120*(8), 934–955.

Liou, C. P. (1998). Pipeline leak detection by impulse response extraction. *Journal of Fluids Engineering, 120*(4), 833–838.

Meniconi, S., Brunone, B., Ferrante, M., & Massari, C. (2011a). Transient tests for locating and sizing illegal branches in pipe systems. *Journal of Hydroinformatics, 13*(3), 334–345.

Meniconi, S., Brunone, B., Ferrante, M., & Massari, C. (2011b). Small amplitude sharp pressure waves to diagnose pipe systems. *Water Resources Management, 25*(1), 79–96.

Misiunas, D., Vítkovskỳ, J., Olsson, G., Simpson, A., & Lambert, M. (2005). Pipeline break detection using pressure transient monitoring. *Journal of Water Resources Planning and Management, 131*(4), 316–325.

Mpesha, W., Gassman, S. L., & Chaudhry, M. H. (2001). Leak detection in pipes by frequency response method. *Journal of Hydraulic Engineering, 127*(2), 134–147.
Nash, G. A., & Karney, B. W. (1999). Efficient inverse transient analysis in series pipe systems. *Journal of Hydraulic Engineering, 125*(7), 761–764.
Nixon, W., Ghidaoui, M. S., & Kolyshkin, A. A. (2006). Range of validity of the transient damping leakage detection method. *Journal of hydraulic engineering, 132*(9), 944–957.
Nixon, W., & Ghidaoui, M. S. (2007). Numerical sensitivity study of unsteady friction in simple systems with external flows. *Journal of Hydraulic Engineering, 133*(7), 736–749.
Puust, R., Kapelan, Z., Savic, D., & Koppel, T. (2010). A review of methods for leakage management in pipe networks. *Urban Water Journal, 7*(1), 25–45.
Sattar, A. M., et al. (2008). Partial blockage detection in pipelines by frequency response method. *Journal of Hydraulic Engineering, 134*(1), 76–89.
Shamloo, H., & Haghighi, A. (2009). Leak detection in pipelines by inverse backward transient analysis. *Journal of Hydraulic Research, 47*(3), 311–318.
Soares, A. K., Covas, D. I., & Reis, L. F. R. (2011). Leak detection by inverse transient analysis in an experimental pvc pipe system. *Journal of Hydroinformatics, 13*(2), 153–166.
Stephens, M., Lambert, M., Simpson, A., Vítkovskỳ, J., & Nixon, J. (2004). Field tests for leakage, air pocket, and discrete blockage detection using inverse transient analysis in water distribution pipes. In *2004 World Water and Environmental Resources Congress*.
Stoianov, I., Karney, B., Covas, D., Masksimovic, C., & Graham, N. (2001). Wavelet processing of transient signals for pipeline leak location and quantification.
Taghvaei, M., Beck, S., & Staszewski, W. (2006). Leak detection in pipelines using cepstrum analysis. *Measurement Science and Technology, 17*(2), 367.
Tuck, J et al. (2013). Analysis of transient signlas in simple pipeline systems with an extended blockage. *Journal of Hydraulic Research, 51*(6), 623–633.
Vítkovskỳ, J. P., Lambert, M. F., Simpson, A. R., & Liggett, J. A. (2007). Experimental observation and analysis of inverse transients for pipeline leak detection. *Journal of Water Resources Planning and Management, 133*(6), 519–530.
Vítkovskỳ, J. P., Simpson, A. R., & Lambert, M. F. (2000). Leak detection and calibration using transients and genetic algorithms. *Journal of Water Resources Planning and Management, 126*(4), 262–265.
Vítkovskỳ, J., Lee, P. J., Stephens, M. L., Lambert, M. F., Simpson, A. R., & Liggett, J. A. (2003). Leak and blockage detection in pipelines via an impulse response method. *Pumps Electromechanical Devices and Systems Applied to Urban Water Management, 1*, 423–430.
Wang, X. J., Lambert, M. F., & Simpson, A. R. (2005). Detection and location of a partial blockage in a pipeline using damping of fluid transients. *Journal of Water Resources Planning and Management, 131*(3), 244–249.
Wang, X. J., Lambert, M. F., Simpson, A. R., Liggett, J. A., & Vítkovskỳ, J. P. (2002). Leak detection in pipelines using the damping of fluid transients. *Journal of Hydraulic Engineering, 128*(7), 697–711.
Wang, X. J., Lambert, M. F., Simpson, A. R., Vitkovsky, J. P., et al. (2001). Leak detection in pipelines and pipe networks: a review. In *6th Conference on Hydraulics in Civil Engineering: The State of Hydraulics; Proceedings* (p. 391). Australia: Institution of Engineers.
Wylie, E. B. & Streeter, V. L. (1978). Fluid transients (401 p.). New York, McGraw-Hill International Book Co., 1.

Chapter 3
Numerical Issues and Approximated Models for the Diagnosis of Transmission Pipelines

Zdzisław Kowalczuk and Marek Tatara

Abstract The chapter concerns numerical issues encountered when the pipeline flow process is modeled as a discrete-time state-space model. In particular, issues related to computational complexity and computability are discussed, i.e., simulation feasibility which is connected to the notions of singularity and stability of the model. These properties are critical if a diagnostic system is based on a discrete mathematical model of the flow process. The starting point of the study is determined by the partial differential equations obtained from the momentum and mass conservation laws by using physical principles. A realizable computational model is developed by approximation of the principal equations using the finite difference method. This model is expressed in terms of the recombination matrix A which is the key of the analysis by taking into account its possible singularity and stability. The nonsingularity of the matrix A for nonzero and finite, time and spatial steps is proven by the Lower–Upper decomposition. A feature of the discrete model allows the derivation of a nonsingular aggregated model, whose stability can be analyzed. By considering the Courant–Friedrichs–Lewy condition and data from experimental studies, numerical stability conditions are derived and limitations for the feasible discretized grid are obtained. Moreover, the optimal relationship between the time and space steps which ensures a maximum stability margin is derived. Because the inverse of matrix A, composed of four tridiagonal matrices, is required for the main diagnosis methods, two analytical methods for the inversion are discussed which reduce the system's initialization time and allow designing an accurate and fast diagnosis algorithm. By considering that each inversion method generates its particular structure, two different flow models are generated: one based on auxiliary variables and the other suitable if the stability condition of A is satisfied. The applicability of the two models is shown by considering the norm of the difference between their behaviors for a finer discretization grid. A similarity measure is proposed which considers the number of pipeline segments as well as the ratio between the time and

---

Z. Kowalczuk (✉) · M. Tatara
Gdansk University of Technology (ETI), Narutowicza 11/12, 80-233 Gdansk, Poland
e-mail: kova@pg.gda.pl

C. Verde and L. Torres (eds.), *Modeling and Monitoring of Pipelines and Networks*, Applied Condition Monitoring 7,
DOI 10.1007/978-3-319-55944-5_3

spatial steps. Thus, the system's computational efficiency is improved and satisfactory results are shown with respect to the base model, if a highly dimensional model with the approximated diagonal matrix is considered.

3.1 Introduction

Model-based approaches for leak detection and identification (LDI) require a robust mathematical description of the flow process with the same behavior as the real pipeline. Throughout the years, many modeling methodologies have been developed for the leak detection, monitoring of parameters, and complexity analysis of pipeline installations. The dominant directions are the (linear) observer-based methods for leaks in a single pipeline (Billmann and Isermann 1987; Kowalczuk and Gunawickrama 1998, 2004) or multiple leaks (Torres et al. 2012), as well as the artificial neural network approaches (Belsito et al. 1998). Recently, Reddy et al. (2011) and Verde and Torres (2015) have reported efforts to describe complex pipeline systems.

Behind every discrete-time model derived for LDI purposes, however, there is a concern about the numerical stability and computability of the model. The specific practical issues relate to computational complexity and computability, therefore, the feasibility of simulation is discussed in this chapter. Section 3.2 starts with the principal (physical) equations of the continuous-time nature, derived from the momentum and mass conservation laws. The feasible models are developed by discretization of the principal equations using the finite difference method, and are analyzed considering possible singularity and stability. In Sect. 3.3, by LU decomposition, the proof of the nonsingularity of the recombination matrix for nonzero and finite time and spatial steps is presented. Once the invertibility of the matrix is proved, in Sect. 3.4 a new aggregated model, named nonsingular, is derived and its features are discussed. In Sect. 3.5 the question of the numerical stability of the model is analyzed. By considering the Courant–Friedrichs–Lewy condition and data obtained from experimental studies, one derives a necessary condition for numerical stability that limits the choice of the discretization grid. This allows the search for the optimal relationship between the steps of time and space, which ensures a maximum stability margin.

In Sect. 3.6, two analytical methods for the inversion of matrix A composed of four tridiagonal matrices are introduced, and the utility of the proposed models for diagnostic systems is discussed in Sect. 3.7. The analysis is based on the norm of the difference between both models and considers the cardinality of the pipeline segmentation as well as the ratio between the time and spatial steps.

3.1.1 Matrices' Notations

Symmetric matrix: A matrix M is symmetric if it is square and equal to its transposition: $M = M^T$.

Diagonal matrix: A matrix is diagonal if its elements out of the main diagonal are all zero.
Diagonal of a matrix: The set of elements of a matrix located in the diagonal, i.e., the elements set $m_{i,j}$ with $i = j$.
Subdiagonal of a matrix: The set of elements of a matrix directly under the main diagonal (it is thus the first diagonal under the main diagonal).
Superdiagonal of a matrix: The set of elements of a matrix directly above the main diagonal (it is thus the first diagonal above the main diagonal).
Tridiagonal matrix: A matrix is tridiagonal if it has nonzero elements only on its main diagonal, subdiagonal, and superdiagonal.
Centrosymmetric matrix: A matrix M is called centrosymmetric if its elements satisfy the following condition:

$$m_{i,j} = m_{K-i+1,K-j+1} \qquad \text{for} \qquad i,\ j \in \{1, 2, \ldots, K\}$$

where K is the number of the matrix rows (or columns, since it is square). In other words, such matrix is symmetric around its center.
Row of a matrix: $r_i(M)$ denotes the i-th row of matrix M.

3.2 Base Model of the Flow Process

Consider the mathematical description of the pressure and flow rate for a flow process in transmission pipelines, which is expressed by the two equations obtained from the momentum and mass conservation laws (Billmann and Isermann 1987):

$$\frac{\mathscr{A}}{\nu^2}\frac{\partial p}{\partial t} + \frac{\partial q}{\partial z} = 0 \tag{3.1}$$

$$\frac{1}{\mathscr{A}}\frac{\partial q}{\partial t} + \frac{\partial p}{\partial z} = -\frac{\lambda \nu^2}{2D\mathscr{A}^2}\frac{q|q|}{p} - \frac{g sin\alpha}{\nu^2}p, \tag{3.2}$$

where $\mathscr{A}$ is the cross-sectional area $[m^2]$, ν is the isothermal velocity of the sound in the fluid $[\frac{m}{s}]$, D is the diameter of the pipe $[m]$, q is the mass flow $[\frac{kg}{s}]$, p is the pressure (Pa), t is the time $[s]$, z is the spatial coordinate $[m]$, λ is the dimensionless generalized friction factor, α is the inclination angle $[rad]$, and g is the gravitational acceleration $[\frac{m}{s^2}]$.

Since the practical operation of a model-based algorithm for pipeline diagnosis requires emulation of the underlying process behavior, the presented set of equations is discretized for the numerical implementation. The idea behind the discretization is to divide the pipeline into N segments of equal size, each one of length Δz, where the pressure at the end of each odd segment and the flow rate at the end of each even segment characterize the flow process. It is assumed that both variables, flow and pressure, are measured at the inlet and outlet of the pipeline. Such a discretization scheme is illustrated in Fig. 3.1.

The discrete-time model is simply obtained by introducing low-order central difference schemes:

$$\frac{\partial x}{\partial t} = \frac{3x_d^{k+1} - 4x_d^k + x_d^{k-1}}{2\Delta t} \tag{3.3}$$

$$\frac{\partial x}{\partial z} = \frac{x_{d+1}^{k+1} - x_{d-1}^{k+1} + x_{d+1}^k - x_{d-1}^k}{4\Delta z}, \tag{3.4}$$

where Δz is a spatial step, and Δt is a time step, subscripts and superscripts denote the number of the pipeline's segment and discrete-time step index, respectively.

The substitution of (3.3) and (3.4) into the composed model (3.1) and (3.2) gives the following discretized set of equations for the flow process in the pipeline:

$$ap_d^{k+1} - b\left(q_{d-1}^{k+1} - q_{d+1}^{k+1}\right) = \frac{a}{3}\left(4p_d^k - p_d^{k-1}\right) + b\left(q_{d-1}^k - q_{d+1}^k\right) \tag{3.5}$$

$$b\left(p_{d+1}^{k+1} - p_{d-1}^{k+1}\right) + cq_d^{k+1} = b\left(p_{d-1}^k - p_{d+1}^k\right) + Y_d p_d^k + \left(\frac{4c}{3} + F_d^k\right) q_d^k - \frac{c}{3} q_d^{k-1} \tag{3.6}$$

with physical coefficients

$$a = \frac{3\mathscr{A}}{2\nu^2 \Delta t}, \quad b = \frac{1}{4\Delta z}, \quad c = \frac{3}{2\mathscr{A}\Delta t}, \quad Y_d = \frac{-gsin\alpha_d}{\nu^2}$$

where α_d denotes the inclination angle of a d-th segment. The nonlinear function is approximated by

$$F_d^k \simeq -\frac{\lambda\nu^2}{D\mathscr{A}} \frac{|q_d^k|}{p_{d-1}^k + p_{d+1}^k}$$

since pressure is monitored at the ends of odd segments only. At the inlet and outlet ($d = 0$ and $d = N$) of the pipeline, the approximation is not required, because the pressure is known or measured, that is $F_d^k = -\frac{\lambda\nu^2}{2D\mathscr{A}} \frac{|q_d^k|}{p_d^k}$.

Thus, (3.5) and (3.6) can be represented by the compact state-space model

$$A\hat{x}^k = B\hat{x}^{k-2} + C\left(\hat{x}^{k-1}\right)\hat{x}^{k-1} + Du^{k-1} + Eu^k, \tag{3.7}$$

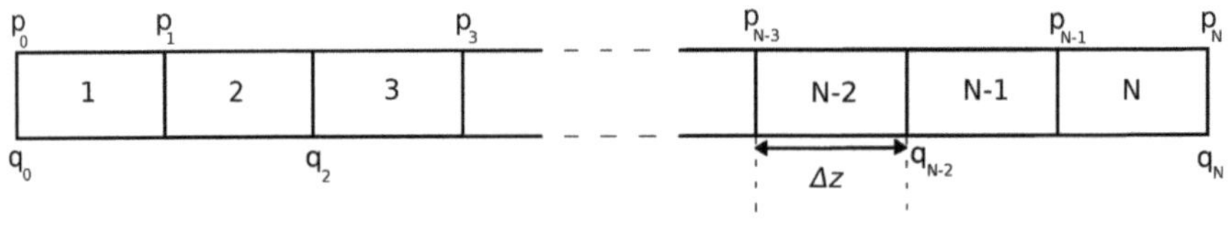

Fig. 3.1 Discretization scheme of a pipeline with N even segments

where B and $C(\hat{x}^k - 1)$ are associated with the nonlinear dynamic of the state

$$\hat{x}^k = \left[\, q_0^k\ q_2^k\ q_4^k \cdots q_N^k \,\middle|\, p_1^k\ p_3^k\ p_5^k \cdots p_{N-1}^k \,\right]^T \in \mathbb{R}^{N+1}$$

and matrices D and E are associated with the input $u^k = \left[\, p_0^k\ p_N^k \,\right]^T \in \mathbb{R}^2$ (Gunawickrama 2001). In particular, the matrices are given by

$$B = \frac{1}{3}\left[\begin{array}{ccccc|ccccc}
-c & 0 & \cdots & 0 & 0 & & & & & \\
0 & -c & \cdots & 0 & 0 & & & & & \\
\vdots & & \ddots & & \vdots & & & 0_{(\frac{N}{2}+1)\times(\frac{N}{2})} & & \\
0 & 0 & \cdots & -c & 0 & & & & & \\
0 & 0 & \cdots & 0 & -c & & & & & \\
\hline
 & & & & & -a & 0 & \cdots & 0 & 0 \\
 & & & & & 0 & -a & \cdots & 0 & 0 \\
 & & 0_{(\frac{N}{2})\times(\frac{N}{2}+1)} & & & \vdots & & \ddots & & \vdots \\
 & & & & & 0 & 0 & \cdots & -a & 0 \\
 & & & & & 0 & 0 & \cdots & 0 & -a
\end{array}\right] \in \mathbb{R}^{(N+1)\times(N+1)} \tag{3.8}$$

$$C(\hat{x}^{k-1}) = \left[\begin{array}{ccccc|ccccc}
H_0^k & 0 & \cdots & 0 & 0 & 2b & 0 & \cdots & 0 & 0 \\
0 & H_2^k & \cdots & 0 & 0 & \Gamma_{2+} & \Gamma_{2-} & \cdots & 0 & 0 \\
\vdots & & \ddots & & \vdots & \vdots & & \ddots & & \vdots \\
0 & 0 & \cdots & H_{N-2}^k & 0 & 0 & 0 & \cdots & \Gamma_{(N-2)+} & \Gamma_{(N-2)-} \\
0 & 0 & \cdots & 0 & H_N^k & 0 & 0 & \cdots & 0 & +2b \\
\hline
b & -b & 0 & \cdots & 0 & \frac{4a}{3} & 0 & \cdots & 0 & 0 \\
0 & b & -b & \cdots & 0 & 0 & \frac{4a}{3} & \cdots & 0 & 0 \\
\vdots & & \ddots & & \vdots & \vdots & & \ddots & & \vdots \\
0 & \cdots & b & -b & 0 & 0 & 0 & \cdots & \frac{4a}{3} & 0 \\
0 & \cdots & 0 & b & -b & 0 & 0 & \cdots & 0 & \frac{4a}{3}
\end{array}\right] \in \mathbb{R}^{(N+1)\times(N+1)} \tag{3.9}$$

where $\Gamma_{d\pm} = \frac{Y_d}{2} \pm b$, $H_0^k = \frac{4c}{3} - \frac{\lambda v^2 \hat{x}^k_{(d/2+1)}}{2D\mathscr{A}u_1^k}$, $H_N^k = \frac{4c}{3} - \frac{\lambda v^2 \hat{x}^k_{(d/2+1)}}{2D\mathscr{A}u_2^k}$,

$$H_d^k = \frac{4c}{3} - \frac{\lambda v^2 \hat{x}^k_{(d/2+1)}}{D\mathscr{A}\left(\hat{x}^k_{(d/2+1+N/2)} + \hat{x}^k_{(d/2+2+N/2)}\right)} \quad \text{for } d = 2, 4, \ldots, N-2$$

.

$$D = \begin{bmatrix} Y_0 + 2b & 0 \\ 0 & 0 \\ \vdots & \vdots \\ 0 & 0 \\ 0 & Y_N - 2b \\ \hline \multicolumn{2}{c}{0_{(\frac{N}{2})\times 2}} \end{bmatrix} \in \mathbb{R}^{(N+1)\times 2}, \qquad E = \begin{bmatrix} 2b & 0 \\ 0 & 0 \\ \vdots & \vdots \\ 0 & 0 \\ 0 & -2b \\ \hline \multicolumn{2}{c}{0_{(\frac{N}{2})\times 2}} \end{bmatrix} \in \mathbb{R}^{(N+1)\times 2} \tag{3.10}$$

One can see from (3.7) that the singularity of A affects the model. This description is called the *singular state-space* model. The term *singular* emphasizes the specific form of the model in which A may be not invertible in general. The matrix A itself is named recombination matrix, and it is written by

$$A = \left[\begin{array}{c|c} A_1 & A_2 \\ \hline A_3 & A_4 \end{array}\right] = \left[\begin{array}{c|c} D_{\frac{N}{2}+1}(c) & \begin{matrix} \searrow 2b & 0 & \cdots & 0 & 0 \\ -b & \searrow b & \cdots & 0 & 0 \\ \vdots & & \ddots & & \vdots \\ 0 & 0 & \cdots & \searrow b & 0 \\ 0 & 0 & \cdots & -b & \searrow b \\ 0 & 0 & \cdots & 0 & -2b \end{matrix} \\ \hline \begin{matrix} -b & b & 0 & \cdots & 0 & 0 & 0 \\ 0 & -b & b & \cdots & 0 & 0 & 0 \\ \vdots & & & \ddots & & & \vdots \\ 0 & 0 & 0 & \cdots & -b & b & 0 \\ 0 & 0 & 0 & \cdots & 0 & -b & b \end{matrix} & D_{\frac{N}{2}}(a) \end{array}\right] \in \mathbb{R}^{(N+1)\times(N+1)}, \tag{3.11}$$

where $D_W(\theta) \in \mathbb{R}^{W\times W}$ denotes a diagonal matrix with θ on the diagonal. Note that the upper right submatrix is non-square and belongs to $\mathbb{R}^{(\frac{N}{2}+1)\times(\frac{N}{2})}$.

3.3 Assessment of the Model's Singularity

According to Kowalczuk and Tatara (2013), to determine the effect of the selected discretization grid on the singularity in the model, the determinant of the recombination matrix A can be calculated by using the matrix LU factorization. To achieve this, the augmented recombination matrix is defined as

$$\bar{A} = \left[\, I_{N+1} \,\vdots\, A \,\right] = \left[\, L \,\vdots\, U \,\right], \tag{3.12}$$

where $I_{N+1} \in \mathbb{R}^{(N+1)\times(N+1)}$ is the proper identity matrix. The key to obtaining the determinant of $\bar{A}$ is the calculation of the lower and upper triangular matrices on the left-hand and right-hand sides of the augmented recombination matrix, respectively. Thus, the determinant is reduced to the product of the diagonal values of the matrices L and U (Kreyszig 2006):

$$\det(A) = \det(L)\ \det(U) \tag{3.13}$$

To obtain the lower triangular matrix on the left-hand side of matrix

$$\bar{A} = \left[\begin{array}{c|cccccccccc} & c & 0 & \cdots & 0 & 0 & 2b & 0 & \cdots & 0 & 0 \\ & 0 & c & \cdots & 0 & 0 & -b & b & \cdots & 0 & 0 \\ & \vdots & & \ddots & & \vdots & \vdots & & \ddots & & \vdots \\ & 0 & 0 & \cdots & c & 0 & 0 & 0 & \cdots & -b & b \\ & 0 & 0 & \cdots & 0 & c & 0 & 0 & \cdots & 0 & -2b \\ \mathbb{I}_{N+1} & -b & b & 0 & \cdots & 0 & a & 0 & \cdots & 0 & 0 \\ & 0 & -b & b & \cdots & 0 & 0 & a & \cdots & 0 & 0 \\ & \vdots & & \ddots & & \vdots & \vdots & & \ddots & & \vdots \\ & 0 & \cdots & -b & b & 0 & 0 & 0 & \cdots & a & 0 \\ & 0 & \cdots & 0 & -b & b & 0 & 0 & \cdots & 0 & a \end{array}\right] \tag{3.14}$$

elementary row transformations are used.

The first elementary transformations of $\bar{A}$ are obtained by adding row $r_1\left(\bar{A}\right)$ multiplied by $d = \frac{b}{c}$ to the row $r_{(\frac{N}{2}+2)}\left(\bar{A}\right)$ and then by subtracting row $r_2\left(\bar{A}\right)$ multiplied by d from row $r_{(\frac{N}{2}+2)}\left(\bar{A}\right)$. This iterative operation should be repeated by increasing the rows' indexes each time by 1, until they reach $\frac{N}{2}$. Thus, the equivalent matrix can be written by

$$\bar{A} = \left[\begin{array}{ccccc|c|ccccc} & & & & & & 2b & 0 & \cdots & 0 & 0 \\ & & & & & & -b & b & \cdots & 0 & 0 \\ & I_{\frac{N}{2}+1} & & 0_{\left(\frac{N}{2}+1\right)\times\frac{N}{2}} & & D_{\frac{N}{2}+1}(c) & \vdots & & \ddots & & \vdots \\ & & & & & & 0 & 0 & \cdots & -b & b \\ & & & & & & 0 & 0 & \cdots & 0 & -2b \\ \hline l_{n,1} & \cdots & 1 & \cdots & 0 & & a+3bd & -bd & \cdots & 0 & 0 \\ & & & & & & -bd & a+2bd & -bd & \cdots & 0 \\ \vdots & & \ddots & & \vdots & 0_{\frac{N}{2}} & \vdots & & \ddots & & \vdots \\ l_{N,1} & l_{N,2} & \cdots & 1 & 0 & & 0 & \cdots & -bd & a+2bd & -bd \\ l_{N+1,1} & l_{N+1,2} & \cdots & l_{N+1,N} & 1 & & 0 & 0 & \cdots & -bd & a+3bd \end{array}\right], \tag{3.15}$$

where the coefficients $l_{j,k}$ for $j - \frac{N}{2}+2, \frac{N}{2}+3, \ldots, N+1$ and $k = 1, 2, \ldots, N$ are irrelevant for the determinant of A.

On the other hand, the upper triangular matrix on the right-hand side can be obtained by removing the coefficients $-bd$ of the subdiagonal on the right-hand matrix by starting with $n = \frac{N}{2} + 2$ and by ending at $n = N$. This is achieved by adding to the row $r_{n+1}\left(\bar{A}\right)$ the row $r_n\left(\bar{A}\right)$ multiplied by $\frac{bd}{f_n}$, with f_n being the n-th coefficient of the main diagonal of the right-hand side matrix. Thus, (3.15) is equivalent to

$$\bar{A} = \left[\begin{array}{ccccc|c|ccccc}
 & & & & & & 2b & 0 & \cdots & 0 & 0 \\
 & & & & & & -b & b & \cdots & 0 & 0 \\
 & I_{\frac{N}{2}+1} & & 0_{(\frac{N}{2}+1)\times\frac{N}{2}} & & D_{\frac{N}{2}+1}(c) & \vdots & & \ddots & & \vdots \\
 & & & & & & 0 & 0 & \cdots & -b & b \\
 & & & & & & 0 & 0 & \cdots & 0 & -2b \\
\hline
l'_{n,1} & \cdots & 1 & \cdots & 0 & & f_n & u_{n,n+1} & \cdots & 0 & 0 \\
 & & & & & & 0 & f_{n+1} & u_{n+1,n+2} & \cdots & 0 \\
\vdots & & \ddots & & \vdots & 0_{\frac{N}{2}} & \vdots & & \ddots & & \vdots \\
l'_{N,1} & l'_{N,2} & \cdots & 1 & 0 & & 0 & \cdots & 0 & f_N & u_{N,N+1} \\
l'_{N+1,1} & l'_{N+1,2} & \cdots & l'_{N+1,N} & 1 & & 0 & 0 & \cdots & 0 & f_{N+1}
\end{array}\right] \tag{3.16}$$

The coefficients f_j for $j = \frac{N}{2} + 2, \frac{N}{2} + 3, \ldots, N + 1$ are obtained recursively with the initial condition $f_{\frac{N}{2}+2} = a + 3bd$ and the final condition $f_{N+1} = a + 3bd - \frac{b^2d^2}{f_N}$. The rest of coefficients are obtained with

$$f_j = a + 2bd - \frac{b^2d^2}{f_{j-1}} \quad \text{for} \quad j = \frac{N}{2} + 3, \frac{N}{2} + 4, \ldots, N$$

Note that coefficients $l'_{j,k}$ for $j = \frac{N}{2} + 2, \frac{N}{2} + 3, \ldots N + 1$ and $k = 1, 2, \ldots, N$ under the main diagonal of the left-hand side matrix have no effect on the $\det(\bar{A})$. Similarly, coefficients $u_{j,k}$ for $j = n, \ldots, N$ and $k = j + 1$ above the main diagonal of the right-hand side matrix have no effect on the $\det(\bar{A})$. As an example, if $N = 2$, there exists only one coefficient $f_3 = a + 4bd$.

As a consequence, the $\det(A)$ is given by

$$\det(A) = \prod_{j=1}^{N+1} l_{j,j} u_{j,j} \tag{3.17}$$

which is generated by the diagonal coefficients of L and U, respectively. Since the diagonal elements of L are equal to 1, and U consists of $\frac{N}{2} + 1$ diagonal elements with value c and $\frac{N}{2}$ diagonal elements of value f_j, the expression (3.17) is reduced to

$$\det(A) = (c)^{\frac{N}{2}+1} \prod_{j=\frac{N}{2}+2}^{N+1} f_j \tag{3.18}$$

Table 3.1 Determinants of the recombination matrix for selected segmentation N

N	Determinant of matrix A
2	$ac^2 + 4b^2c$
4	$a^2c^3 + 6ab^2c^2 + 8b^4c$
6	$a^3c^4 + 8a^2b^2c^3 + 19ab^4c^2 + 12b^6c$

and the general expression for a dimension N is written by

$$\det(A) = \sum_{i=0}^{\frac{N}{2}} C_i^N a^{\frac{N}{2}-i} b^{2i} c^{\frac{N}{2}-i+1} \tag{3.19}$$

with boundary values $C_0^N = 1$, for $N = 0, 2, 4, \ldots$ and $C_{\frac{N}{2}}^N = 2N$, for $N = 2, 4, 6, \ldots$. The coefficients C_i^N can be recursively calculated as

$$C_i^N = 2C_{i-1}^{N-2} + C_i^{N-2}, \quad \text{for } N = 4, 6, 8, \ldots \text{ and } i = 1 \tag{3.20}$$

$$C_i^N = 2C_{i-1}^{N-2} + C_i^{N-2} - C_{i-2}^{N-4}, \quad \text{for } N = 4, 6, 8, \ldots; \text{ and } i = 2, 3 \ldots, \frac{N}{2} - 1 \tag{3.21}$$

The determinant values for $N = 2, 4, 6$ can be obtained by substituting the values of f_j into (3.18) and are reported in Table 3.1.

By analyzing (3.20) and (3.21), one can see that all of the coefficients are positive for $i > 0$ and $N > 0$. Moreover, since (3.19) depends on a, b, and c, with even exponentials, the only way to get a zero determinant is that any of them is equal to zero. By considering that the cross-section and the sound velocity are always positive, the only way to achieve singularity of the recombination matrix is if $\Delta z = \infty$ or $\Delta t = \infty$. The former is equivalent to an infinitely long pipeline, and the latter can be neglected by maintaining the numerical stability of the algorithm discussed in Sect. 3.5.

Note that the time and spatial steps are positive, and negative spatial steps could not affect the nonsingularity of A, since b always has an even exponent. Negative time steps also do not appear in the practical flow process. As a consequence, one can establish that the recombination matrix A is nonsingular for every finite time and spatial steps, and thus, invertible.

3.4 Aggregated Model

Once the invertibility of A has been shown, the base model (3.7) can be represented in the nonsingular state-space form

$$\hat{x}^k = A^{-1}\left(B\hat{x}^{k-2} + C\left(\hat{x}^{k-1}\right)\hat{x}^{k-1} + Du^{k-1} + Eu^k\right) \tag{3.22}$$

According to Kowalczuk and Tatara (2013), by defining an aggregated state vector $\tilde{x}^k = \left[\hat{x}^{k\,T}\ \hat{x}^{k-1\,T}\right]^T$ and an augmented input vector, $\tilde{u}^k = \left[u^{k\,T}\ u^{k-1\,T}\right]^T$, the flow process model can be rewritten in the form of state-space equation:

$$\tilde{x}^k = A_c\tilde{x}^{k-1} + B_c\tilde{u}^k, \tag{3.23}$$

where

$$A_c = \left[\begin{array}{c|c} A^{-1}C\left(\hat{x}^{k-1}\right) & A^{-1}B \\ \hline I & 0 \end{array}\right] \tag{3.24}$$

$$B_c = \left[\begin{array}{c|c} A^{-1}E & A^{-1}D \\ \hline 0 & 0 \end{array}\right] \tag{3.25}$$

Note that matrix A_c is a function of the state vector $\tilde{x}^{k-1}$, and matrix B_c is dependent on the friction factor λ. Thus, A_c must be recalculated during the operation of the algorithm. Since (3.23) represents a regular state-space form, it can be analyzed with well-known methods from control theory, but the recombination matrix A must be inverted at each Δt. For this task, the following matrix inversion lemma (MIL) can be applied at each time step (Brogan 1991):

$$A^{-1} = \left[\begin{array}{c|c} A_1' & A_2' \\ \hline A_3' & A_4' \end{array}\right] = \left[\begin{array}{c|c} (A_1 - A_2A_4^{-1}A_3)^{-1} & -A_1^{-1}A_2(A_4 - A_3A_1^{-1}A_2)^{-1} \\ \hline -A_4^{-1}A_3(A_1 - A_2A_4^{-1}A_3)^{-1} & (A_4 - A_3A_1^{-1}A_2)^{-1} \end{array}\right] \tag{3.26}$$

3.5 Selection of the Discretization Grid

The numerical stability problem in discretized differential equations is connected with the choice of discretization grid, which determines information's propagation speed between the nodes of the algorithm. According to Strikwerda (2007), the Courant–Friedrichs–Lewy condition (CFL) establishes that the propagation speed of information in a numerical algorithm $\frac{\Delta z}{\Delta t}$ must be greater or equal to the information exchange velocity in the corresponding differential equations for maintaining stability. This is only a necessary condition, but it is not sufficient and gives only a lower boundary for the velocity. At the same time, discretization must allow online simulation. Since the highest speed present in the flow process is the sound velocity ν, the following inequality must be satisfied:

$$\frac{\Delta z}{\Delta t} \geqslant \nu \tag{3.27}$$

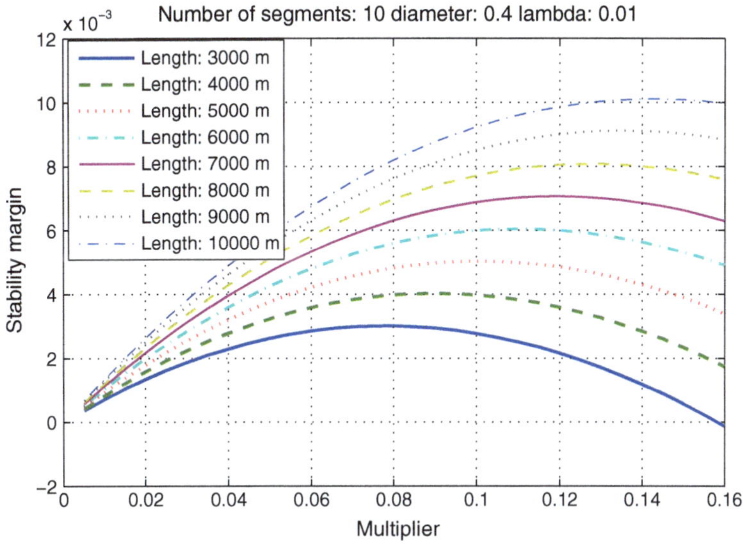

Fig. 3.2 Distribution of the stability margin versus coefficient μ for eight lengths of the pipeline shown in $[m]$. Experimental setup: $N = 10$, $D = 0.4$ $[m]$, $\lambda = 0.01$, $\nu = 304$ $[\frac{m}{s}]$, $p_{inlet} = 3.2$ $[MPa]$, and $p_{outlet} = 3.0$ $[MPa]$

In addition, rewriting it as an equality, one obtains

$$\Delta t = \mu \frac{\Delta z}{\nu}, \tag{3.28}$$

where μ is a coefficient within the range $(0, 1\rangle$, binding the discretization steps.

According to Kowalczuk and Tatara (2016), there exists a coefficient μ_{opt}, which maximizes the stability margin s_m for the discrete-time system on the z-plane (domain of the Z-transform) for the specific physical parameters of a pipeline. The stability margin is calculated as

$$s_m = 1 - e_{max}, \tag{3.29}$$

where e_{max} is the largest absolute eigenvalue of the state transition matrix A_c.

The distribution of the stability margin versus the *stability tuner*, μ, for a few values of a pipeline's length is shown in Fig. 3.2.

The eigenvalues are calculated after the transition response to diminish the influence of initial values on the stability margin. As one can see from Fig. 3.2, the stability margin is in the magnitude order of 10^{-2}, which emphasizes the need for a detailed numeric stability analysis. Such a system can be easily destabilized, and special attention must be paid during the discretization phase.

To find a maximum stability margin for a physical flow parameter, such as pipeline dimension or fluid properties, the influence of each of them on the estimation of μ_{opt} must be studied. In particular, numerical optimization was performed by using the Hooke and Jeeves (1961) algorithm to find the values of μ_{opt} (the dependent variable) for specified flow parameters. Six experiments were conducted where one of the parameters was an independent variable, and the others were fixed.

The toolbox *curve fitting tool* (cftool) of MathWorks (2012) was used to fit the curve to the experimental data. By assuming the common fitting curve

$$\mu_{opt} = C_1 p_p^{C_2}, \tag{3.30}$$

where C_1 and C_2 are the coefficients to be optimized and p_p is the physical flow parameter which could be: N, D, L, λ, the mean pressure p_m in the pipeline and the pressure drop p_m along the pipeline. The quality of the fitting is evaluated via the R-squared coefficient of determination R^2. The closer it is to 1, the better the curve fits the experimental data (Walpole et al. 2012). The adjusted functions are presented in Fig. 3.3, where each subplot refers to a different physical parameter as the independent variable.

By interpreting the identified parameters of (3.30) for a global case as follows: (1) the term $p_p^{C_2}$ combined into the single parameter ξ which includes all the physical parameters with a common power $\frac{1}{2}$; and (2) the individual scaling factor C_1 which is united in a global scaling coefficient C_3; the simple optimal stability adjustment is proposed

$$\mu_{opt} = C_3 \sqrt{\frac{p_d L \lambda}{p_m D} \frac{1}{N}} = C_3 \xi, \tag{3.31}$$

where C_3 is the sought coefficient. For this task, the MATLAB's cftool is used to determine the value of the global scaling coefficient C_3 in (3.31).

The data set for the experiment is generated by 10 pipelines with randomly produced parameters. The fitting result is shown in Fig. 3.4, where the experimental data, aggregated to the single parameter ξ, are indicated by the symbol $\times$. The fitted value of $C_3 = 0.36$ is characterized by the coefficient of determination $R^2 = 0.97$. Note in Fig. 3.4 that some experimental data are slightly displaced with respect to the fitting curve. The reason for this should be attributed to the most 'severe' mathematical operation consisting in rounding the power coefficient 0.44 (the case reflected in Fig. 3.3c) to 0.5 (the case implemented by Eq. (3.31)). In both cases, the issue of fitting the μ_{opt} to the pressure drop along the pipeline is considered. Nevertheless, the relatively high value of the computed determination coefficient R^2 shows that the curve fits the experimental data satisfactorily, and thus the obtained expression can be practically used for approximating μ_{opt}.

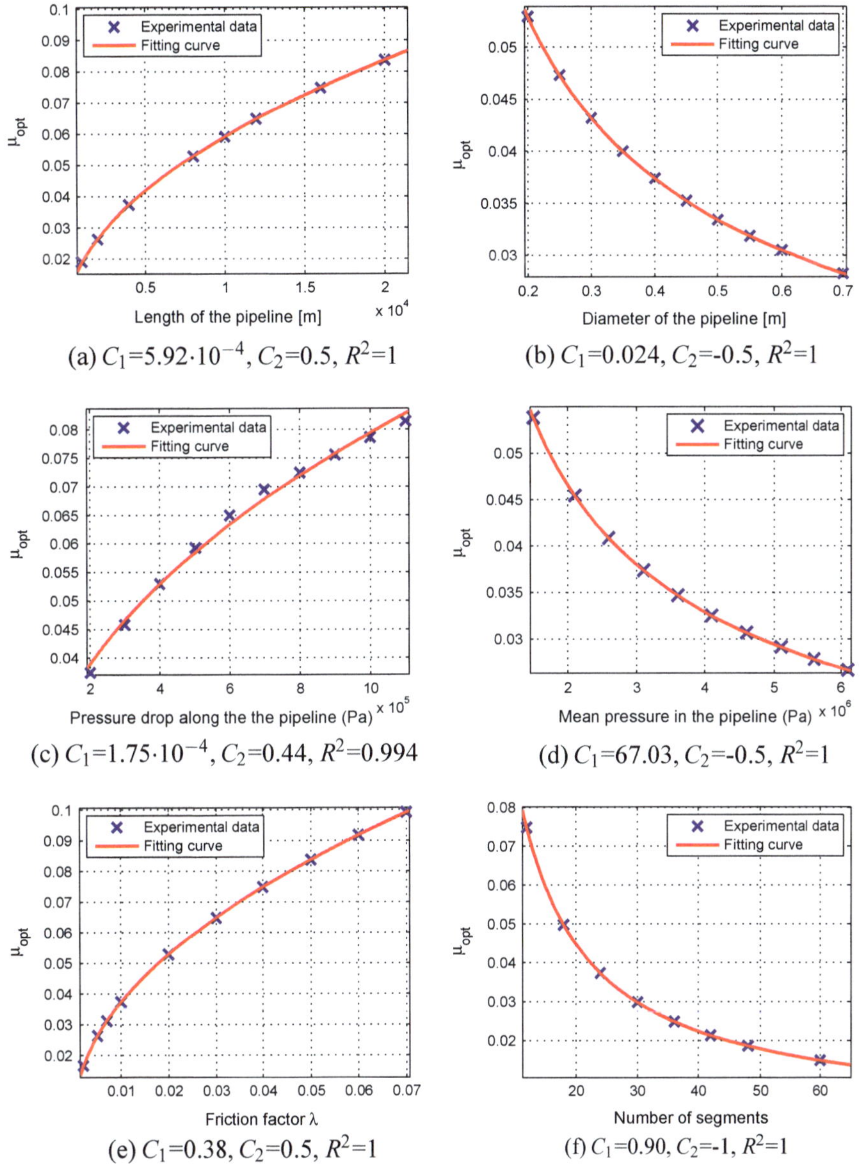

(a) C_1=5.92·10^{-4}, C_2=0.5, R^2=1

(b) C_1=0.024, C_2=-0.5, R^2=1

(c) C_1=1.75·10^{-4}, C_2=0.44, R^2=0.994

(d) C_1=67.03, C_2=-0.5, R^2=1

(e) C_1=0.38, C_2=0.5, R^2=1

(f) C_1=0.90, C_2=-1, R^2=1

Fig. 3.3 Results of adjusted function (3.30) for μ_{opt} by varying the following: **a** length of the pipeline, **b** diameter of the pipeline, **c** difference between the inlet and the outlet pressure, **d** mean pressure in the pipeline, **e** friction factor and **f** number of the pipeline's segments. The used fixed parameters (other than the applied argument) are: N = 12, D = 0.4 [m], λ = 0.01, v = 304 [$\frac{m}{s}$], p_{inlet} = 4.1 [MPa], and p_{outlet} = 3.9 [MPa]

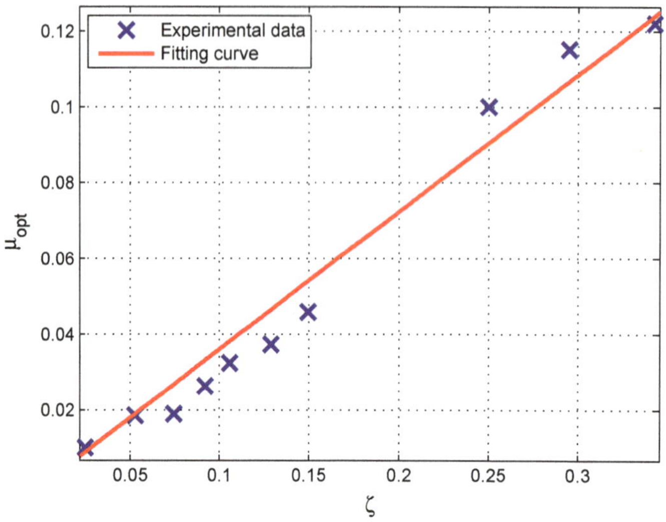

Fig. 3.4 Adjusting of (3.31) to the experimental data

Thus the formula connecting the time and spatial steps with the maximal stability margin is

$$\mu_{opt} = 0.36\sqrt{\frac{p_d L\lambda}{p_m D}\frac{1}{N}} \tag{3.32}$$

This stability margin is applicable to models (3.7) and (3.23). For models constructed with other assumptions, (3.32) does not guarantee stability; however, the optimization process can be reproduced analogically, obtaining the result fitted to the analyzed model.

3.6 Analytic Inversion of the Recombination Matrix

In the case of flow process model, there are two main reasons for applying efficient computational analytic inversion matrix methods. The first relates to the Gauss–Jordan elimination method, which has the time complexity $O(n^3)$, which implies that the computation time strongly increases with the dimension of the model. The second is caused by a property of the recombination matrix which has a ratio of its largest to its smallest singular values of order 10^9, and as a consequence matrix A is very sensitive to numerical errors.

The above facts motivated the inverse problem formulation within the framework of the tridiagonal matrix with the matrix A divided into four submatrices

$$A = \left[\begin{array}{c|c} A_1 & A_2 \\ \hline A_3 & A_4 \end{array}\right] = \left[\begin{array}{cccccc|cccccc} & & & & & & 2b & 0 & \cdots & 0 & 0 \\ & & & & & & -b & b & \cdots & 0 & 0 \\ & & D_{\frac{N}{2}+1}(c) & & & & \vdots & & \ddots & & \vdots \\ & & & & & & 0 & 0 & \cdots & -b & b \\ & & & & & & 0 & 0 & \cdots & 0 & -2b \\ \hline -b & b & 0 \cdots & 0 & 0 & 0 & & & & & \\ 0 & -b & b \cdots & 0 & 0 & 0 & & & & & \\ \vdots & & \ddots & & & \vdots & & & D_{\frac{N}{2}}(a) & & \\ 0 & 0 & 0 \cdots & -b & b & 0 & & & & & \\ 0 & 0 & 0 \cdots & 0 & -b & b & & & & & \end{array}\right] \tag{3.33}$$

Since matrix A consists of two diagonal submatrices and two submatrices having elements only on two diagonals, by (3.26) the matrix A^{-1} can be represented as

$$A^{-1} = \left[\begin{array}{c|c} A_1 & A_2 \\ \hline A_3 & A_4 \end{array}\right]^{-1} = \left[\begin{array}{c|c} A_1' & A_2' \\ \hline A_3' & A_4' \end{array}\right] \tag{3.34}$$

with

$$A_1' = \left(D_{\frac{N}{2}+1}(c) - A_2 D_{\frac{N}{2}}(a^{-1}) A_3\right)^{-1} \tag{3.35}$$

$$A_4' = \left(D_{\frac{N}{2}}(a) - A_3 D_{\frac{N}{2}+1}\left(c^{-1}\right) A_2\right)^{-1} \tag{3.36}$$

$$A_2' = D_{\frac{N}{2}+1}\left(-c^{-1}\right) A_2 A_4' \tag{3.37}$$

$$A_3' = D_{\frac{N}{2}}\left(-a^{-1}\right) A_3 A_1' \tag{3.38}$$

From the set of matrices, one can see that A_2' and A_3' are dependent on matrices A_4' and A_1', which have to be determined first. Thus, (3.35) and (3.36) can be directly calculated which results in the following tridiagonal matrices:

$$A_1' = \begin{bmatrix} c + \frac{2b^2}{a} & -\frac{2b^2}{a} & 0 & \cdots & 0 & 0 & 0 \\ -\frac{b^2}{a} & c + \frac{2b^2}{a} & -\frac{b^2}{a} & \cdots & 0 & 0 & 0 \\ \vdots & & & \ddots & & & \vdots \\ 0 & 0 & 0 & \cdots & -\frac{b^2}{a} & c + \frac{2b^2}{a} & -\frac{b^2}{a} \\ 0 & 0 & 0 & \cdots & 0 & -\frac{2b^2}{a} & c + \frac{2b^2}{a} \end{bmatrix}^{-1} \tag{3.39}$$

$$A_4' = \begin{bmatrix} a+\frac{3b^2}{c} & -\frac{b^2}{c} & 0 & \cdots & 0 & 0 & 0 \\ -\frac{b^2}{c} & a+\frac{2b^2}{c} & -\frac{b^2}{c} & \cdots & 0 & 0 & 0 \\ \vdots & & & \ddots & & & \vdots \\ 0 & 0 & 0 & \cdots & -\frac{b^2}{c} & a+\frac{2b^2}{c} & -\frac{b^2}{c} \\ 0 & 0 & 0 & \cdots & 0 & -\frac{b^2}{c} & a+\frac{3b^2}{c} \end{bmatrix}^{-1} \tag{3.40}$$

3.6.1 Tridiagonal Matrix Inversion Method

Da Fonseca and Petronilho (2001) determined the analytical expression of the inverse matrix for a general tridiagonal matrix with the general structure

$$T = \begin{bmatrix} \alpha_1 & \beta_1 & & & 0 \\ \gamma_1 & \alpha_2 & \beta_2 & & \\ & \gamma_2 & \ddots & \ddots & \\ & & \ddots & \ddots & \beta_{n-1} \\ 0 & & & \gamma_{n-1} & \alpha_n \end{bmatrix} \tag{3.41}$$

The authors showed that the coefficients of T^{-1} are given by

$$T^{-1}(i,j) = t_{i,j} = \begin{cases} \frac{(-1)^{i+j}\beta_i\cdots\beta_{j-1}\theta_{i-1}\phi_{j+1}}{\theta_n} & \text{for } i < j \\ \frac{\theta_{i-1}\phi_{j+1}}{\theta_n} & \text{for } i = j \\ \frac{(-1)^{i+j}\gamma_j\cdots\gamma_{i-1}\theta_{j-1}\phi_{i+1}}{\theta_n} & \text{for } i > j \end{cases} \tag{3.42}$$

with the parameters θ_i and ϕ_i calculated recursively as

$$\theta_i = \alpha_i\theta_{i-1} - \beta_{i-1}\gamma_{i-1}\theta_{i-2} \quad \text{for } i = 2, 3, \ldots, n; \quad \theta_0 = 1, \quad \theta_1 = \alpha_1 \tag{3.43}$$

$$\phi_i = \alpha_i\phi_{i+1} - \beta_i\gamma_i\phi_{i+2} \quad \text{for } i = n-1, n-2, \ldots, 1; \quad \phi_{n+1} = 1, \quad \phi_n = \alpha_n \tag{3.44}$$

Because the matrix (3.39) has a centrosymmetric structure, $\phi_i = \theta_{n+1-i}$, and its inverse is also centrosymmetric, thus, only half of the coefficients of A_1' must be calculated by using (3.42). Therefore, the general matrix is obtained:

$$A'_1 = \begin{bmatrix} a'_{1,1} & a'_{1,2} & \cdots & a'_{1,n-1} & a'_{1,n} \\ a'_{n-1,n} & a'_{2,2} & & a'_{2,n-1} & a'_{2,n} \\ \vdots & & \ddots & & \vdots \\ a'_{2,n} & a'_{2,n-1} & & a'_{2,2} & a'_{n-1,n} \\ a'_{1,n} & a'_{1,n-1} & & a'_{1,2} & a'_{1,1} \end{bmatrix} \tag{3.45}$$

Similar reasoning applies to (3.40). Moreover, this matrix is symmetric not only centrosymmetric. Thus, the number of elements to be calculated is only $n^2/4$. Hence, matrix A'_4 has the following form

$$A'_4 = \begin{bmatrix} a''_{1,1} & a''_{1,2} & \cdots & a''_{1,n-1} & a''_{1,n} \\ a''_{1,2} & a''_{2,2} & & a''_{2,n-1} & a''_{1,n-1} \\ \vdots & & \ddots & & \vdots \\ a''_{1,n-1} & a''_{2,n-1} & & a''_{2,2} & a''_{1,2} \\ a''_{1,n} & a''_{1,n-1} & \cdots & a''_{1,2} & a''_{1,1}, \end{bmatrix} \tag{3.46}$$

where the values of $a''_{i,j}$ are calculated with (3.42). Once the matrices A'_1 and A'_4 are evaluated, one can use them to calculate (3.37) and (3.38), resulting in

$$A'_2 = \frac{b}{c} \left[\begin{array}{c|c|c|c|c} -2a''_{1,1} & -2a''_{1,2} & \cdots & -2a''_{1,n-1} & -2a''_{1,n} \\ a''_{1,1} - a''_{1,2} & a''_{1,2} - a''_{2,2} & \cdots & a''_{1,n-1} - a''_{2,n-1} & a''_{1,n} - a''_{1,n-1} \\ \vdots & & \ddots & & \vdots \\ a''_{1,n-1} - a''_{1,n} & a''_{2,n-1} - a''_{1,n-1} & \cdots & a''_{2,2} - a''_{1,2} & a''_{1,2} - a''_{1,1} \\ 2a''_{1,n} & 2a''_{1,n-1} & \cdots & 2a''_{1,2} & 2a''_{1,1} \end{array} \right] \tag{3.47}$$

$$A'_3 = \frac{b}{a} \left[\begin{array}{c|c|c|c|c} a'_{1,1} - a'_{n-1,n} & a'_{1,2} - a'_{2,2} & \cdots & a'_{1,n-1} - a'_{2,n-1} & a'_{1,n} - a'_{2,n} \\ a'_{n-1,n} - a'_{n-2,n} & a'_{2,2} - a'_{n-2,n-1} & \cdots & a'_{2,n-1} - a'_{3,n-1} & a'_{2,n} - a'_{3,n} \\ \vdots & & \ddots & & \vdots \\ a'_{3,n} - a'_{2,n} & a'_{3,n-1} - a'_{2,n-1} & \cdots & a'_{n-2,n-1} - a'_{2,2} & a'_{n-2,n} - a'_{n-1,n} \\ a'_{2,n} - a'_{1,n} & a'_{2,n-1} - a'_{1,n-1} & \cdots & a'_{2,2} - a'_{1,2} & a'_{n-1,n} - a'_{1,1} \end{array} \right] \tag{3.48}$$

Matrices A'_2 and A'_3 are anti-centrosymmetric, which reduces again the number of the required mathematical operations. Thus, the four submatrices which form matrix A^{-1} are known and can be assembled into one. The state-space model expressed in term of the A^{-1} will be called the analytical model of band matrix inversion (AMBMI). A disadvantage of this method is that for high-order systems, the inverse cannot be computed because some θ_i and ψ_i have values higher than the computer's numerical representation range (over 10^{300}). Therefore, this model is recommended only for a lower number of segments.

3.6.2 Diagonal Approximation Model

To avoid the disadvantage of the analytical inversion matrix, Kowalczuk and Tatara (2016) suggested approximating the tridiagonal matrices (3.39) and (3.40) with their diagonal counterparts. To make such approximation feasible, the coefficients' values on the subdiagonal and superdiagonal must be significantly lower than the ones on the main diagonal. This sufficient condition satisfies if

$$|c| \gg \left|4\frac{b^2}{a}\right| \tag{3.49}$$

By substituting physical parameters of the pipeline in (3.49), the inequality

$$|\frac{3}{2\mathscr{A}\,\Delta t}| \gg |4\frac{1}{16\Delta z^2}\frac{2\nu^2\Delta t}{3\mathscr{A}}| \tag{3.50}$$

is obtained. Since Δt, Δz and ν are positive, the above equation can be simplified to

$$\Delta t^2 \ll 9\frac{\Delta z^2}{\nu^2} \tag{3.51}$$

which is equivalent to

$$\Delta t \ll 3\frac{\Delta z}{\nu} \tag{3.52}$$

By incorporating the CFL condition (3.27) in (3.52), the restriction on μ is reduced to

$$\mu \ll 3 \tag{3.53}$$

By considering the practical assumption that μ must be at least two orders of magnitude fewer than the value obtained with (3.53), one suggests the following stability tuner's condition

$$\mu < 0.03 \tag{3.54}$$

By comparing inequality (3.54) with the Courant–Friedrichs–Lewy result discussed by Dick (2012), who says that μ must be lower than 0.90 to 0.95 for stationary flows (depending on flow parameters), it is clear that (3.54) is more restrictive because it has been obtained by considering an approximated model. Moreover, the CFL criterion solely defines a necessary condition; thus the values may not assure numerical stability.

By considering (3.54), the two submatrices of the inverted recombination matrix can be approximated as

$$A_1' \approx \breve{A}_1' = \begin{bmatrix} c+\frac{2b^2}{a} & 0 & 0\cdots 0 & 0 & 0 \\ 0 & c+\frac{2b^2}{a} & 0\cdots 0 & 0 & 0 \\ \vdots & & \ddots & & \vdots \\ 0 & 0 & 0\cdots 0 & c+\frac{2b^2}{a} & 0 \\ 0 & 0 & 0\cdots 0 & 0 & c+\frac{2b^2}{a} \end{bmatrix}^{-1} \tag{3.55}$$

and

$$A_4' \approx \breve{A}_4' = \begin{bmatrix} a+\frac{3b^2}{c} & 0 & \cdots\ 0\ 0 & 0 & \\ 0 & a+\frac{2b^2}{c} & 0\ \cdots 0 & 0 & 0 \\ \vdots & & \ddots & & \vdots \\ 0 & 0 & 0\ \cdots 0 & a+\frac{2b^2}{c} & 0 \\ 0 & 0 & 0\ \cdots 0 & 0 & a+\frac{3b^2}{c} \end{bmatrix}^{-1} \tag{3.56}$$

The inverses of these matrices are obtained by substituting their reciprocals on the main diagonal. Therefore, the inverse recombination matrix

$$A^{-1} \approx \left[\begin{array}{cccccc|cccccc} \frac{a}{\sigma} & 0 & \cdots & & 0 & 0 & -\frac{2b}{3b^2+ca} & 0 & \cdots & & 0 & 0 \\ 0 & \frac{a}{\sigma} & & & 0 & 0 & \frac{b}{3b^2+ca} & -\frac{b}{\sigma} & 0 & & 0 & 0 \\ & & & & & & 0 & \frac{b}{\sigma} & -\frac{b}{\sigma} & & 0 & 0 \\ \vdots & & \ddots & & & \vdots & \vdots & & \ddots & & & \vdots \\ & & & & & & 0 & 0 & & \frac{b}{\sigma} & -\frac{b}{\sigma} & 0 \\ 0 & 0 & & & \frac{a}{\sigma} & 0 & 0 & 0 & & 0 & \frac{b}{\sigma} & -\frac{b}{3b^2+ca} \\ 0 & 0 & \cdots & & 0 & \frac{a}{\sigma} & 0 & 0 & \cdots & & 0 & \frac{2b}{3b^2+ca} \\ \hline \frac{b}{\sigma} & -\frac{b}{\sigma} & 0 & \cdots & 0 & 0 & \frac{c}{3b^2+ca} & 0 & \cdots & & 0 & 0 \\ 0 & \frac{b}{\sigma} & -\frac{b}{\sigma} & & 0 & 0 & 0 & \frac{c}{\sigma} & & & 0 & 0 \\ 0 & 0 & \frac{b}{\sigma} & & 0 & 0 & & & & & & \\ \vdots & & \ddots & & & \vdots & \vdots & & \ddots & & & \vdots \\ 0 & 0 & & -\frac{b}{\sigma} & 0 & 0 & & & & & & \\ 0 & 0 & & \frac{b}{\sigma} & -\frac{b}{\sigma} & 0 & 0 & 0 & & & \frac{c}{\sigma} & 0 \\ 0 & 0 & \cdots & 0 & \frac{b}{\sigma} & -\frac{b}{\sigma} & 0 & 0 & \cdots & & 0 & \frac{c}{3b^2+ca} \end{array}\right] \tag{3.57}$$

is obtained by using (3.26) with $\sigma = 2b^2 + ca$. This matrix constitutes an explicit form of the inverted recombination matrix. Since this inversion is performed after the estimation of the friction factor which could change with the operation condition, or during the initialization, the inversion algorithm consumes short computational time. Since (3.57) is now a sparse matrix, another advantage is that the online computation of (3.24) and (3.25), and the premultiplication by a sparse matrix is not time-consuming. The model obtained with this inverse method for the recombination matrix is called the analytic model of diagonal approximation (AMDA). Certainly, the model is an approximation of the base model; thus its accuracy and calculation speed must be examined.

3.6.2.1 Note on the Model's Dimension

By considering the maximum stability margin given in (3.32) with the CFL condition (3.54) for the AMDA model, one deduces that

$$\mu_{opt} = 0.36\sqrt{\frac{p_d L\lambda}{p_m D}\frac{1}{N}} < 0.03 \tag{3.58}$$

Thus, if the above condition is satisfied, the AMDA model is a good approximation for the base model (3.7), and the stability tuner μ_{opt} for AMDA can be computed by using (3.32). The formula (3.58) can be rewritten in a practical form, which gives us information about the minimal cardinality of the AMDA model as

$$N > 12\sqrt{\frac{p_d L\lambda}{p_m D}} \tag{3.59}$$

As a consequence, if a properly designed AMDA's dimension is selected beyond its minimum, one can apply (3.32) to calculate μ_{opt}, which, in turn, satisfies (3.58) and assures a maximum stability margin.

Note that μ_{opt} differs slightly from model to model. When the accuracy of the inversion improves, however, the optimal value of the coefficient for the AMDA model approaches the value for the principal base model; certainly, the segmentation (3.59) must be rounded up to the nearest even integer.

3.7 Analysis of the Models

Based on the two methods AMBMI and AMDA for inverting the recombination matrix which have been presented in the previous section, one can say that the former method generates values beyond the acceptable computer representation range, making it impossible to compare accuracy and convergence with other models from

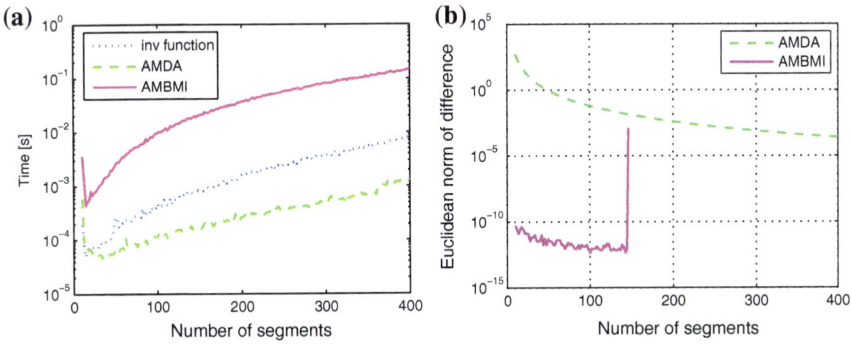

Fig. 3.5 Comparison of AMDA and AMBMI models: **a** inversion time; continuous line is used for the AMBMI method, the discontinuous line for the AMDA model, and the dots line for the MATLAB inv function; **b** accuracy measure, the continuous line denotes the norm of the difference between the AMBMI and inv function, and dashed line denotes the difference between the functions AMDA and inv

a practical point of view. On the other hand, the AMDA model for lower (rough) segmentation produces an inverse which is not similar to the numerically inverted matrix considering the Euclidean norm of the variables as a measure. For this reason, computation accuracy and speed will be assessed in detail only for the AMDA model, if one compares it with the base model. The inversion time for the three methods has been evaluated with respect to the model's dimension, and the results are shown in Fig. 3.5a. The inversion accuracy is assessed in terms of the Euclidean norm of the difference between the matrices inverted by the AMDA or AMBMI methods and the MATLAB inv function. The results are shown in Fig. 3.5b. The experimental data used are $L = 4000$ [m], $\nu = 304.23$ [$\frac{m}{s}$], $D = 0.4$ [m], and $\lambda = 0.01$. Each calculation was repeated 10 times, and the results are averaged.

A more detailed analysis has been conducted for the AMDA model as compared to the inv function. The models are analyzed in terms of the Euclidean norm of the difference between the two inverted matrices (which should ideally be equal to zero). The inversion time of the matrix gained from the two methods has been presented as a function of μ and the number of segments N. Finally, the models have been compared in Fig. 3.6 in terms of the mean squared error between the mass flow estimates (at the inlet and outlet) obtained by both of them.

Note that in Fig. 3.6 the axes in subplots (a, c, e,) have a log–log scale with N ranging from 10 to 400. For subplots (b, d, f) N=100 and μ is calculated by using (3.32). Subplots a and b describe with a continuous line the Euclidean norm of the difference between the inverted recombination matrices by AMDA and by inv functions; the dashed line is the Euclidean norm of the difference between state transition matrices obtained by AMDA and inv functions. The subplots c and d describe the numerical inversion time results with continuous line for the inv function,

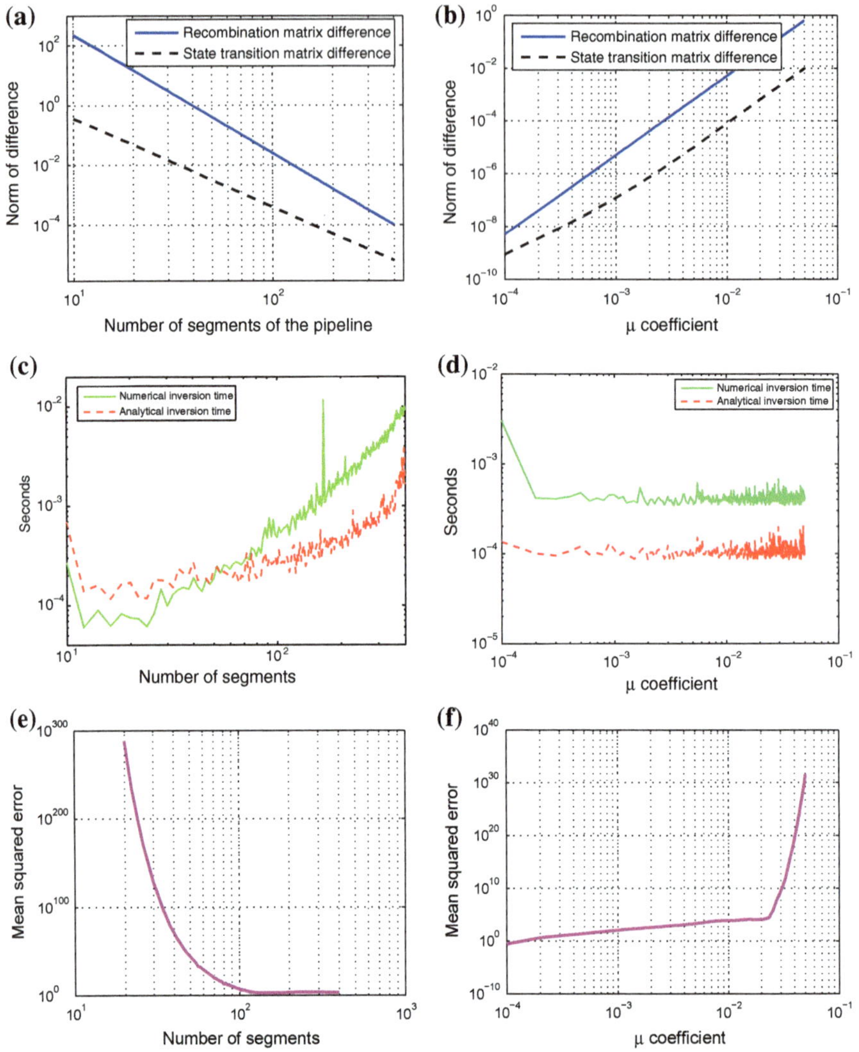

Fig. 3.6 Numerical results of the AMDA inversion method as compared to the MATLAB inv function for a pipeline with L=2000 [m], v=304.23 [$\frac{m}{s}$], D=0.4 [m], p_{inlet}=3.2 [MPa], p_{outlet}=3.0 [MPa], and $\lambda = 0.1$

and dashed line for the analytic inversion time (AMDA). Finally, subplots e and f show the result of the mean quadratic difference between mass flow estimated by the two analyzed models. All the presented results were obtained with mean values of 10 runs.

3.8 Conclusions

In the case of inversion time, the AMDA outperforms other methods for almost each segmentation, while AMBMI is the slowest one. Note, however, that the MATLAB `inv` function is effectively optimized, which may not be available in a field computer running the diagnostic algorithm. Note also that for $N \approx 140$ segments, the AMBMI model is not computable at all, though, because of the lower segmentation it results in a better approximation of inversion. The AMBMI method is, however, limited to a lower order due to the high values of the auxiliary variables. The upper boundary for the use of this model depends on the physical parameters of the flow; therefore it is difficult to give a number. A safe upper boundary seems to be $N \approx 50$, however.

As seen in Figs. 3.6a, b, the norm difference between the two recombination matrices decreases by decreasing the coefficient μ or by increasing the order of the model. Since the state transition matrices depend on the recombination matrix, the norm difference is smaller in every case. The inversion time is almost independent from the coefficient μ for both methods; however, increasing the model's order increases the time necessary for inversion. For lower segmentation, the analytical approach is slower, but for the number of segments above approximately 50, the analytic approach outperforms the numerical one. Figures 3.6e, f show the mean quadratic difference between the two models. Clearly, for lower segmentation, the difference level is significant and decreases when one increases segmentation. Above 100 segments, the error line becomes flat, and this region seems to constitute the preferred operation range for the model. In the case of the coefficient μ, the error is almost flat for $\mu < 0.02$ and rapidly increases with μ above this limit.

In summary, the recommendations for using the AMDA model are that the number of segments should be higher than 100, for the following reasons: to benefit from the low quadratic error of the mass flow estimates and from faster inversion, while the coefficient μ should remain below 0.02. The model should also be tested in an actual LDI algorithm, however, to assess the resulting accuracy of the target estimates of leak parameters.

References

Belsito, S., Lombardi, P., Andreussi, P., & Banerjee, S. (1998). Leak detection in liquefied gas pipelines by artificial neural networks. *AIChE*, *44*(12), 2675–2688.

Billmann, L., & Isermann, R. (1987). Leak detection methods for pipelines. *Automatica*, *23*(3), 381–385.

Brogan, W. (1991). *Modern control theory*. Boston: Prentice Hall.

Da Fonseca, C., & Petronilho, J. (2001). Explicit inverses of some tridiagonal matrices. *Linear Algebra and its Applications*, *325*, 7–21.

Dick, M. (2012). Stabilization of the gas flow in networks: Boundary feedback stabilization of quasilinear hyperbolic systems on networks. Ph.D. thesis. Erlangen-Nürnberg: Friedrich-Alexander-Universität.

Gunawickrama, K. (2001). Leak detection methods for transmission pipelines. Ph.D. thesis. Gdańsk: Gdańsk Univeristy of Technology.

Hooke, R., & Jeeves, T. A. (1961). Direct search solution of numerical and statistical problems. *Journal of the Association for Cumputing Machinery*, *8*(2), 212–229.

Kowalczuk, Z., & Gunawickrama, K. (1998). Detection of leakages in industry pipelines using a cross-correlation approach. *Pomiary Automatyka Kontrola*, *44*(4), 140–146.

Kowalczuk, Z., & Gunawickrama, K. (2004). Detecting and locating leaks in transmission pipelines. In J. Korbicz, J. M. Koœcielny, Z. Kowalczuk, & W. Cholewa (Eds.), *Fault Diagnosis. Models, Artificial Intelligence, Applications*, chapter 21 (pp. 821–864). Berlin, Heidelberg, New York: Springer.

Kowalczuk, Z. & Tatara, M. (2013). Analytical modeling of flow processes: Analysis of computability of a state-space model. In *XI International Conference on Diagnostics of Processes and Systems* (pp. 74.1–12). £agów, Lubuski.

Kowalczuk, Z. & Tatara, M. (2016). Approximate models and parameter analysis of the flow process in transmission pipelines. In Z. Kowalczuk (Ed.), *Advanced and Intelligent Computations in Diagnosis and Control. Advances in Intelligent Systems and Computing* (Vol. AISC 386, pp. 209–220). Cham, Heidelberg, New York, Dordrecht, London: Springer. doi:10.1007/978-3-319-23180-8_17

Kreyszig, E. (2006). *Advanced engineering mathematics* (Vol. 9). Columbus: John Wiley and Sons Inc.

MathWorks. (2012). *MATLAB and Curve Fitting Toolbox Release 2012b*. Matlab, Natick, Massachusetts, United States: Technical Report.

Reddy, H., Narasimhan, S., Bhallamudi, S. M., & Bairagi, S. (2011). Leak detection in gas pipeline networks using an efficient state estimator. Part-I: Theory and simulations. *Computers and Chemical Engineering*, *35*(4), 651–661.

Strikwerda, J. (2007). Finite difference schemes and partial differential equations. SIAM.

Torres, L., Besançon, G., & Verde, C. (2012). Leak detection using parameter identification. In *8th IFAC symposium SAFEPROCESS-2012*. Mexico City, Mexico.

Verde, C., & Torres, L. (2015). Referenced model based observers for locating leaks in a branched pipeline. In *9th International Federation of Automatic Control (IFAC) Symposium SAFEPROCESS* (pp. 1066–1071). Paris: IFAC.

Walpole, R., Myers, R., Myers, S., & Ye, K. (2012). *Probability and statistics for engineers and scientists* (9th ed.). Boston: Prentice Hall.

Chapter 4
One-Dimensional Modeling of Pipeline Transients

Jean François Dulhoste, Marcos Guillén, Gildas Besançon and Rafael Santos

Abstract This chapter summarizes the one-dimensional modeling of transients in a pipeline, commonly used for detection and location of faults (such as leaks and obstructions) by means of model-based methods. The modeling starts with the discretization via finite-difference method of classical water hammer equations. The result of such a discretization is a system of ordinary differential equations, which is considered together with boundary conditions that represent faults and pipeline accessories. Some illustrative results are finally given based on a test bench.

4.1 Introduction

Fault detection and isolation (FDI) in real pipelines remains an important challenge for proper distribution of fluids, specifically in all aspects related to the opportune diagnosis of leaks and obstructions. Many of the FDI techniques rely on models which are themselves based on the fluid dynamics in the pipeline. Some of these techniques have been presented in Billman and Isermann (1984); Benkherouf and Allidina (1988); Verde (2005); Besançon et al. (2007) to cite a few. In the present chapter, the purpose is to summarize the modeling approach commonly used for the design of FDI techniques based on dynamical models. First, *water hammer* equations which represent the fluid transient phenomena in pipelines (Roberson et al.

J.F. Dulhoste (✉) · M. Guillén · R. Santos
Grupo Termofluidodinámica, Dpto. de Ciencias Térmicas, Escuela de Ing. Mecánica, Universidad de Los Andes Mérida, Mérida, Venezuela
e-mail: djean@ula.ve

M. Guillén
e-mail: marcosg@ula.ve

R. Santos
e-mail: rsantosl@ula.ve

G. Besançon
Université Grenoble Alpes, CNRS, GIPSA-Lab, Grenoble 38000, France
e-mail: gildas.besancon@grenoble-inp.fr

C. Verde and L. Torres (eds.), *Modeling and Monitoring of Pipelines and Networks*, Applied Condition Monitoring 7,
DOI 10.1007/978-3-319-55944-5_4

1997) are recalled in Sect. 4.2. In Sect. 4.3, the friction modeling is discussed, which is important since most of the FDI algorithms are based on the assumption of a constant friction coefficient, while friction in a pipeline in fact depends on the flow (Potter and Wiggert 2001). While various numerical methods can be considered for solving equations like water hammer ones (Fletcher 2006), Sect. 4.4 focuses on the discretization by means of finite-difference method. Section 4.5 subsequently introduces two types of pipeline faults: leaks and obstructions. For each of them, a model based on hydrodynamics is developed (Çengel and Cimbala 2006). In Sect. 4.6, the possibility of adding some accessory equations to the model is considered for more accuracy representing hydraulic transients in the presence of a pipe fault (Guillén et al. 2014). Finally, some application examples of pipeline modeling with faults are given in Sect. 4.7, before concluding in Sect. 4.8.

4.2 Water Hammer Equations

For incompressible fluid flows, the classical model that describes the nonstationary phenomena in a pipeline is expressed by the well-known water hammer (WH) equations, which are a couple of nonlinear hyperbolic partial differential equations that represent the continuity and the movement of the fluid in a pipeline. The WH equations are given by

$$
\begin{aligned}
&\frac{\partial H(z,t)}{\partial t} + \frac{c^2}{gA}\frac{\partial Q(z,t)}{\partial z} = 0 \\
&\frac{1}{A}\frac{\partial Q(z,t)}{\partial t} + g\frac{\partial H(z,t)}{dz} + J(Q(z,t)) = 0,
\end{aligned} \tag{4.1}
$$

where z denotes the one-dimensional spatial coordinate $[m]$, t is the time $[s]$, $H(z,t)$ is the hydraulic head $[m]$, $Q(z,t)$ is the fluid flow $[m^3/s]$, A is the cross-sectional area $[m^2]$, g is the gravity acceleration $[m/s^2]$, c is the fluid pressure wave speed $[m/s]$, and J is the friction losses affecting the fluid dynamics within the pipe.

Classically, J is expressed under a stationary form given by

$$
J = J_s = \frac{fQ|Q|}{2DA^2}, \tag{4.2}
$$

where f corresponds to the Darcy-Weisbach friction coefficient and D to the diameter.

4.3 Friction Modeling

In most of model-based fault detection and isolation approaches, coefficient f is considered as a constant, even if sometimes it is updated when a leak is detected. This coefficient, however, is actually known to depend on the so-called Reynolds number Re and the rugosity coefficient of the pipe e. The implicit Colebrook equation describes this coefficient value for a pipe with a circular section of diameter D as

$$\frac{1}{\sqrt{f}} = 0.86 \ln \left(\frac{1}{3.7} \frac{e}{D} + \frac{2.51}{Re\sqrt{f}} \right) \tag{4.3}$$

where the Reynolds number can be calculated with

$$Re = \frac{\rho V D}{\mu} = \frac{4\rho Q}{\pi D \mu} \tag{4.4}$$

for ρ as the fluid density and μ as the fluid viscosity.

This equation cannot easily be implemented. Therefore, an approximate explicit formulation, known as the Swamee–Jain equation, can be used instead:

$$f = 1.325 \left\{ ln \left[0.27 \left(\frac{e}{D} \right) + 5.74 \left(\frac{1}{Re} \right)^{0.9} \right] \right\}^{-2}. \tag{4.5}$$

This equation is valid for $10^{-8} < e/D < 0.01$ and $5000 < Re < 10^8$. For an even more complete friction modeling, some *non-stationary* friction losses J_u can also be added (namely $J = J_s + J_u$ in (4.1)) according to the model below:

$$J_u = \frac{k}{2A} \left(\frac{\partial Q}{\partial t} + c\Phi_A \left| \frac{\partial Q}{\partial z} \right| \right), \tag{4.6}$$

where $\Phi_A = sgn(Q)$ and k denotes the Brunone coefficient, which can be calculated with

$$\begin{aligned} k &= \sqrt{0.0476}/2, \quad \text{or} \\ k &= \left(\sqrt{\frac{7.41}{Re^{\log(14.3/Re^{0.05})}}} \right) /2 \end{aligned} \tag{4.7}$$

for laminar flow and turbulent flow, respectively.

With the two terms of friction losses, pipeline Eq. (4.1) becomes

$$\begin{aligned} \frac{\partial H}{\partial t} &= -\frac{c^2}{gA} \frac{\partial Q}{\partial z} \\ \frac{\partial Q}{\partial t} &= \frac{1}{(2+k)} \left(-2gA \frac{\partial H}{dz} - c\Phi_A k \left| \frac{\partial Q}{\partial z} \right| - \frac{fQ|Q|}{DA} \right), \end{aligned} \tag{4.8}$$

where coefficients f and k can be calculated at any time. Omitting nonstationary friction losses simply means $k = 0$.

Dulhoste et al. (2011) have shown that for leak detection and isolation (LDI), the better formulation is the use of a variable friction but without nonstationary losses. The model with a constant friction coefficient is limited to nearly constant flow applications, and therefore it is not recommended for real systems, while the non-stationary formulation makes the LDI structure more complex, without significantly improving leak estimations. For this reason, in the rest of the chapter we will use the recommended variable friction. More precisely, we use the following:

$$\begin{aligned}\frac{\partial H}{\partial t} &= -\frac{c^2}{gA}\frac{\partial Q}{\partial z}\\ \frac{\partial Q}{\partial t} &= -gA\frac{\partial H}{dz} - \frac{fQ|Q|}{2DA}.\end{aligned} \tag{4.9}$$

4.4 Finite-Difference Discretization

There are several forms of finite-difference discretization. Here explicit discretization is considered (Chaudhry 2014): the pipe length (which is continuous) is approximated using explicit discrete spatial sections.

To perform this discretization the pipe is divided into N sections, as shown in Fig. 4.1,

and approximations are made as follows:

- $H(z,t) \cong H_i(t)$: pressure at space abscissa z_i;
- $Q(z,t) \cong Q_i(t)$: flow at space abscissa z_i;
- $\dfrac{\partial H(z,t)}{\partial t} \cong \dot{H}_i(t)$: temporal variation of the pressure at space abscissa z_i; and
- $\dfrac{\partial Q(z,t)}{\partial t} \cong \dot{Q}_i(t)$: temporal variation of the flow at space abscissa z_i.

Partial derivatives with respect to z are approximated by first-order finite differences, either backward or forward, according to the chosen boundary conditions:

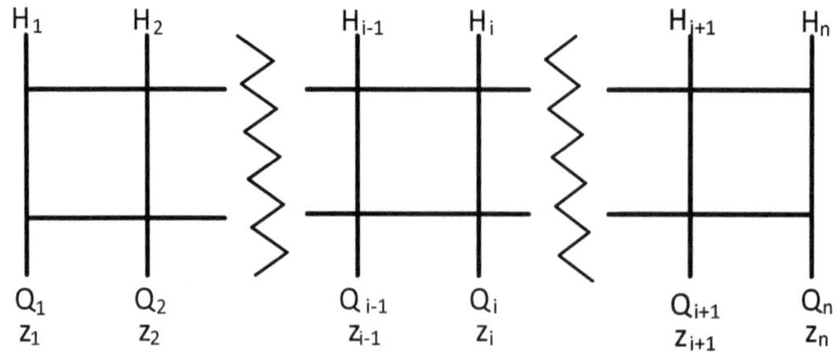

Fig. 4.1 Scheme of finite-difference discretization where index i refers to pipe section i

$$\frac{\partial H(z,t)}{\partial z} \approx \frac{\Delta H_i(t)}{\Delta z_i} = \frac{H_{i+1} - H_i}{\Delta z_i} \text{ or } \frac{H_i - H_{i-1}}{\Delta z_i}$$
$$\frac{\partial Q(z,t)}{\partial z} \approx \frac{\Delta Q_i(t)}{\Delta z_i} = \frac{Q_{i+1} - Q_i}{\Delta z_i} \text{ or } \frac{Q_i - Q_{i-1}}{\Delta z_i}. \tag{4.10}$$

Regarding the boundary conditions, they can be chosen among the following: inlet pressure H_{in}, outlet pressure H_{out}, inlet flow Q_{in}, and outlet flow Q_{out}.

Notice that other discretization schemes could be considered, such as the collocation-based one (Torres et al. 2008), which are not discussed here.

4.5 Fault Models

In this section, some fault models in pipelines, as they can be used in FDI applications, are presented. Leaks and obstructions are more particularly considered.

4.5.1 Leak Modeling

A mathematical expression for a leak can be deduced from Bernoulli's equation, which relates the pressure difference between the pipeline inside and outside. Bernoulli's equation applies under the following considerations (Çengel and Cimbala 2006):

- Nonviscous flow,
- Permanent or continuous flow $\left(\frac{\partial V}{\partial t} = 0\right)$,
- Along the power line $\left(a_s = V\frac{\partial V}{\partial s}\right)$,
- Constant density $\left(\frac{\partial \rho}{\partial s} = 0\right)$.

where V is the velocity of the fluid, a_s is the acceleration of the particle of the fluid in the direction s, and ρ is the density of the fluid.

Let us consider a hole in a pipe as shown in Fig. 4.2. Bernoulli's equation between points 1 and 2 is given by

$$\frac{P_1}{\gamma} + \frac{V_1^2}{2g} + Z_1 = \frac{P_2}{\gamma} + \frac{V_2^2}{2g} + Z_2, \tag{4.11}$$

where P is the height or pressure of the fluid load, V is the speed of the fluid, and Z is the elevation.

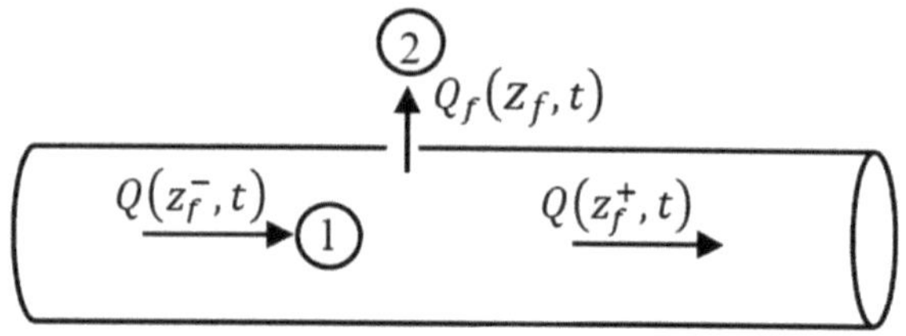

Fig. 4.2 Scheme of a hole in a pipe

By taking into account that $Z_1 - Z_2 = 0$, $P_2 = P_{atm} = 0$ and $H = \frac{P_1}{\gamma} + \frac{V_1{}^2}{2g}$, (4.11) reduces to

$$V_2 = \sqrt{2gH}. \tag{4.12}$$

This represents the theoretical speed of the fluid outside the hole. Since the flow at such a point is given by the product of the output velocity and the hole area, we get

$$Q_t = A\sqrt{2gH}, \tag{4.13}$$

where Q_t is the theoretical flow rate through the hole and A is the hole area.

In practice, the fluid leakage also generates speed losses, which can be taken into account via a multiplicative factor known as the discharge coefficient c_d, giving a flow rate of the form

$$Q = c_d A\ \sqrt{2gH}. \tag{4.14}$$

To simplify the expression, constants are gathered into a single global coefficient of discharge F expressed as follows:

$$F = c_d A\sqrt{2g}. \tag{4.15}$$

Finally, the expression for the flow of a leak in a pipe at a point z_f is given by

$$Q_f(z_f, t) = F\sqrt{H(z_f, t)} \tag{4.16}$$

with $F \geq 0$.

4.5.2 *Obstruction Modeling*

Blockages are also common faults in pipes and pipeline networks. They could be formed by the accumulation of the transported fluid or they can be caused by the partial closure of a valve. Blockages reduce the efficiency of pipeline systems, but can also cause serious damage in the security of the plant. Hereafter three types of blockages in a pipeline are presented.

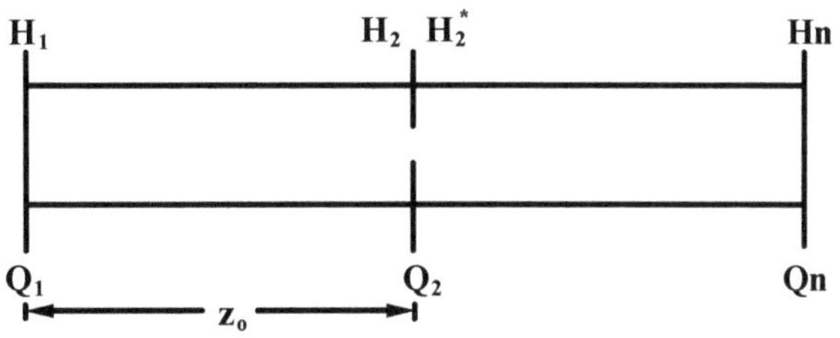

Fig. 4.3 Partial blockage

4.5.2.1 Modeling of a Partial Blockage

Figure 4.3 represents a blockage that occurs in the central area of a pipe. Pressure H_2 changes because of the blockage and such a changed pressure are given by H_2^*. If Bernoulli's and continuity equations are both applied between point 2 (before the blockage) and point 2* (after the blockage), one has

$$H_2 + \frac{(V_2)^2}{2g} + Z_2 + h_W - h_L = H_2^* + \frac{(V_2^*)^2}{2g} + Z_2^* \tag{4.17}$$

where H is the pressure head of the fluid, V is the fluid flow speed, ρ is the density of the fluid, Z is the elevation at points 2 and 2*, h_W represents energy inputs by pumps or turbines, and h_L represents pressure losses in the pipeline section. As levels Z_2 and Z_2^* are equal, and energy inputs by pumps and/or turbines are zero (in this case), then one gets

$$H_2 + \frac{(V_2)^2}{2g} - h_L = H_2^* + \frac{(V_2^*)^2}{2g}. \tag{4.18}$$

By continuity of conservation of mass,

$$\rho_2 V_2 A = \rho V_2^* A_o, \tag{4.19}$$

where A is the cross-sectional area of the pipe and A_o is the cross-sectional area at the blockage. Therefore, $V_2^* = \frac{V_2\ A}{A_o}$. Using expression (4.18) in conjunction with (4.19), one gets

$$H_2^* = H_2 + \frac{(V_2)^2}{2g}\left(1 - \left(\frac{A}{A_o}\right)^2\right) - h_L, \tag{4.20}$$

which, expressed in terms of the volumetric flow $Q_2 = V_2 A$, becomes

$$H_2^* = H_2 + \frac{Q_2^2}{2gA^2}\left(1 - \left(\frac{A}{A_o}\right)^2\right) - h_L. \tag{4.21}$$

It may be difficult to analytically obtain h_L, since it is a function of the flow and the blockage geometry. Therefore, for practical purposes, the following expression that includes a discharge coefficient C_d can be used:

$$H_2^* = H_2 + \frac{Q_2^2}{2gC_dA^2}\left(1 - \left(\frac{A}{A_o}\right)^2\right). \tag{4.22}$$

Both coefficients C_d and h_L require additional information to be numerically determined. This additional information could be, for example, the shape and length of geometry changes.

4.5.2.2 Modeling of a Blockage Stretch

A blocked stretch is a reduced cross-sectional area of the pipe with significant length l_o. This is illustrated in Fig. 4.4.

To model the blocked stretch, the blockage area, denoted by A_o, is quantified as a percentage of the pipeline area A. Therefore, constant A must be replaced by A_o in the equations of the section that is blocked. This is done as follows:

$$\begin{aligned}
\dot{H}_0 &= -\frac{c^2}{gA_o}\frac{Q_{0+1} - Q_0}{l_o} \\
\dot{Q}_{0+1} &= -A_o g\frac{H_{0+1} - H_0}{l_o} - \frac{f_{0+1}}{2DA_o}Q_{0+1}\left|Q_{0+1}\right| \\
\dot{H}_{0+1} &= -\frac{c^2}{gA}\frac{Q_{0+2} - Q_{0+1}}{z_{0+2} - z_{0+1}} \\
\dot{Q}_{0+2} &= -Ag\frac{H_{0+2} - H_{0+1}}{z_{0+2} - z_{0+1}} - \frac{f_{0+2}}{2DA}Q_{0+2}\left|Q_{0+2}\right|.
\end{aligned} \tag{4.23}$$

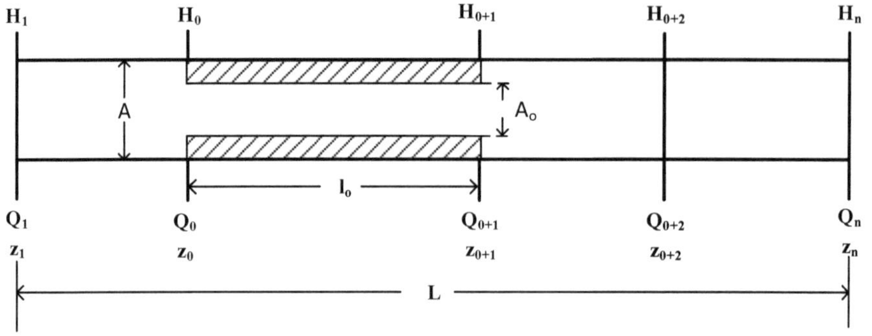

Fig. 4.4 Scheme of a blocked stretch

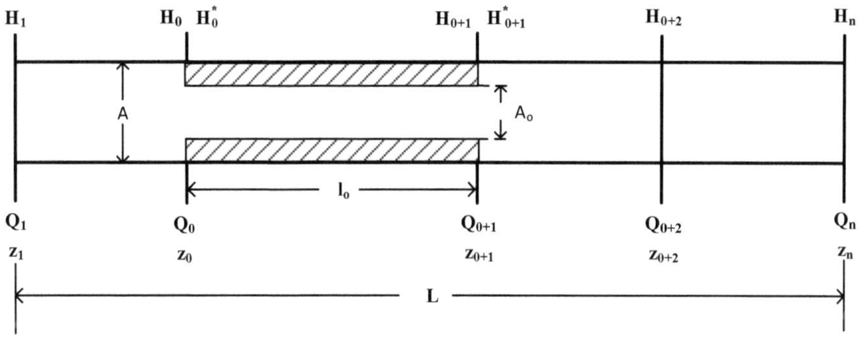

Fig. 4.5 Scheme of the contraction and expansion phenomena

4.5.2.3 Modeling of a Blockage with Contraction and Expansion Phenomena

This model takes into account the two phenomena occurring at the two edges of the blocked section: a contraction appearing upstream of the blockage (from A to A_o) and an expansion occurring downstream (from A_o to A), as shown in Fig. 4.5.

The following equations describe the dynamics of pressures and flows before blockage (H_0, Q_{0+1}) and after it (H_{0+1}, Q_{0+2}) respectively:

$$\begin{aligned}
\dot{H}_0 &= -\frac{c^2}{gA}\frac{Q_{0+1}-Q_0}{l_o}\\
\dot{Q}_{0+1} &= -Ag\frac{H_{0+1}-H_0^*}{l_o} - \frac{f_{0+1}}{2DA}Q_{0+1}\left|Q_{0+1}\right|\\
\dot{H}_{0+1} &= -\frac{c^2}{gA}\frac{Q_{0+2}-Q_{0+1}}{z_{0+2}-z_{0+1}}\\
\dot{Q}_{0+2} &= -Ag\frac{H_{0+2}-H_{0+1}^*}{z_{0+2}-z_{0+1}} - \frac{f_{0+2}}{2DA}Q_{0+2}\left|Q_{0+2}\right| ,
\end{aligned} \tag{4.24}$$

where H_0^* and H_{0+1}^* are the pressures at the beginning and just after the blockage, respectively, which are calculated using Bernoulli's and continuity equation

$$\begin{aligned}
H_0^* &= H_0 + \frac{Q_0^2}{2gA^2}\left[1-\left(\frac{A}{A_o}\right)^2\right]\\
H_{0+1}^* &= H_{0+1} + \frac{Q_{0+1}^2}{2gA_o^2}\left[1-\left(\frac{A_o}{A}\right)^2\right] .
\end{aligned} \tag{4.25}$$

4.6 Boundary Conditions

The solution of discretized models, by any method, requires defining some boundary conditions, which represent known values of the variables at the border of the studied domain. Water hammer equations require two boundary conditions. They can only be pressures, only flows, or a combination of both. The choice of boundary conditions changes the structure of the models and can even vary the number of equations in which the discretized model is subdivided.

Figure 4.6 represents the possible boundary conditions in a scheme of a discretized pipeline.

The figure clearly shows that when boundary conditions are of the same variable type (pressure or flow) on both sides, there will be fewer components to be computed for that type of variable in the discretized scheme than for the other one. Conversely, if the boundary conditions are made from two variables of different types, then the discretized model will include the same number of components for both types of variables.

The selection of the boundary condition in some cases can be based on the model that one wants to attain; in other cases, it can be based on known or measurable variables in the physical system. In real systems, boundary conditions are given by the elements available in the hydraulic system.

For example, let us consider the pipeline scheme shown in Fig. 4.7.

As can be observed in the figure, this pipeline has a pump that raises the pressure from the tank input to the pump output, and then has a series of accessories (valves,

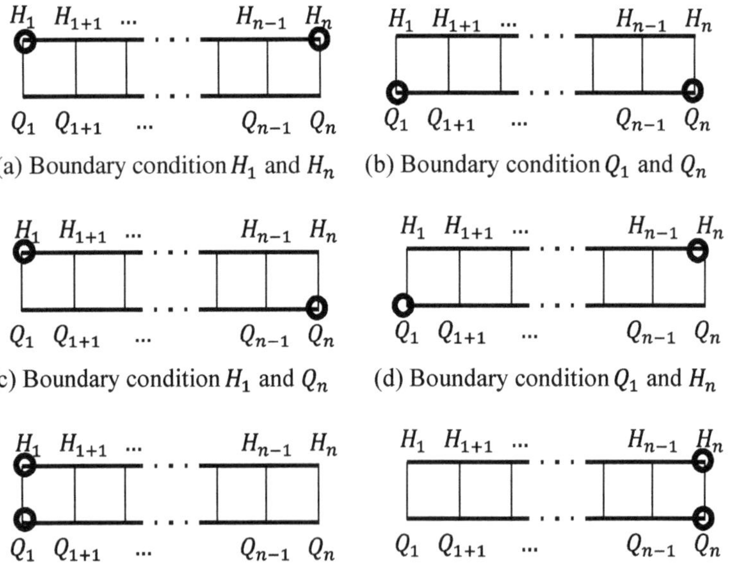

(a) Boundary condition H_1 and H_n (b) Boundary condition Q_1 and Q_n

(c) Boundary condition H_1 and Q_n (d) Boundary condition Q_1 and H_n

(e) Boundary condition H_1 and Q_1 (f) Boundary condition Q_n and H_n

Fig. 4.6 Possible boundary conditions

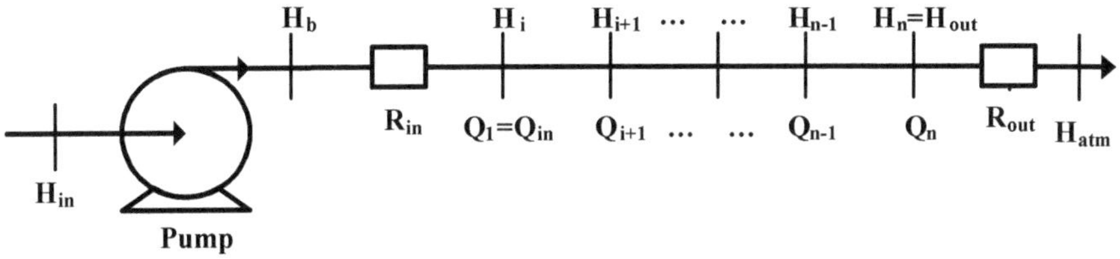

Fig. 4.7 Pipeline system

elbows, reductions, etc.) that act as a restriction to the flow. Adding these accessories allows the simulation of the behavior of the inflow and outflow of the pipe when a fault is present in the pipe.

The pump equation can be approximated as follows:

$$H_b - H_{in} = C_2 Q_1^2 + C_1 Q_1 + C_0, \tag{4.26}$$

where $H_b - H_{in} = H_M$ is the pressure head of the pump, H_b is the pressure at the exit of the pump, H_{in} is the pressure at the pump suction, Q_1 is the inlet flow of the pipe, and C_0, C_1, C_2 are constant coefficients that depend on the pump.

The hydraulic constraint at the input of the pipe can be modeled by the following expression:

$$R_{in} {Q_1}^2 = H_b - H_i, \tag{4.27}$$

where R_{in} is the restriction coefficient and depends on the loss of pressure that occurs between the download of the pump and the start of the pipeline. When the pump described by (4.26) as well as the hydraulic restriction represented by (4.27) is added to the pipe model, the boundary condition varies depending on the pressure. In the case when the flow (Q_1) is considered as an upstream boundary condition, this gives the following expression:

$$Q_1 = \frac{-C_1 + \sqrt{{C_1}^2 - 4\,(R_{in} + C_0)\,(-C_2 + H_1 - H_{in})}}{2(R_{in} + C_0)}, \tag{4.28}$$

where H_{in} depends on the conditions of the tank, while H_1 is calculated at each step in simulation.

The hydraulic restriction at the outlet of the pipe can be similarly modeled as the one at the input:

$$R_{out} Q_n^2 = H_n - H_{atm}, \tag{4.29}$$

where R_{out} is the coefficient of restriction and depends on the pressure loss occurring at the outlet of the pipe. In the case when H_n is used as a boundary condition, it must be calculated according to the following expression:

$$H_n = H_{atm} + R_{out} Q_n^2 . \quad (4.30)$$

Now H_{atm} is the new condition of border, and H_n is calculated at each simulation step.

4.7 Application Examples

To illustrate the formerly presented possible modeling approaches, three examples are given here: a model based on pressure head boundary conditions in both pipe sides, a model with flow rate boundary condition upstream and pressure head downstream, and a third one including pump and restriction models. In each case, a 50-section discretized model is used for simulations, and three leaks appearing at three different times and three different positions along the pipe are considered.

The simulated behaviors of boundary pressures and flows are compared with experimental ones, obtained on a test bench with the following parameters: pipe length $L = 85\,[\mathrm{m}]$, sound speed $c = 407.75\,[\mathrm{m/s}]$, and diameter $D = 0.0635\,[\mathrm{m}]$. This test bench is available at the CINVESTAV Guadalajara, Mexico, and its full technical description can be found in the work of Begovich et al. (2012).

In all plots, simulated boundary pressures are denoted by $H_{in},\ H_{out}$ and $Q_{in},\ Q_{out}$ and experimental ones by $h_{in},\ h_{out}$ and $q_{in},\ q_{out}$.

4.7.1 *Example 1: Modeling with Two Pressure Boundary Conditions*

In this subsection, let us present a model with upstream pressure $H_1 = H_{in}$ and downstream pressure $H_n = H_{out}$ as the boundary conditions, considering upstream flow Q_1 and downstream flow Q_n as outputs; see Fig. 4.8.

This model has been used by several authors for the detection and location of leaks because of its simplicity (Torres et al. 2011; Castro Burgos and Valdés-González 2009; Besançon et al. 2007; Verde 2005), and since it complies with certain conditions for observability. The equations of such a model are

$$\dot{H}_{i+1} = -\frac{c^2}{gA}\frac{(Q_{i+1} - Q_i)}{(z_{i+1} - z_i)} \quad \forall i = 1, ..., n-2 \quad (4.31)$$

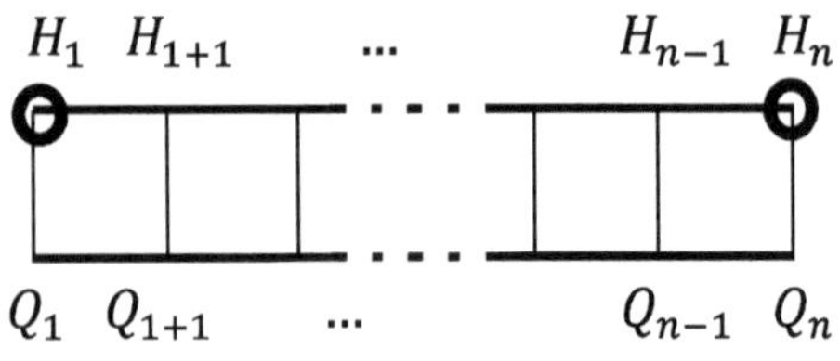

Fig. 4.8 Discretization scheme for the model of example 1

$$\dot{Q}_i = -gA\frac{(H_{i+1} - H_i)}{(z_{i+1} - z_i)} - \frac{f}{2DA}Q_{i+1}\left|Q_{i+1}\right| \quad \forall i = 1, ..., n\,. \tag{4.32}$$

In order to consider a leak in the pipe, the leak expression should be added to the equation of mass conservation (4.31):

$$Q_i = Q_i^* + F\sqrt{H_i}\,. \tag{4.33}$$

Here Q_i represents the flow at point i, Q_i^* represents the flow after the leak, and F represents the coefficient that allows determining the leak flow depending on the internal pressure. Thus, the pipe model taking into account the leak effect is

$$\dot{H}_{i+1} = -\frac{c^2}{gA}\frac{\left(Q_{i+1} - (Q_i - F\sqrt{H_i})\right)}{(z_{i+1} - z_i)} \quad \forall i = 1, ..., n-2 \tag{4.34}$$

$$\dot{Q}_i = -gA\frac{(H_{i+1} - H_i)}{(z_{i+1} - z_i)} - \frac{f}{2DA}Q_{i+1}\left|Q_{i+1}\right| \quad \forall i = 1, ..., n. \tag{4.35}$$

The final model will depend on the chosen section number, or equivalently, the node number denoted by n. For instance, if one takes $n = 5$, three equations are obtained for H, and five for Q, but with only four of them being independent. Hence, if we introduce the leak term in the node 3, we obtain a system as

$$\begin{aligned}
\dot{H}_2 &= -\frac{c^2}{gA}\frac{(Q_2 - Q_1)}{(z_2 - z_1)}\\
\dot{H}_3 &= -\frac{c^2}{gA}\frac{(Q_3 - Q_2)}{(z_3 - z_2)}\\
\dot{H}_4 &= -\frac{c^2}{gA}\frac{\left(Q_4 - (Q_3 - F\sqrt{H_3})\right)}{(z_4 - z_3)}\\
\dot{Q}_1 &= -gA\frac{(H_2 - H_1)}{(z_2 - z_1)} - \frac{f}{2DA}Q_2\left|Q_2\right|\\
\dot{Q}_2 &= -gA\frac{(H_2 - H_2)}{(z_3 - z_2)} - \frac{f}{2DA}Q_3\left|Q_3\right|\\
\dot{Q}_3 &= -gA\frac{(H_4 - H_3)}{(z_4 - z_3)} - \frac{f}{2DA}Q_4\left|Q_4\right|\\
\dot{Q}_4 &= -gA\frac{(H_5 - H_4)}{(z_5 - z_4)} - \frac{f}{2DA}Q_5\left|Q_5\right|\,.
\end{aligned} \tag{4.36}$$

Notice that the equation for Q_5 cannot be used because there is no value for Q_6, which is necessary in the considered forward difference. If one would like to write an equation for Q_5, a change to *backward difference* would be needed, giving rise to the same equation as that of Q_4.

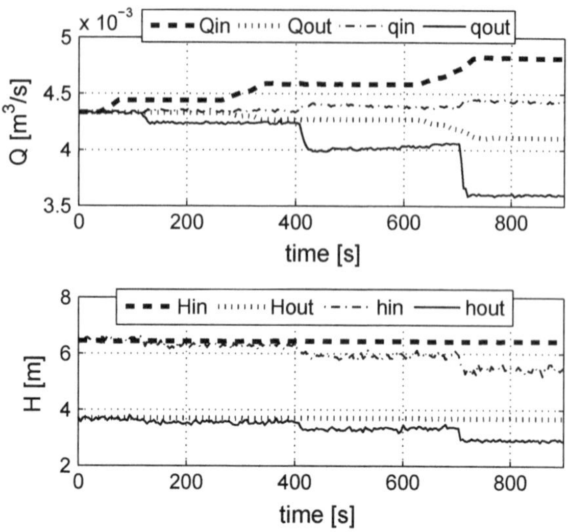

Fig. 4.9 Simulation and experiment results for example 1

Notice also that the expression for the leak can be included in any of the nodes of the discretization, and depending on the number of nodes, one may place the leak in any position.

Finally, the sizes of the sections do not have to be identical in such a model. For this example, if we take the same size for the four sections, the leak at node 3 is placed at the center of the pipe. If the size of the first two sections is half of the last two though, the leak at node 3 is in fact placed at one-third of the pipe's length.

As explained in the section introduction, to illustrate the behavior of a model built in this way, we consider the pipe test bench of Begovich et al. (2012), with three leaks along the pipeline, and a 50-section model. The obtained behavior for pressures and flows at the boundaries both in simulation (notations H, Q) and in real experiments (notations h, q) are presented in Fig. 4.9. In this figure, one can appreciate how Q_{out} is decreasing as leaks occur in the pipeline. In addition, a growth can be seen in Q_{in}, which is because one of the boundary conditions is the inlet pressure. Changes in simulated pressures are not observed, because boundary conditions are introduced as constants, which is a behavior that, however, deviates from the real behavior of a pipe.

4.7.2 Example 2: Modeling with Flow-Pressure Boundary Conditions

In this second example, we present a model with upstream flow rate $Q_1 = Q_{in}$ and downstream pressure head $H_n = H_{out}$ as boundary conditions, leaving upstream pressure head H_1 and downstream flow rate Q_n as outputs (Fig. 4.10).

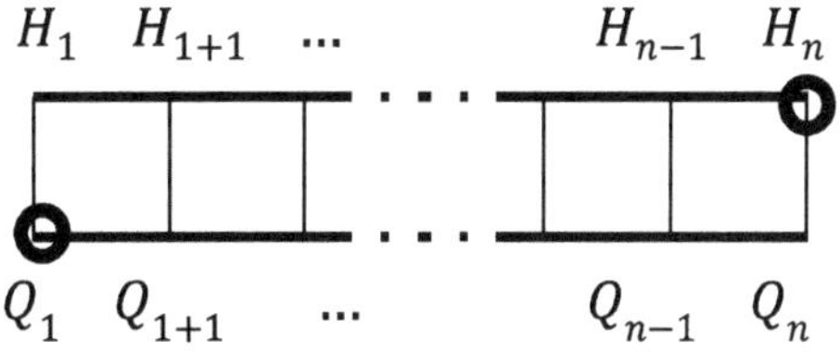

Fig. 4.10 Discretization scheme for model of example 2

Because of the selected boundary conditions, the first element Q_1 of the flow and the last element H_n of pressure are known, and we thus must consider a forward discretization for H equation and a backward discretization for Q equation. With the leak expression at each node of the pipe, the discretized system would be then as follows:

$$\frac{\partial H_i}{\partial t} = -\frac{c^2}{gA}\frac{\left(Q_{i+1} - (Q_i - F\sqrt{H_i})\right)}{\left(z_{i+1} - z_i\right)} \quad \forall i = 1, ..., n-1 \tag{4.37}$$

$$\frac{\partial Q_i}{\partial t} = -Ag\frac{(H_i - H_{i-1})}{(z_i - z_{i-1})} - \frac{f_i}{2DA} Q_i \left|Q_i\right| \quad \forall i = 2, ..., n, \tag{4.38}$$

with Q_1 and H_n as entries or known signals and H_1 and Q_n as outputs of the system; n is still the number of nodes in the pipe and subscript i refers to the node position in the pipe.

In the same way as in example 1, the final model then depends on the chosen number of nodes n, and if we take $n = 5$ for instance, we obtain four equations for H and four equations for Q. With a leak in node 3, we obtain a model as follows:

$$\begin{aligned}
\frac{\partial H_1}{\partial t} &= -\frac{c^2}{gA}\frac{(Q_2 - Q_1)}{(z_2 - z_1)} \\
\frac{\partial H_2}{\partial t} &= -\frac{c^2}{gA}\frac{(Q_3 - Q_2)}{(z_3 - z_2)} \\
\frac{\partial H_3}{\partial t} &= -\frac{c^2}{gA}\frac{\left(Q_4 - (Q_3 - F\sqrt{H_3})\right)}{(z_4 - z_3)} \\
\frac{\partial H_4}{\partial t} &= -\frac{c^2}{gA}\frac{(Q_5 - Q_4)}{(z_5 - z_4)} \\
\frac{\partial Q_2}{\partial t} &= -Ag\frac{(H_2 - H_1)}{(z_2 - z_1)} - \frac{f_2}{2DA} Q_2 \left|Q_2\right| \\
\frac{\partial Q_3}{\partial t} &= -Ag\frac{(H_3 - H_2)}{(z_3 - z_2)} - \frac{f_3}{2DA} Q_3 \left|Q_3\right| \\
\frac{\partial Q_4}{\partial t} &= -Ag\frac{(H_4 - H_3)}{(z_4 - z_3)} - \frac{f_4}{2DA} Q_4 \left|Q_4\right| \\
\frac{\partial Q_5}{\partial t} &= -Ag\frac{(H_5 - H_4)}{(z_5 - z_4)} - \frac{f_5}{2DA} Q_5 \left|Q_5\right| .
\end{aligned} \tag{4.39}$$

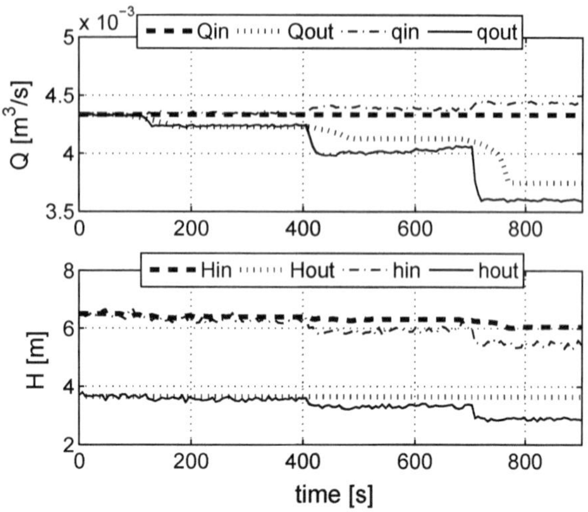

Fig. 4.11 Simulation and experiment results for example 2

Figure 4.11 shows the results obtained in this example, when simulated with 50 sections, and the same parameters as in example 1. Here one can again appreciate how the output flow Q_{out} diminishes as leaks occur, which is an expected result and natural behavior. On the other hand, the inlet pressure H_{in} decreases as a result of the leaks. Constant input flow and output pressure, however, corresponding to simulated boundary conditions, deviate from the real behavior in the pipe, which should present a flow increase at the entrance and pressure decrease at the outlet.

4.7.3 Example 3: Modeling with Flow-Pressure Boundary Conditions and Pump-Restriction Models

In this third example, we again present a model with upstream flow rate $Q_1 = Q_{in}$ and downstream pressure head $H_n = H_{out}$ as boundary conditions, with H_1 and Q_n as outputs (as in Fig. 4.10). We now include, however, an equation for the pump and the input and output restrictions, as shown in Fig. 4.7. This gives the following equations:

$$\frac{\partial \dot{H}_1}{\partial t} = -\frac{c^2}{gA}\frac{\left(Q_2 - \left(\frac{-C_1+\sqrt{C_1{}^2-4(R_{in}+C_0)(-C_2+H_1-H_{in})}}{2(R_{in}+C_0)}\right)\right)}{(z_2 - z_1)} \tag{4.40}$$

$$\frac{\partial H_i}{\partial t} = -\frac{c^2}{gA}\frac{\left(Q_{i+1} - (Q_i - F\sqrt{H_i})\right)}{\left(z_{i+1} - z_i\right)} \quad \forall i = 2, ..., n-1 \tag{4.41}$$

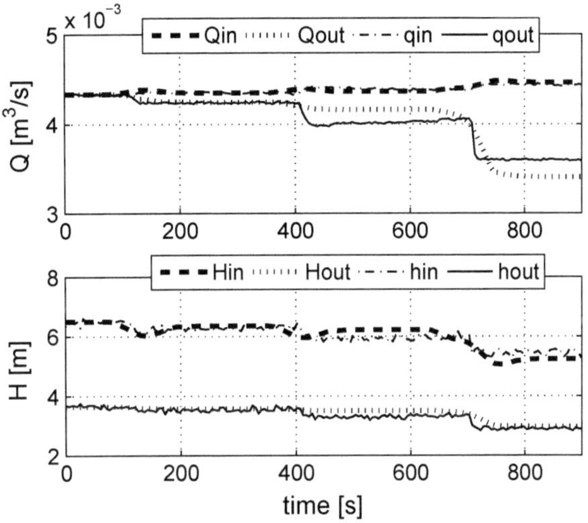

Fig. 4.12 Simulation and experiment results for example 3

$$\frac{\partial Q_i}{\partial t} = -Ag\frac{(H_i - H_{i-1})}{(z_i - z_{i-1})} - \frac{f_i}{2DA}Q_i\,|Q_i|\ \forall i = 2, ..., n-1 \tag{4.42}$$

$$\frac{\partial Q_n}{\partial t} = -Ag\frac{\left(\left(R_{sal}Q_n^2 + H_{atm}\right) - H_{n-1}\right)}{(z_n - z_{n-1})} - \frac{f_n}{2DA}Q_n\,|Q_n|\ . \tag{4.43}$$

As in the former examples, the final model depends on the number of nodes n, and again with $n = 5$ for instance, we obtain four equations for H and four equations for Q. Finally with a leak in node 3, we obtain the following model:

$$\begin{aligned}
\frac{\partial H_1}{\partial t} &= -\frac{c^2}{gA}\frac{\left(Q_2 - \left(\frac{-C_1+\sqrt{C_1{}^2-4(R_{in}+C_0)(-C_2+H_1-H_{in})}}{2(R_{in}+C_0)}\right)\right)}{(z_2 - z_1)}\\
\frac{\partial H_2}{\partial t} &= -\frac{c^2}{gA}\frac{(Q_3 - Q_2)}{(z_3 - z_2)}\\
\frac{\partial H_3}{\partial t} &= -\frac{c^2}{gA}\frac{\left(Q_4 - (Q_3 - F\sqrt{H_3})\right)}{(z_4 - z_3)}\\
\frac{\partial H_4}{\partial t} &= -\frac{c^2}{gA}\frac{(Q_5 - Q_4)}{(z_5 - z_4)}\\
\frac{\partial Q_2}{\partial t} &= -Ag\frac{(H_2 - H_1)}{(z_2 - z_1)} - \frac{f_2}{2DA}Q_2\,|Q_2|\\
\frac{\partial Q_3}{\partial t} &= -Ag\frac{(H_3 - H_2)}{(z_3 - z_2)} - \frac{f_3}{2DA}Q_3\,|Q_3|
\end{aligned} \tag{4.44}$$

$$\frac{\partial Q_4}{\partial t} = -Ag\frac{(H_4 - H_3)}{(z_4 - z_3)} - \frac{f_4}{2DA} Q_4 \left| Q_4 \right|$$
$$\frac{\partial Q_5}{\partial t} = -Ag\frac{\left(\left(R_{sal} Q_5^2 + H_{atm}\right) - H_4\right)}{(z_5 - z_4)} - \frac{f_5}{2DA} Q_5 \left| Q_5 \right|.$$

Figure 4.12 presents the results for this example with the 50-section model and the corresponding experimental data. One can now see how there are changes in Q and H both at the inlet and the outlet of the pipe. This behavior correctly matches the real system behavior.

4.8 Conclusion

In this chapter, the modeling of water flow dynamics in pipelines has been discussed, with the particular purpose of including effects of faults, such as leaks and obstructions. The physical modeling has been recalled, as well as the finite-difference discretization-based implementation. Illustrative examples have finally been given.

References

Begovich, O., Pizano, A., & Besançon, G. (2012). On-line implementation of a leak isolation algorithm in a plastic pipeline prototype. *Latin American Applied Research, 57*, 131–141.

Benkherouf, A. & Allidina, A. (1988). Leak detection and location in gas pipielines. *IEE Proceedings. Part D. Control Theory Applications, 135*(2), 142–148.

Besançon, G., Georges, D., Begovich, O., Verde, C., & Aldana, C. (2007). Direct observer design for leak detection and estimation in pipelines. *European Control Conference* (pp. 5666–5670). Greece: Kos.

Billman, L., & Isermann, R. (1984). Leak detection methods for pipelines. *9th IFAC World Congress* (pp. 1813–1818). Hungary: Budapest.

Castro Burgos, L., & Valdés-González, H. (2009). Detección de pérdidas en tuberías de agua: Propuesta basada en un banco de filtros. *Revista Chilena de Ingeniería, 17*(3), 375–385.

Çengel, Y. & Cimbala, J. (2006). *Fluid Mechanics*. New York: McGraw-Hill.

Chaudhry, M. H. (2014). *Applied Hydraulic Transients*. New York: Springer.

Dulhoste, J., Besançon, G., Torres, L., Begovich, O., & Navarro, A. (2011). About friction modeling for observer-based leak estimation in pipelines. In *50th IEEE Conference on Decision and Control and European Control Conference*, 4413–4418.

Fletcher, C. A. (2006). *Computational Galerkin Methods*. Heidelberg: Springer.

Guillén, M., Dulhoste, J. F., Santos, R., Scola, I. R., Besançon, G., & Georges, D. (2014). Modelo dinámico para la detección y localización de obstrucciones parciales en tuberías. In *XII International Congress on Numerical Methods in Engineering and Applied Sciences*, Sartenajas, Venezuela.

Potter, M. & Wiggert, D. (2001). *Mechanics of Fluids*. New York: CL Engineering.

Roberson, J., Cassidy, J., & Chaudhry, M. (1997). *Hydraulic Engineering*. New York: Wiley.

Torres, L., Besançon, G., & Georges, D. (2008). A collocation model for water hammer dynamics with application to leak detection. In *47th IEEE Conference on Decision and Control*, Cancun, Mexico.

Torres, L., Besançon, G., Navarro, A., Begovich, O., & Georges, D. (2011). Examples of pipeline monitoring with nonlinear observers and real-data validation. In *8th IEEE International Multi-Conference on Signals Systems and Devices, Sousse, Tunisia.*

Verde, C. (2005). Accommodation of multi-leak location in a pipeline. *Control Engineering Practice*, *13*(8), 1071–1078.

Chapter 5
Observer Tools for Pipeline Monitoring

Gildas Besançon

Abstract This chapter discusses how the problem of fault monitoring in pipelines can be addressed by state observer tools. In short, the approach relies on a dynamical modeling of water flow dynamics in the pipeline subject to fault effects, and on this basis fault parameters are directly estimated by observer techniques. Motivated by typical pipeline models and faults, possible observer tools are recalled and illustrated with some application examples.

5.1 Introduction

The problem of pipeline monitoring corresponds to detecting and locating various possible faults affecting the flow in the pipe. The most typical fault is the occurrence of a *leak*, which results in losses of the product transported within the pipe, possible environmental damage depending on the effect of the product spread around, energy waste related to the pumps used to feed the pipe, etc. The problem of Leak Detection and Isolation (LDI) has generated a lot of work, with various possible approaches. Among them, one can find methods based on observer design, in the usual sense of systems' theory: an auxiliary dynamical system that estimates the internal (state) variables of a process, from which is fed with available data. In short, observer approaches can be used in two ways: either as a residual generator, allowing the detection of faults and their indirect isolation (e.g., as in Verde (2001)), or as a direct fault estimator, which is supposed to simultaneously detect and isolate possible faults (as originally proposed for LDI in Besançon et al. (2007)). This turns the monitoring problem into a state observer one. Depending on the system dynamics, various observer designs are available, and the purpose of the present chapter is to review some possible candidates for addressing various possible faults. In addition to leaks, one can also manage *obstructions*, which reduce the flow in the pipe, either locally or all along a piece of the pipe. One may also monitor faults affecting flow

G. Besançon (✉)
University of Grenoble Alpes, CNRS, GIPSA-Lab, 38000 Grenoble, France
e-mail: gildas.besancon@grenoble-inp.fr

C. Verde and L. Torres (eds.), *Modeling and Monitoring of Pipelines and Networks*, Applied Condition Monitoring 7,
DOI 10.1007/978-3-319-55944-5_5

sensors or even a pump feeding the pipe. This list is not exhaustive, and examples in the chapter will mostly be based on the leak case, even though some results also exist for other cases. In the same way, the presented possible observers do not claim to be exhaustive either. In fact, the entire presentation is mostly based on the author's own experience, and his collaborations with people listed in the acknowledgement section.

The remainder of the chapter is as follows: Sect. 5.2 first gives the main ideas for observer-based approaches in pipeline monitoring, and Sect. 5.3 then presents some examples of possible observers for possible fault identifications. Section 5.4 finally concludes the chapter.

5.2 Principle for Observer-Based Pipeline Monitoring

5.2.1 Model-Based Approach

Considering a simple pipeline of length L as depicted in Fig. 5.1, let $H(z,t)$ and $Q(z,t)$, respectively, denote the pressure head and the flow rate in the pipeline at position $z \in [0, L]$ and time $t \geq 0$. The dynamical behavior of water flow inside the pipe is then classically described by a couple of partial differential equations of the following form (Chaudry 1979):

$$\begin{aligned} &\frac{\partial H}{\partial t}(z,t) + \frac{c^2}{ga}\frac{\partial Q}{\partial z}(z,t) = 0 \\ &\frac{1}{a}\frac{\partial Q}{\partial t}(z,t) + g\frac{\partial H}{\partial z}(z,t) + \frac{f\,Q(z,t)|Q(z,t)|}{2Da^2} = 0, \end{aligned} \tag{5.1}$$

where c stands for the wave speed in the fluid, g is the gravitational acceleration, D is the diameter of the pipe, a is its cross-sectional area, and f is the friction coefficient. In general, this coefficient may depend on various elements, including flow rate or temperature for example, as in Delgado-Aguiñaga et al. (2015), but this will be neglected in the present discussion.

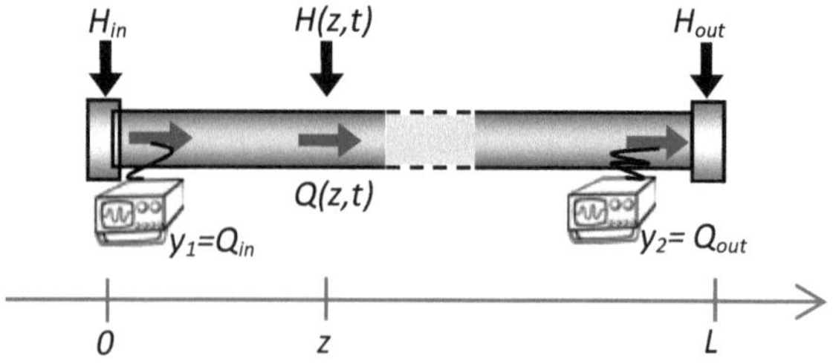

Fig. 5.1 Schematic view of a pipeline

The fluid dynamics are finally fully defined by initial conditions $Q(z, 0)$, $H(z, 0)$ for $z \in [0, L]$ and a couple of boundary conditions on the flow rate Q and pressure head H, for $z \in \{0, L\}$.

One can typically assume that pressure heads are fixed at each end of the pipe ($H_{in}(t)$ at $z = 0$ and $H_{out}(t)$ at $z = L$).

Monitoring means that some measurements are available on this pipe, and those variables are typically flow rates at each end of the pipe given by

$$y(t) = (Q(0,t) \quad Q(L,t))^T = (Q_{in}(t) \quad Q_{out}(t))^T,$$

together with the knowledge of pressure heads at those points.

The monitoring problem then consists of detecting and locating possible faults affecting the flow dynamics in the pipeline, from only the knowledge of available measurements. In model-based approaches, this is done using a model derived from (5.1), as well as candidate models for possible faults.

5.2.2 *Model Discretization*

Model (5.1) is infinite-dimensional, which makes it not easily tractable in general. One could still directly use it for purposes of observer designs (e.g., as in Hauge, Aamo and Godhavn (2007)), but this will not be discussed any further here. Instead let us reduce the problem to a finite-dimensional one by considering some *space discretization*. Once again, various techniques could be used, but here we will focus on the very basic finite-difference approach; see e.g., Verde (2001) for an early use of observer-based LDI. Notice finally that discretization in time can also be employed, but since we are focusing on observer tools mainly, we will stick to space discretization only.

By considering discretization points z_1 up to z_n along the pipe, with $z_1 = 0$, $z_n = L$ (as in Fig. 5.2), and forward finite differences for the derivative of H, backward differences for the derivative of Q (see Chap. 4 of the present book for a discussion about possible choices), one obtains

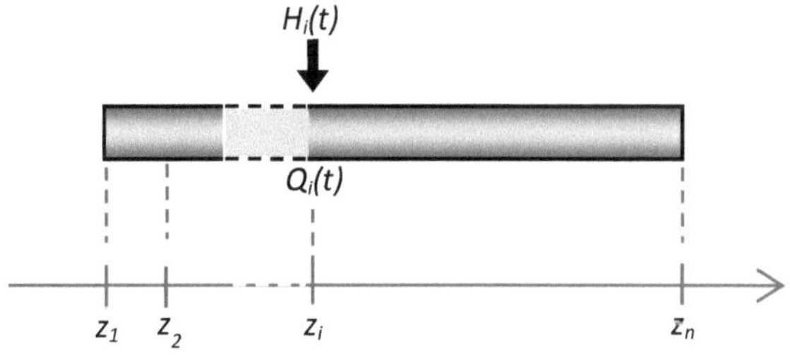

Fig. 5.2 Schematic view of pipeline discretization

$$
\begin{aligned}
\frac{dH_i}{dt} &= -\frac{c^2}{ga}\frac{Q_i(t)-Q_{i-1}(t)}{z_i-z_{i-1}} \\
\frac{dQ_i}{dt} &= -ag\frac{H_{i+1}(t)-H_i(t)}{z_{i+1}-z_i} - \frac{fQ_i(t)|Q_i(t)|}{2Da},
\end{aligned}
\tag{5.2}
$$

where H_i, Q_i stand for approximate solutions for H, Q of (5.1) at point z_i. This discretization is appropriate for boundary conditions on H_1, H_n typically, with $i = i_0, ..n-1$, starting with $i_0 = 2$ for the pressure head H, and with $i_0 = 1$ for the flow rate Q).

For the purpose of fault detection, one must to enhance the model with the effect of possible faults to be monitored. In the case of a leak, one just has to use the balance equation:

$$
Q(z_{leak}^-, t) = Q(z_{leak}^+, t) + Q_{leak}(t),
$$

where z_{leak}^- is the position just before the point where the leak occurs and z_{leak}^+ is the one just after, while Q_{leak} denotes the leakage flow. From the Bernoulli's law, the leak out flow rate can be expressed as

$$
Q_{leak}(t) = F_{leak}\sqrt{H(z_{leak}, t)},
$$

where F_{leak} is related to the size of the leak orifice. This latter coefficient can typically be assumed to be zero when no leak is present, and different from zero (almost constantly) when a leak occurs.

By considering such a possible leak in each section of the considered finite-difference approximation, model (5.2) then becomes

$$
\begin{aligned}
\frac{dH_i}{dt} &= -\frac{c^2}{ga}\frac{Q_i(t)+F_i\sqrt{H_i}-Q_{i-1}(t)}{z_i-z_{i-1}} \\
\frac{dQ_i}{dt} &= -ag\frac{H_{i+1}(t)-H_i(t)}{z_{i+1}-z_i} - \frac{fQ_i(t)|Q_i(t)|}{2Da}.
\end{aligned}
\tag{5.3}
$$

Notice that on the basis of a similar approximation, one could adapt the model to other types of faults, like obstruction; see Chap. 4 of the present book. For obstruction, though, it was shown that finite-difference approximation is not very appropriate (Rubio-Scola 2015; Guillén 2016), and for this reason, we will here focus here on *leak* faults.

5.2.3 *Observer Formulation*

From a model of the form (5.3), one easily gets a state-space description of water flow dynamics, with H_i, Q_i's as state variables of a vector x_0, boundary conditions as known input variables of a vector u, and measured variables as output variables of a vector y. In addition, fault variables, such as leak magnitudes F_i's and related positions z_i's in (5.3), can be gathered in a fault vector d. The model then takes the following form:

$$\begin{aligned} \dot{x}_0(t) &= f_0(x_0(t), u(t), d(t)) \\ y(t) &= h_0(x_0(t)), \end{aligned} \tag{5.4}$$

for the appropriate known functions f_0, h_0.

Observers are then typical tools for recovering the knowledge on state x_0, from that of y, u, f_0, h_0; see e.g., Besançon (2007).

An observer generally reads as follows:

$$\begin{aligned} \dot{x}_{obs}(t) &= f_{obs}(x_{obs}(t), u(t), y(t)) \\ \hat{x}_0(t) &= h_{obs}(x_{obs}(t), u(t), y(t)), \end{aligned} \tag{5.5}$$

and should be such that $\hat{x}_0(t) - x_0(t)$ decays to zero when t grows to infinity, for appropriate functions f_{obs}, h_{obs}.

Assuming that $d = d_0$ represents the fault-free operation, an observer can take the form

$$\dot{\hat{x}}_0(t) = f_0(\hat{x}_0(t), u(t), d_0) - k(t)[h_0(\hat{x}_0(t)) - y(t)], \tag{5.6}$$

where k is appropriately designed. In that case, $r(t) = h_0(\hat{x}_0(t)) - y(t)$ can be chosen as a so-called *residual*, in the sense of a fault indicator, being (asymptotically) zero in a fault-free operation and deviating from zero when a fault occurs. This requires that this residual be indeed sensitive to faults, and if so, it will in general only reveal whether a fault has appeared or not.

One may instead try to directly estimate the faults after their occurrence. This can be done by extending model (5.4) by some fault model characterizing dynamics of d: $\dot{d}(t) = D(d(t), t)$ for some known D (typically $\dot{F}_i = 0,\ \dot{z}_i = 0$ for leaks in (5.3)).

In that case, one can consider some extended state vector $x := \begin{pmatrix} x_0 \\ d \end{pmatrix}$ to yield an extended model such as

$$\begin{aligned} \dot{x}(t) &= f(x(t), u(t)) \\ y(t) &= h(x(t)), \end{aligned} \tag{5.7}$$

for functions f, h directly deduced from f_0, D, h_0 and extended vector x.

In addition, one is brought to an observer issue for this extended model. This is the approach discussed further in next section.

5.3 Examples of Observer Tools for Pipeline Monitoring

In this section, various tools for an observer approach to fault detection in pipelines are presented, with an increasing *"complexity"*, from linear to nonlinear approaches.

More details about the related observer background can be found for instance in Besançon (2007).

5.3.1 *Linear Approaches*

5.3.1.1 Luenberger Observer

Model (5.3) resulting in (5.7) is in general nonlinear, and a very first approach could be to consider an approximate linearization around some fixed operation regime, giving rise to a standard linear time-invariant system of the form

$$\begin{aligned} \dot{x}(t) &= Ax(t) + Bu(t) \\ y(t) &= Cx(t). \end{aligned} \tag{5.8}$$

Under this form, it is well known that a simple observability condition on A, C allows guarantees that a so-called *Luenberger observer* (Luenberger 1964) can solve for the state reconstruction issue, such as

$$\dot{\hat{x}}(t) = A\hat{x}(t) - K(C\hat{x}(t) - y(t)) + Bu(t), \tag{5.9}$$

for any matrix K such that $A - KC$ is Hurwitz (i.e., has all its eigenvalues with strictly negative real parts).

In fact, because the great sensitivity of the operation point with respect to faults like leaks or obstructions, this approach is not successful in general. See e.g., Guillén (2016) for more details.

Instead one can rather rely on linearization around a varying point, updated according to the current estimate of the state; this corresponds to the well-known *extended Kalman filter* approach (Gelb 1974). An early use for leak detection can for instance be found in Besançon et al. (2007). The corresponding equations are recalled in next subsection.

5.3.1.2 Kalman Observer

The extended Kalman filter (or *EKF*) is an extension of original Kalman approach (Kalman and Bucy 1960) developed for linear time-varying systems of the form

$$\begin{aligned} \dot{x}(t) &= A(t)x(t) + B(t)u(t) \\ y(t) &= C(t)x(t). \end{aligned} \tag{5.10}$$

Under the *uniform complete observability* condition, which is stated as

$$\int_{t-T}^{t} \Phi^T(\tau, t-T)C^T(\tau)C(\tau)\Phi(\tau, t-T) \geq \alpha Id; \quad T>0, \quad \alpha>0, \tag{5.11}$$

where Id is the identity matrix and Φ satisfies

- $\dfrac{d\Phi(\tau,t)}{d\tau} = A(\tau)\Phi(\tau,t)$,
- $\Phi(t,t) = Id$,

one has then an observer for system (5.10) as follows:

$$\begin{aligned} \dot{\hat{x}}(t) &= A(t)\hat{x}(t) + B(t)u(t) - K(t)(C(t)\hat{x}(t) - y(t)) \\ K(t) &= M(t)C^T(t)W^{-1} \\ \dot{M}(t) &= M(t)A(t) + A^T(t)M(t) - M(t)C^T(t)W^{-1}C(t)M(t) + V. \end{aligned} \tag{5.12}$$

If W and V are, respectively, the measurement noise and state noise covariance matrices, this yields the optimal state estimate in the sense of least square minimization. In addition, under the above observability condition, if V is positive definite matrix, or replaces the matrix γM for a large enough γ, the estimation error exponentially decays to zero.

On this basis, the EKF for a system of the form (5.7) reads as follows:

$$\begin{aligned} \dot{\hat{x}}(t) &= f(\hat{x}(t), u(t)) - K(t)(h(\hat{x}(t)) - y(t)) \\ K(t) &= M(t)C^T(t)W^{-1} \\ \dot{M}(t) &= M(t)A(t) + A^T(t)M(t) - M(t)C^T(t)W^{-1}C(t)M(t) + V; \; M(0) > 0, \end{aligned} \tag{5.13}$$

where $A(t) := \dfrac{\partial f}{\partial x}(\hat{x}(t), u(t)), \quad C(t) := \dfrac{\partial h}{\partial x}(\hat{x}(t), u(t))$.

The strong interest in such an approach is that it can directly be applied to the physical model, when extended to (5.7), and it works fairly well in practice. It is for instance the underlying approach in Besançon et al. (2007). The following cases have been reported using this approach. If the instrumentation issue is considered, results are reported by Guillén et al. (2014). The scenario of multiple leaks' detection and isolation has been solved by Delgado-Aguiñaga et al. (2016b). The pipeline model calibration has been validated by Navarro et al. (2017). Chapter 12 of the present book shows the performance of the filter in a real application of a Mexican aqueduct of Guadalajara City. This observer can also be considered for obstruction detection and isolation in some cases (for example as it is shown in Besançon et al. (2013)).

Notice that from a theoretical point of view, even if this EKF approach appears to be more efficient than the Luenberger extension for nonlinear systems for the considered monitoring issue, one may only expect *local* convergence of the estimation error. Furthermore, to be formally guaranteed, the latter in general needs stronger conditions in the model. This will be discussed in the next subsection.

5.3.2 Nonlinear Approaches

5.3.2.1 Luenberger Extension

The most famous extension of the Luenberger observer is the so-called *high-gain observer* (Gauthier et al. 1992), which applies to systems (5.7) when it is rewritten under the form

$$\begin{aligned} \dot{\xi}(t) &= A_0\xi(t) + \varphi(\xi(t), u(t)) \\ y(t) &= C_0\xi(t), \end{aligned} \tag{5.14}$$

where $A_0 = \begin{pmatrix} 0 & I_p & & 0 \\ & \ddots & \ddots & \\ & & & I_p \\ 0 & \dots & & 0 \end{pmatrix}$, $C_0 = \begin{pmatrix} I_p & 0 & \cdots & 0 \end{pmatrix}$ and $\varphi(\xi, u)$ satisfies the Lipschitz condition in ξ uniformly in u and can be described by

$$\varphi(\xi, u) = \begin{pmatrix} \varphi_1(\xi_1, u), \\ \varphi_2(\xi_1, \xi_2, u) \\ \vdots \\ \varphi_{q-1}(\xi_1, \cdots \xi_{q-1}, u) \\ \varphi_q(\xi, u) \end{pmatrix} \text{ for } \xi = \begin{pmatrix} \xi_1 \\ \xi_2 \\ \vdots \\ \xi_q \end{pmatrix}$$

with $\xi_i \in I\!R^p$, if $y \in I\!R^p$, and I_p stands for the $p \times p$ identity matrix.

This is possible under the *uniform observability* property (Gauthier and Bornard 1981) and an observer then takes the form

$$\begin{aligned} \dot{\hat{\xi}}(t) &= A_0\hat{\xi}(t) + \varphi(\hat{\xi}(t), u(t)) - \Lambda(\lambda)K_0[C_0\hat{\xi}(t) - y(t)] \\ \Lambda(\lambda) &= \begin{pmatrix} \lambda I_p & 0 & \cdots & 0 \\ 0 & \lambda^2 I_p & & \\ \vdots & & \ddots & 0 \\ 0 & \cdots & 0 & \lambda^q I_p \end{pmatrix}, \end{aligned} \tag{5.15}$$

where K_0 is such that $A_0 - K_0C_0$ is Hurwitz, and λ is to be chosen large enough to guaranty convergence, simultaneously allowing the tuning of the convergence rate.

This approach needs the effort of rewriting the physical model into the form (5.14), but it provides guarantees for exponential convergence of the estimation error. It has been applied successfully to leak detection for instance in Torres et al. (2011), but also in Torres et al. (2014) for liquefied petroleum gas transport monitoring.

Let us recall how such an approach can be applied to leak position and magnitude estimation in a similar way as in Torres et al. (2011). To illustrate, by considering the model (5.3) reduced to two sections, that is Q_1, H_2, Q_2 with Q_1 and Q_2 measured variables and as state variables x_1, x_2, x_3. If this state is extended with a single leak at unknown position z_2 (defined as state variable x_4) and the unknown coefficient

F_2 (defined as state variable x_5), and subject to inputs $u_1 = H_1, u_2 = H_3$, we are brought to a state-space representation under the form (5.7) as follows:

$$\begin{aligned} \dot{x}_1 &= -a_1 \frac{x_2 - u_1}{x_4} - \rho f x_1 |x_1| \\ \dot{x}_2 &= -a_2 \frac{x_3 + x_5\sqrt{x_2} - x_1}{x_4} \\ \dot{x}_3 &= -a_1 \frac{u_2 - x_2}{L - x_4} - \rho f x_3 |x_3| \\ \dot{x}_4 &= 0 \\ \dot{x}_5 &= 0 \\ y &= \begin{pmatrix} x_1 \\ x_3 \end{pmatrix}, \end{aligned} \tag{5.16}$$

where

$$a_1 = ag,\ a_2 = \frac{c^2}{ag}, \quad \text{and} \quad \rho = \frac{1}{2Da}. \tag{5.17}$$

Let us consider an operation under constant inputs. A representation of the form (5.14) can then be obtained by considering the transformation

$$\xi = \begin{pmatrix} x_1 \\ x_3 \\ -a_1 \frac{x_2 - u_1}{x_4} \\ -a_1 \frac{u_2 - x_2}{L - x_4} \\ a_1 a_2 \frac{x_5\sqrt{x_2}}{x_4^2} \\ -a_1 a_2 \frac{x_5\sqrt{x_2}}{x_4(L - x_4)} \end{pmatrix}.$$

In that case, one can indeed check that the system is turned into (5.14) with

$$\varphi(\xi, u) = \begin{pmatrix} -\rho f \xi_1 |\xi_1| \\ -\rho f \xi_2 |\xi_2| \\ \frac{a_1 a_2}{p(\xi)^2}(\xi_2 - \xi_1) \\ -\frac{a_1 a_2}{p(\xi)(L - p(\xi))}(\xi_2 - \xi_1) \\ -\frac{a_1 \xi_5}{2(a_1 u_1 - p(\xi)\xi_3)} \left(\frac{\xi_5 p(\xi)}{a_1} + \frac{a_2(\xi_2 - \xi_1)}{p(\xi)} \right) \\ -\frac{a_1 \xi_6}{2(a_1 u_1 - p(\xi)\xi_3)} \left(\frac{\xi_5 p(\xi)}{a_1} + \frac{a_2(\xi_2 - \xi_1)}{p(\xi)} \right) \end{pmatrix},$$

where ξ_i denotes component i of ξ and $p(\xi) = \frac{a_1(u_1 - u_2) - L\xi_4}{\xi_3 - \xi_4}$.

For the system in those new coordinates, an observer (5.15) can be designed to obtain an estimate for ξ, and in particular for $p(\xi)$, which gives the estimate of leak position x_4 (the leak magnitude being given by $x_5 = \frac{\xi_5 p(\xi)^2}{a_2\sqrt{a_1}\sqrt{a_1 u_1 - \xi_3 p(\xi)}}$).

For the illustration of a concrete pipeline, let us further recall in Fig. 5.3, the experimental results reported in Torres et al. (2011) obtained with a similar approach

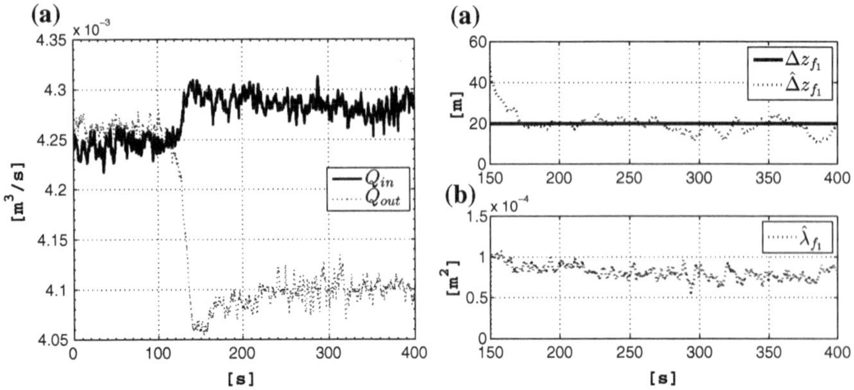

Fig. 5.3 Experimental results for leak identification with observer of type (5.15) (Torres et al. 2011). *Left* input and output measured flow rates; *right* **a** estimated versus actual position, **b** orifice magnitude estimation

and based on a CINVESTAV prototype of about 85 [m] as described in Begovich et al. (2012). In this experiment, a leak has been generated at a point approximately 20 [m] away from the inlet point, and it can be seen how the observer properly estimates it.

Notice that even if the tuning parameter λ in such a high-gain approach opens the possibility of an arbitrarily fast asymptotic convergence, the use of discontinuous correction terms in the observer design, as in *sliding mode* approaches (Levant 1998; Moreno and Osorio 2008), even allows us to achieve *finite time* convergence. An application for leak detection in pipelines can be found in Delgado-Aguiñaga et al. (2016a).

5.3.2.2 Kalman Extension

In some cases, one may possibly rewrite the system in a linear form in the state variable, with only multiplicative and additive terms depending on known signals, such as

$$\begin{aligned} \dot{\xi}(t) &= A(v(t))\xi(t) + B(v(t)) \\ y(t) &= C\xi(t), \end{aligned} \tag{5.18}$$

where v can be reduced to known input u or can also include measured signal y, for instance.

In such a case, an observer can be given as a Kalman-like one

$$\begin{aligned} &\dot{\hat{\xi}}(t) = A(v(t))\hat{\xi}(t) + B(v(t)) - K(t)[C\hat{\xi}(t) - y(t)] \\ &K(t) = M(t)C^T W^{-1} \\ &\dot{M}(t) = M(t)A^T(v(t)) + A(v(t))M(t) - M(t)C^T W^{-1} C + V + \gamma M(t), \end{aligned} \tag{5.19}$$

where W is a positive definite matrix, and either V is also definite positive matrix, or $V \geq 0$ and γ is large enough.

In that condition, the convergence requires that v be exciting enough so that the uniform complete observability property required for the Kalman observer holds.

Such a model for instance arises when considering leak monitoring under measurement of only *one* boundary flow rate, as considered in Rubio-Scola et al. (2013). In such a case, when a problem reduced to leak detection and magnitude identification is considered, model (5.2) can be considered over a single section only, with the introduction of a leak effect either related to position z_2 (as in (5.3)) or to z_1 as follows:

$$\begin{aligned} \dot{H}_2 &= -\frac{c^2}{ga}\frac{Q_2 + F\sqrt{H_1} - Q_1}{L} \\ \dot{Q}_1 &= -ag\frac{H_2 - H_1}{L} - f\frac{Q_1|Q_1|}{2Da}. \end{aligned} \tag{5.20}$$

By considering that $y = Q_1$ is measured in the latter, and $u_1 = H_1$, $u_2 = Q_2$ are input variables, the model can be rewritten with state variables $x_1 := Q_1$, $x_2 := H_2$ and notations (5.17) as follows:

$$\begin{aligned} \dot{x}_1 &= -a_1\frac{x_2 - u_1}{L} - \rho f x_1|x_1| \\ \dot{x}_2 &= -a_2\frac{u_2 + F\sqrt{u_1} - x_1}{L} \\ y &= x_1. \end{aligned} \tag{5.21}$$

Extended with $x_3 := F$, $\dot{x}_3 = 0$, the model easily takes a form (5.18) with $A(v) = A(u_1)$ and $B(v) = B(u_1, u_2, y)$. Notice that the problem can even be extended to the case when Q_2 is constant and also to be estimated, keeping a similar structure (with $x_4 := Q_2$ and $\dot{x}_4 = 0$).

In that case, a special care is to be taken on the pipeline driving pressure u_1 to ensure convergence of the estimation; see e.g., Rubio-Scola et al. (2013).

Such a special condition on the input becomes even more demanding when model (5.18) further includes unmeasured state variables in additive nonlinearities. In that case, the following structure is needed for a possible observer design:

$$\begin{aligned} \dot{\xi}(t) &= A_0(v(t))\xi(t) + \varphi(\xi(t), u(t)) \\ y(t) &= C_0\xi(t), \end{aligned} \tag{5.22}$$

with C_0, φ as in (5.14),

and $A_0(v)$ here of the form $A_0(v) = \begin{pmatrix} 0 & A_{12}(v) & & 0 \\ & \ddots & \ddots & \\ & & & A_{q-1,q}(v) \\ 0 & \dots & & 0 \end{pmatrix}$

for $A_{ii+1} \in I\!R^{n_i \times n_{i+1}}$, $\xi_i \in I\!R^{n_i}$, $n_1 = p$, if $y \in I\!R^p$.

For such systems, an appropriate excitation condition (roughly of the form (5.11) for arbitrarily short T, and αId replaced by $\frac{\alpha}{T}\Lambda^2(T)$), yields an exponential observer as in Besançon and Ţiclea (2007) or the extension provided in Torres et al. (2012):

$$\begin{aligned}
\dot{\hat{\xi}}(t) &= A_0(v(t))\hat{\xi}(t) + \varphi(\hat{\xi}(t), u(t)) - \Lambda(\lambda)K(t)[C\hat{\xi}(t) - y(t)] \\
\Lambda(\lambda) &= \begin{pmatrix} \lambda I_{n_1} & 0 & \cdots & 0 \\ 0 & \lambda^2 I_{n_2} & & \\ \vdots & & \ddots & 0 \\ 0 & \cdots & 0 & \lambda^q I_{n_q} \end{pmatrix} \\
K(t) &= M(t)C^T W^{-1} \\
\dot{M}(t) &= \lambda(M(t)A^T(v(t)) + A(v(t))M(t) - M(t)C^T W^{-1} C + V + \gamma M(t)) \\
M(0) &> 0,
\end{aligned} \tag{5.23}$$

for V, W, γ as in a Kalman observer, A corresponding to A_0 (or extended to $A_0 + \frac{1}{\lambda}\Lambda^{-1}\frac{\partial\varphi}{\partial\xi}(\hat{\xi}, u)\Lambda$ (Torres et al. 2012)), and a sufficiently large $\lambda = \frac{1}{T}$. Here I_{n_i} means $n_i \times n_i$ identity.

Such a model arises for instance when considering the problem of

- leak estimation with unknown friction coefficient to be additionally estimated, as considered in Torres et al. (2012); or
- simultaneous leaks in the pipeline, i.e., more than one nonzero F_i in model (5.3) as studied in Torres et al. (2009); Torres (2011)

Let us recall some details of Torres et al. (2012) for the sake of illustration: coming back to model (5.16), the considered problem here is that of detecting and locating possible leaks under an uncertain friction coefficient. In other words, the problem can be addressed as an estimation one for leak parameters θ_1 :=leak position, θ_2 :=leak magnitude (x_4, x_5 in (5.16)), but also friction parameter $\theta_3 := f$, which can be defined as an additional state variable x_6.

By considering the following new variables,

$$\xi = \begin{pmatrix} x_1 \\ x_6 \\ \frac{1}{x_4} \\ \frac{x_2}{x_4} \\ \frac{x_5\sqrt{x_2}}{x_4^2} \end{pmatrix}$$

it can be checked that the model indeed takes the form (5.22) with

$$A_0 = \begin{pmatrix} 0 & -\rho y_1|y_1| & a_1 u_1 & -a_1 & 0 \\ 0 & 0 & 0 & 0 & 0 \\ 0 & 0 & 0 & 0 & 0 \\ 0 & 0 & 0 & 0 & -a_2 \\ 0 & 0 & 0 & 0 & 0 \end{pmatrix}$$

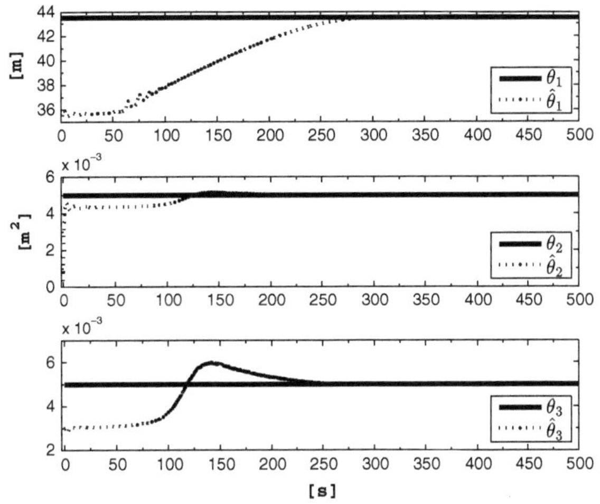

Fig. 5.4 Simulation results for leak and friction identification with observer of type (5.15) (Torres et al. 2012). Top graphic, position estimation; middle graphic, leak orifice estimation; bottom, friction estimation

$$\varphi = \begin{pmatrix} 0 \\ 0 \\ 0 \\ a_2\xi_3^2(y_1 - y_2) \\ a_2\frac{\xi_3\xi_5}{2\xi_4}(\xi_3(y_1 - y_2) - \frac{\xi_5}{\xi_3}) \end{pmatrix}.$$

For this representation, an observer (5.23) can now be designed, and some related simulation results as obtained in Torres et al. (2012) are recalled in Fig. 5.4. In this formulation, the driving input u_1 has been chosen as a pulse signal.

Experimental results with the same prototype as in Begovich et al. (2012) and similar approaches can also be found in Torres (2011).

5.4 Conclusions

In this chapter, it has been emphasized how pipeline monitoring issues can be reformulated as observer design problems and how various possible observers can be used. Future work on validation or implementation of such tools is still in progress, as well on their extension to various general situations (type of faults, number and occurrence times of faults, pipeline configurations, etc.).

Acknowledgements The author would like to thank Drs. Delgado, Guillén, Rubio Scola, Navarro, for their contributions in the field via their recent PHD studies, Dr. Begovich and Prof. Verde for early collaborations on this topic, Dr. Torres for first—and continuous—significant developments in that respect, Prof. Georges, Prof. Dulhoste and Prof. Santos for fruitful interactions in the collaborations formerly mentioned.

References

Begovich, O., Pizano, A., & Besançon, G. (2012). Online implementation of a leak isolation algorithm in a plastic pipeline prototype. *Latin American Applied Research, 57*, 131–141.
Besançon, G. (2007). *Nonlinear observers and applications*. Berlin: Springer.
Besançon, G., & Ţiclea, A. (2007). An immersion-based observer design for rank-observable nonlinear systems. *IEEE Transaction on Automatic Control, 52*(1), 83–88.
Besançon, G., Georges, D., Begovich, O., Verde, C., & Aldana, C. (2007). Direct observer design for leak detection and estimation in pipelines. In *European Control Conference, Kos, Greece*.
Besançon, G., Rubio-Scola, I., Guillen, M., Dulhoste, J., Santos, R., & Georges, D. (2013). Observer-based detection and location of partial blockages in pipelines. In *Internantional Conference on Control and Fault Tolerant Systems, Nice, France*.
Chaudhry, M. H. (1979). *Applied hydraulic transients*. New York: Van Nostrand Reinhold Co.
Delgado-Aguiñaga, J., Besançon, G., & Begovich, O. (2015). Leak isolation based on extended Kalman filter in a plastic pipeline under temperature variations with real-data validation. In *Proceedings of IEEE Mediterranean Control Conference, Torremolinos, Spain*.
Delgado-Aguiñaga, J., Begovich, O., & Besançon, G. (2016a). Exact-differentiation-based leak detection and isolation in a plastic pipeline under temperature variations. *Journal of Process Control, 42*, 114–124.
Delgado-Aguiñaga, J., Besançon, G., Begovich, O., & Carvajal, J. (2016b). Multi-leak diagnosis in pipelines based on extended Kalman filter. *Control Engineering Practice, 49*, 139–148.
Gauthier, J., & Bornard, G. (1981). Observability for any u(t) of a class of nonlinear systems. *IEEE Transaction on Automatic Control, 26*(5), 922–926.
Gauthier, J. P., Hammouri, H., & Othman, S. (1992). A simple observer for nonlinear systems: Applications to bioreactors. *IEEE Transactions on Automatic Control, 37*(6), 875–880.
Gelb, A. (1974). Applied optimal estimation. Technical report, M.I.T.
Guillén, M. (2016). *Sistema de detección y localización de fallas en tuberías basado en observadores de estado*. PhD. thesis, Universidad de Los Andes.
Guillén, M., Dulhoste, J., Besançon, G., Rubio-Scola, I., Santos, R., & Georges, D. (2014). Leak detection and location based on improved pipe model and nonlinear observer. In *Proceedings of 13th European Control Conference, Strasbourg, France*.
Hauge, E., Aamo, O., & Godhavn, J. (2007). Model-based pipeline monitoring with leak detection. In *Proceedings of 7th IFAC symposium on nonlinear control systems, Pretoria, South Africa*.
Kalman, R., & Bucy, R. S. (1960). New results in linear filtering and prediction theory. *Journal of Basic Engineering, 82*(D), 35–40.
Levant, A. (1998). Robust exact differentiation via sliding mode technique. *Automatica, 34*(3), 379–384.
Luenberger, D. (1964). Observing the state of a linear system. *IEEE Transactions on Military Electronics, 8*, 74–80.
Moreno, J., & Osorio, M. (2008). A Lyapunov approach to second-order sliding mode controllers and observers. In *Proceedings of 47th IEEE Conference on Decision and Control, Cancun, Mexico*.
Navarro, A., Begovich, O., Sanchez-Torres, J., & Besançon, G. (2017). Real-time leak isolation based on state estimation with fitting loss coefficient calibration in a plastic pipeline. *Asian Journal of Control, 19*, 1–11.
Rubio-Scola, I. (2015). *Contributions à l observation par commande d observabilité et à la surveillance de pipelines par observateurs*. PhD. thesis, Université de Grenoble.
Rubio-Scola, I., Besançon, G., & Georges, D. (2013). Input optimization for observability of state affine systems. In *5th IFAC symposium systems structure and control, Grenoble, France*.
Torres, L. (2011). *Modèles et observateurs pour systèmes d écoulement sous pression - extension aux systèmes chaotiques*. PhD. thesis, Université de Grenoble.
Torres, L., Besançon, G., & Georges, D. (2009). Multi-leak estimator for pipelines based on an orthogonal collocation model. In *Proceedings of the 48th IEEE Conference on Decision and Control, Shanghai, China*.

Torres, L., Besançon, G., & Georges, D. (2012). EKF-like observer with stability for a class of triangular nonlinear systems. *IEEE Transactions on Automatic Control, 57*(6), 1570–74.

Torres, L., Besançon, G., Navarro, A., Begovich, O., & Georges, D. (2011). Examples of pipeline monitoring with nonlinear observers and real-data validation. In *8th International IEEE Multi-Conference on Systems, Signals and Devices, Sousse, Tunisia.*

Torres, L., Verde, C., Besançon, G., & González, O. (2014). High gain observers for leak detection in subterranean pipelines of liquefied petroleum gas. *International Journal of Robust and Nonlinear Control, 24*(6), 1127–1141.

Verde, C. (2004). Minimal order nonlinear observer for leak detection. *ASME J Dynamic Systems, Measurement and Control, 126*(3), 467–472.

Chapter 6
Auxiliary Signal Design and Liénard-type Models for Identifying Pipeline Parameters

Javier Jiménez, Lizeth Torres, Ignacio Rubio and Marco Sanjuan

Abstract This chapter presents the implementation of optimization algorithms to build auxiliary signals that can be injected as inputs into a pipeline in order to estimate—by using state observers—physical parameters such as the friction or the wave speed. For the design of the state observers, we propose to incorporate the parameters to be estimated into the state vector of Liénard-type models of a pipeline such that the observers can be constructed from the modified models. The proposed optimization algorithms guarantee a prescribed observability degree of the modified models by building an optimal input for the identification. The optimality of the input is defined with respect to the minimization of the input energy, whereas the observability through a lower bound for the observability Gramian, which is constructed from the Liénard-type models of the pipeline. The proposed approach to construct auxiliary inputs was experimentally tested by using real data obtained from a laboratory pipeline.

6.1 Introduction

Despite a good maintenance plan for fault prevention, leaks in pipelines are unfortunately very common events that must be found and identified in time to avoid irretrievable losses. Hence, fluid distribution systems must be under constant surveillance and should be diagnosed at least by a leak detection and identification (LDI) technique.

Among the most used LDI techniques are those based on measurements of flow and pressure together with a model of the fluid dynamics. To use these techniques,

J. Jiménez (✉) · M. Sanjuan
Universidad del Norte, Barranquilla, Colombia
e-mail: cabasa@uninorte.edu.co

L. Torres
Universidad Nacional Autónoma de México,
Cuidad de México, Mexico

I. Rubio
Centro Federal de Educação Tecnológica de Minas Gerais,
Divinópolis, MG, Brazil

C. Verde and L. Torres (eds.), *Modeling and Monitoring of Pipelines and Networks*, Applied Condition Monitoring 7,
DOI 10.1007/978-3-319-55944-5_6

measurements at appropriate points of the pipe are required as well as a model that properly represents the behavior of the pipeline under diagnosis. Having an appropriate model means that the parameters of the model equations, which govern the dynamic behavior of the pipeline, should be correct; pipelines, however, are characterized by different parameters that can change significantly from their design values because of the manufacturing execution, the installation process, and the aging deterioration. Therefore, in order to design algorithms based on the physics of the pipeline, employing or developing available methodologies is necessary for continuously updating the physical parameters of the pipeline model.

The updating of the parameters of a process is a task known as parameter identification, and many methodologies for such a task have been developed over time. We recommend consulting the works analyzed and summarized by Sinha and Kuszta (1983); Isermann and Münchhof (2010) to get a better perspective on the subject. For the particular case of the parameter identification of pipelines, some good approaches have been presented by Yang et al. (2008) and Zecchin et al. (2013), each one with advantages and disadvantages to be evaluated according to the characteristics of the problem.

In particular, Torres et al. (2015) introduced a parameter identification methodology based on state observers and Liénard-type models, and a validation of such a methodology with experimental data was presented by Torres et al. (2016). In this chapter, we have incorporated this approach with important theoretical improvements for its implementation.

The methodology starts with the converting of the hyperbolic partial differential equations that represent the fluid dynamics in a pipeline into Liénard equations. Note that one of the most important benefits of having Liénard-type models is their suitable structure for the design of state observers.

For the parameter estimation, the Liénard-type models are extended with the incorporation of the parameters to be estimated into their state vectors. From the extended Liénard-type models, exponential state observers may then be constructed to identify the parameters. The exponential convergence of the estimations is guaranteed under an appropriate excitation condition on the inputs, which means that the observability of the parameters depend on the applied inputs.

For simple systems represented by simple models, such inputs could be found analytically; however, for more complex systems, this may not be feasible. A common solution is to heuristically look for an input signal that guarantees the observability condition.

In this work, the approach presented by Scola et al. (2013) was adopted and modified: an optimization procedure to find an optimal sequence that guarantees the observability. In other words, we propose to build an auxiliary signal that guarantees the identification of the parameters of a pipeline by using state observers based on Liénard-type models.

In this work, by using the approach presented by Scola et al. (2013), the observability of the Liénard-type models is characterized through the so-called observability Gramian, which is in short a measure of the energy in the output signal. Furthermore, a degree of observability is defined from the eigenvalues of the Gramian, in particular the lowest of those values.

The design of an adequate input is crucial for the successful identification of parameters. In fact, there is extensive literature on optimal input design generated in the 1960's (Litman and Huggins 1963; Levadi 1966; Mehra 1974a): for linear systems (Mehra 1974b), for distributed systems (Rafajłowicz 1989), for nonlinear systems (Kalaba and Spingarn 1977), for specific applications (Shafer 1984; Jauberthie et al. 2006; Lu 2010), and for fault detection (Campbell and Nikoukhah 2015). The efforts of this work are aimed at estimating the parameters of a pipeline that can be modeled as a nonlinear Liénard-type system.

The organization of the chapter is as follows. Section 6.2 presents some background of observability, whereas Sect. 6.3 is dedicated to recollections of the Liénard equation. In Sect. 6.4, Liénard-type models for a pipeline are introduced. Section 6.5 describes the optimization algorithm proposed by Scola et al. (2013) for the design of optimal inputs and a new algorithm based on it. Section 6.6 presents some simulations and experimental tests to show the input design procedure as well as the estimation of some parameters by using the synthesized inputs. Finally, Sect. 6.7 concludes this chapter.

6.2 Recalls on Observability

Observability for linear systems is a global property that can be checked from the rank of the observability Gramian matrix for time-varying systems or from the rank of the observability matrix for time-invariant systems. Contrary to the linear case, nonlinear system observability is determined locally around a given equilibrium point. This section presents a brief description of the observability characterization of a class of state affine systems with output injection of the form

$$\begin{aligned} \dot{x}(t) &= A_c(u(t), y(t))\, x(t) + B_c(u(t), y(t)), \\ y(t) &= C_c x(t) \end{aligned} \tag{6.1}$$

where $x(t) \in \mathbb{R}^n$ denotes the state vector, $u(t) \in \mathbb{R}^m$ the control input vector, and $y(t) \in \mathbb{R}$ the measured output. System (6.1) can be discretized through a single matrix exponential as follows (Van Loan 1978):

$$\begin{bmatrix} A(k) & B(k) \\ 0 & I \end{bmatrix} = \exp\left(T_s \begin{bmatrix} A_c(u(k), y(k)) & B_c(u(k), y(k)) \\ 0 & 0 \end{bmatrix} \right),$$

where k is any sampling instant, $u(k) \equiv u(kT_s)$, $y(k) \equiv y(kT_s)$, T_s is the sampling time, and 0 and I stand for null and identity matrices of appropriate dimensions.

According to Ţiclea and Besançon (2009), given the initial condition $x(0)$ and the input function, the observation problem of (6.1) can be reduced to the observation of the linear time-varying system

$$x(k+1) = A(k)\,x(k) + B(k)\,,$$
$$y(k) = C(k)\,x(k)\,, \tag{6.2}$$

where the state, input, and output matrices $A(k)$, $B(k)$ and $C(k)$ are assumed to be bounded functions of k.

It is important to address that equation system (6.1) is *not uniformly observable*, i.e., it may admit inputs for which observability is lost (Besançon and Ţiclea 2007). Therefore, characterizing the observability property of the system through the observability Gramian is necessary, which for system (6.2) is defined as

$$\Gamma(u(k), \sigma) = \sum_{l=k-\sigma}^{k} \Phi(l,k)^T\, C_l^T C_l \Phi(l,k)\,, \tag{6.3}$$

where σ is the length of the Gramian windows and

$$\Phi(k, k_0) = A(k)\,A(k-1)\cdots A(k_0) = \prod_{p=k_0}^{k} A(p)$$

is the state transition matrix. All matrices $A(p)$ in the product are assumed invertible to guarantee the existence of an inverse for $\Phi(k, k_0)$. The eigenvalues of Γ, specifically the lowest eigenvalue, can be used to define a *degree of observability*.

As a result, a *regularly persistent input* for the observability of (6.1) can be defined as done by Ţiclea and Besançon (2009).

Definition 6.1 An input sequence $u(k)$ is a *regularly persistent input* for (6.1) if, for any initial condition $x(0)$, the induced linear time-varying representation (6.2) is *completely uniformly observable*: it satisfies the next property:

There exists a fixed natural number σ such that at any sampling instant k we have

$$\Gamma(u(k), \sigma) - \alpha I > 0 \tag{6.4}$$

where I stands for identity matrix.

Notice that given the system (6.1) different kinds of observer structures can be designed for both state and parameter estimation. The extended Kalman filter, for example, has been used as a parameter and state estimator for stochastic and deterministic systems (Ljung 1979; Torres et al. 2011). The Kalman-like observer presented by Besançon et al. (1996) has been widely used as the parameter and state estimator in fluid transmission lines (Besançon et al. 2013; Torres et al. 2015, 2016). The use of classical high-gain observers for the design of parameter identification algorithms has been studied by Torres et al. (2011). An exponential forgetting factor observer in discrete time, presented by Ţiclea and Besançon (2009), was analyzed by Scola et al. (2013) for parameter and state estimation.

It is emphasized that in this work the attention is focused upon the calculation of the *regularly persistent input*, more than in the observer used in the solution of the corresponding nonlinear estimation problem.

6.3 Input Optimization Algorithm

To guarantee the convergence of the observer used in the solution of the corresponding nonlinear estimation problem, the input $u\left(k\right)$ must be regularly persistent (Ţiclea and Besançon 2009). For some really simple systems, a regularly persistent input can be analytically found by calculating the Gramian expressed by (6.3) and by taking into account the observability condition in Definition 6.1 (Besançon 2007; Scola et al. 2013). For complex systems, however, analytically finding an appropriate $u\left(k\right)$ is not so easy. Because of that, Scola et al. (2013) proposed an offline optimization algorithm for finding a periodic persistent input sequence for a given system over a time window of N steps, where the optimization problem under consideration can be summarized as follows:

$$\begin{array}{c} \min\limits_{u_k} \sum_{k=0}^{N-1} \left(u_{ref} - u_k\right)^2 \\ \Gamma\left(U\left(k+i\right), N\right) - \alpha I > 0, \ 0 \leq i \leq N-1 \\ U\left(k\right) = \left[u_0 \, u_1 \, \cdots \, u_{N-2} \, u_{N-1}\right] \\ u^{\min} < u_k < u^{\max}, \ 0 \leq k \leq N-1 \\ \Delta u^{\min} < \left|u_{k+1} - u_k\right| < \Delta u^{\max}, \ 0 \leq k \leq N-2 \\ \Delta u^{\min} < \left|u_0 - u_{N-1}\right| < \Delta u^{\max} \end{array} \tag{6.5}$$

where $\Gamma\left(U\left(k+i\right), N\right)$ represents the observability Gramian over the time window $[0, N-1]$ with input sequence $U\left(k\right)$, since the input sequence is periodic

$$U\left(k+N-1\right) = \left[u_{N-1} \, u_0 \, \cdots \, u_{N-3} \, u_{N-2}\right],$$

where N should be greater than or equal to the system dimension, u_{ref} is a reference signal, $u^{\min}$ and $u^{\max}$ are the minimum and maximum value for each element of the input sequence $U\left(k\right)$ and $\Delta u^{\min}$ and $\Delta u^{\max}$ are the minimum and maximum value for the difference between two consecutive elements of the sequence. In the original algorithm (Scola et al. 2013), only the maximum absolute difference between two consecutive elements of the sequence $\Delta u^{\max}$ was considered. The minimum absolute difference, $\Delta u^{\min}$, was included in this work to guarantee that the actuator resolution limitation was respected. In other words, there is a limitation in the smallest increment or step that can be taken or seen by the device that generates the physical input.

Notice that the algorithm looks for the persistent input with minimal energy, by considering the energy of the difference between the reference signal u_{ref} and the additional persistent input $U\left(k\right)$. Additionally, since the input needs to be defined at any time (for the observer application purpose (Scola et al. 2013)), the sequence $U\left(k\right)$ calculated through the algorithm must be repeated as many times as necessary. The N Gramian constraints guarantee that the observability condition (6.4) holds at any shifted time.

6.4 Recalls on Liénard Equation

In the study of dynamical systems and differential equations, a Liénard system is a second-order differential equation that governs the behavior of second-order mechanical systems, corresponding to the equation (Torres et al. 2015)

$$\ddot{x}(t) + F_0(x(t))\dot{x}(t) + G_0(x(t)) = 0, \tag{6.6}$$

where $\dot{x}(t) = \frac{dx(t)}{dt}$, $\ddot{x}(t) = \frac{d^2x(t)}{dt^2}$ for given functions F_0, G_0 and position $x(t)$. By considering $x(t)$ and $\dot{x}(t)$ as state variables $x_1(t)$, $x_2(t)$, the Liénard-type equation can be rewritten in a state-space form as follows:

$$\begin{aligned} \dot{x}_1(t) &= x_2(t), \\ \dot{x}_2(t) &= -F_0(x_1(t))\, x_2(t) - G_0(x_1(t)). \end{aligned} \tag{6.7}$$

For state and parameter estimation, a more appropriate representation of (6.7) is obtained by the application of the following change of variables (Torres et al. 2015, 2016):

$$\begin{aligned} \begin{pmatrix} \zeta_1(t) \\ \zeta_2(t) \end{pmatrix} &= \Phi \begin{pmatrix} x_1(t) \\ x_2(t) \end{pmatrix}, \\ \Phi : \begin{pmatrix} x_1(t) \\ x_2(t) \end{pmatrix} &\to \begin{pmatrix} x(t) \\ \dot{x}(t) + F(x(t)) \end{pmatrix}, \end{aligned} \tag{6.8}$$

where $F(x) = \int_0^x F_0(\sigma)\, d\sigma$, which transforms (6.7) into

$$\begin{aligned} \dot{\zeta}_1(t) &= \zeta_2(t) - F(\zeta_1(t)), \\ \dot{\zeta}_2(t) &= -G_0(\zeta_1(t)), \\ y(t) &= \zeta_1(t). \end{aligned} \tag{6.9}$$

The representation (6.9) is called the Liénard form, and since now nonlinear functions are decoupled from the unmeasured state $\zeta_2(t)$, it facilitates estimation.

In the particular case in which functions F_0, G_0 are linearly parameterized with respect to some parameter vector θ, a Liénard system can take the following form (Besançon et al. 2010):

$$\begin{aligned} \dot{\zeta}(t) &= A_o\zeta(t) + \Phi(u(t), y(t))\theta + \varphi(u(t), y(t)), \\ y(t) &= C_o(t), \end{aligned} \tag{6.10}$$

where $A_o = \begin{bmatrix} 0 & 1 \\ 0 & 0 \end{bmatrix}$, $C_o = \begin{bmatrix} 1 & 0 \end{bmatrix}$ and Φ, φ are functions that depend exclusively on the inputs and outputs of the system.

By considering the augmented state vector $\xi(t) = \left[\zeta(t)\ \theta\right]^T$, system (6.10) can be rewritten as follows:

$$\dot{\xi}(t) = \underbrace{\begin{bmatrix} A_o & \Phi(u(t), y(t)) \\ 0 & 0 \end{bmatrix}}_{A(u(t),y(t))} \xi(t) + \underbrace{\begin{bmatrix} \varphi(u(t), y(t)) \\ 0 \end{bmatrix}}_{B(u(t),y(t))}$$

$$y(t) = \underbrace{\left[C_o\ 0\right]}_{C} \xi(t) \tag{6.11}$$

where 0 stands for null matrix of appropriate dimensions. Notice that system (6.11) has the general form given by (6.1).

6.5 Liénard-type Models for a Pipelines

6.5.1 *Hydraulic Equations*

By assuming negligible convective changes in velocity, as well as both the liquid density and the cross-sectional area are constant, the momentum and continuity equations governing the dynamics of the fluid in a horizontal pipeline can classically be expressed as (Torres et al. 2015)

$$\begin{aligned} \frac{\partial Q(z,t)}{\partial t} &= -a_1 \frac{\partial H(z,t)}{\partial z} - \mu Q(z,t)\left|Q(z,t)\right|, \\ \frac{\partial H(z,t)}{\partial t} &= -a_2 \frac{\partial Q(z,t)}{\partial z}, \end{aligned} \tag{6.12}$$

where $(z, t) \in (0, L) \times (0, \infty)$ gathers the space (m) and time (s) coordinates respectively, L is the length of the pipe, $H(z, t)$ is the pressure head (m) and $Q(z, t)$ is the flow rate $\left(m^3/s\right)$. The parameters of the model are

$$a_1 = gA_r,\ a_2 = \frac{b^2}{gA_r},\ \mu = \frac{f}{2\phi A_r},$$

where b is the wave speed in the fluid (m/s), g is the gravitational acceleration $\left(m/s^2\right)$, A_r is the cross-sectional area of the pipe $\left(m^2\right)$, ϕ is the inside diameter of the pipe (m), and f is the Darcy–Weisbach friction factor.

To approximate equation system (6.12), two of the following Dirichlet conditions must be imposed at the boundaries of the pipeline: (i) upstream pressure head, $H(0,t) = H_{in}(t)$; (ii) downstream pressure head, $H(L,t) = H_{out}(t)$; (iii) upstream flow rate, $Q(0,t) = Q_{in}(t)$; and (iv) downstream flow rate, $Q(L,t) = Q_{out}(t)$. On one hand, in this work the boundary conditions considered for all the simulations and the discretization of (6.12) are $H_{in}(t)$ and $H_{out}(t)$. On the other hand, measurements of flow and pressure head at the ends of the pipeline are denoted with the same nomenclature of the boundary conditions.

6.5.2 Liénard Representation

The equations system (6.12) can be rewritten into a Liénard-like form as done by Torres et al. (2015), where the so-called *flow-based Liénard form* of the pipeline dynamics was obtained and can be written as

$$\begin{aligned} \frac{\partial Q^a(z,t)}{\partial t} &= Q^b(z,t) - \mu Q^a(z,t)|Q^a(z,t)|, \\ \frac{\partial Q^b(z,t)}{\partial t} &= b^2 \frac{\partial^2 Q^a(z,t)}{\partial^2 z}. \end{aligned} \tag{6.13}$$

By using the finite-difference method, the spatial discrete version of (6.13) is expressed by

$$\begin{aligned} \dot{Q}_i^a(t) &= Q_i^b(t) - \mu Q_i^a(t)|Q_i^a(t)|, \\ \dot{Q}_i{}^b(t) &= b^2 \left[\frac{Q_{i+1}^a(t) - 2Q_i^a(t) + Q_{i-1}^a(t)}{(\Delta z)^2} \right], \end{aligned} \tag{6.14}$$

where Δz is the spatial step. Conversely, Torres et al. (2016) developed the *hybrid Liénard model* (flow rate and pressure head derivative), which takes the following form:

$$\begin{aligned} \frac{\partial Q^a(z,t)}{\partial t} &= Q^b(z,t) - \mu Q^a(z,t)|Q^a(z,t)|, \\ \frac{\partial Q^b(z,t)}{\partial t} &= -gA_r \frac{\partial^2 H(z,t)}{\partial t \partial z}, \end{aligned} \tag{6.15}$$

where $\dot{H}$ stands for $\partial H/\partial t$. If (6.15) is discretized in space, one gets the following ODE system:

$$\begin{aligned} \dot{Q}_i^a(t) &= Q_i^b(z,t) - \mu Q_i^a(z,t)|Q_i^a(z,t)|, \\ \dot{Q}_i{}^b(t) &= a_1 \frac{\dot{H}_i(t) - \dot{H}_{i+1}(t)}{\Delta z}, \end{aligned} \tag{6.16}$$

where Δz is the spatial step.

An advantage of using Liénard-type models for expressing the dynamical behavior of a fluid transmission line is the suitability of these types of models for formulating estimation algorithms (Torres et al. 2015, 2016).

6.5.3 Extension of the Input Optimization Algorithm to Liénard-type Models for Pipelines

In this work, two illustrative examples regarding the application of the proposed optimization algorithm for generating *regularly persistent inputs*, while taking into consideration Liénard-type models, are presented. This has not been reported previously. In the first example, the model and optimization algorithm given by (6.14) and (6.5), respectively, are considered. In the second one, the model used for representing the dynamical behavior of the pipeline is the given by (6.16). In this case the model inputs are the time derivatives of the downstream and upstream pressure head, $\dot{H}_{in}(t)$ and $\dot{H}_{out}(t)$, and additionally it is assumed that such variables are not independent. Let $u^1(t) = H_{in}(t)$ and $u^2(t) = H_{out}(t)$. Since in pipeline systems the flow is usually forced by upstream pumping, H_{in} is considered the independent variable, that is $u^2(t) = g\left(u^1(t)\right)$. For simplicity, a steady-state gain model is used, thereby $u^2(t) = K^{21}u^1(t)$. Because of this, it was necessary to make some additional modifications to the original algorithm given by (6.5) as follows:

$$\begin{array}{c}\min\limits_{u_k} \sum_{k=0}^{N-1} \left(u_{ref} - u_k^1\right)^2 \\ \Gamma\left(U_D^1(k+i), N\right) - \alpha I > 0,\ 0 \le i \le N-1 \\ U_D^1(k) = \left[\frac{u_1^1 - u_0^1}{\Delta t}\ \frac{u_2^1 - u_1^1}{\Delta t}\ \cdots\ \frac{u_{N-1}^1 - u_{N-2}^1}{\Delta t}\ \frac{u_0^1 - u_{N-1}^1}{\Delta t}\right] \\ U_D^2(k) = g\left(U_D^1(k)\right),\ 0 \le k \le N-1 \\ u^{\min} < u_k^1 < u^{\max},\ 0 \le k \le N-1 \\ \Delta u^{\min} < \left|u_{k+1}^1 - u_k^1\right| < \Delta u^{\max},\ 0 \le k \le N-2 \\ \Delta u^{\min} < \left|u_{N-1}^1 - u_0^1\right| < \Delta u^{\max}\end{array} \tag{6.17}$$

where $\Gamma\left(U_D^1(k+i), N\right)$ stands for the observability Gramian over the time window $[0, N-1]$ with input sequence' derivative $U_D^1(k)$, since the input sequence is periodic

$$U_D^1(k+N-1) = \left[\frac{u_0^1 - u_{N-1}^1}{\Delta t}\ \frac{u_1^1 - u_0^1}{\Delta t}\ \cdots\ \frac{u_{N-3}^1 - u_{N-2}^1}{\Delta t}\right],$$

where N should be greater than or equal to the system dimension, u_{ref} is a reference signal for the input sequence $U_D^1(k)$, $u^{\min}$ and $u^{\max}$ are the minimum and maximum value for each element of the sequence $U_D^1(k)$ and $\Delta u^{\min}$ and $\Delta u^{\max}$ are the minimum and maximum value for the difference between two consecutive elements of the sequence $U_D^1(k)$.

The proposed algorithm works on the input sequence $U_D^1(k)$, but at each iteration the derivative of that sequence is approximated by using forward finite differences to calculated the observability Gramians $\Gamma\left(U_D^1(k+i), N\right)$. As in algorithm (6.5), for the observer application purpose the input needs to be defined at any time, so the sequence $U_D^1(k)$ calculated must be repeated as many times as necessary.

6.6 Tests: Parameter Identification in a Pipeline

Two examples regarding the application of the proposed optimization algorithms, taking into consideration Liénard-type models, are presented. In the first example, the model and optimization algorithm given by (6.14) and (6.5), respectively, were contemplated. In the second one, the model used for representing the dynamical behavior of the pipeline was the given by (6.15), and the optimization algorithm was described by (6.17). Note that in both examples it was verified that the *regularly persistent inputs* calculated were independent of the initial conditions $\hat{\xi}(0)$. Additionally, in this section the exponential observer presented by Besançon et al. (1996) was used in the solution of the corresponding nonlinear estimation problem.

6.6.1 *Simulation Test: Estimation of the Friction Coefficient and the Wave Speed*

In the system under consideration, the pipeline is assumed to be connected to a pump that provides the upstream pressure head. The corresponding model to this system is described by the differential equations given by (6.12), and the assumed boundary conditions were the upstream and downstream pressure heads, $H_{in}(t)$, $H_{out}(t)$. In addition, Table 6.1 provides the list of model parameters considered here. In order to compare results, such parameters were taken from Torres et al. (2015).

If the parameters f, b are unknown, the augmented system (as in 6.11) would be

$$\dot{\xi}(t) = \begin{bmatrix} 0 & 1 & -\frac{y(t)|y(t)|}{2\phi A_r} & 0 \\ 0 & 0 & 0 & \rho(t) \\ 0 & 0 & 0 & 0 \\ 0 & 0 & 0 & 0 \end{bmatrix} \xi(t), \tag{6.18}$$

where $\rho(t) = (u_1(t) - 2y(t) + u_2(t))/(\Delta z)^2$, $u_1(t) = Q_{in}(t)$, $u_2(t) = Q_{out}(t)$, $y(t)$ is an intermediate flow measurement and $\xi(t) = \left[\, Q_i^a(t)\; Q_i^b(t)\; f\; b^2 \,\right]^T$. The followings settings were considered for the simulation (Torres et al. 2015)

Table 6.1 Physical parameters

Symbol	Value	Units	Description
g	9.81	m/s^2	Gravitational acceleration
L	200.16	m	Pipeline length
ϕ	0.1016	m	Pipeline diameter
b	1284	m/s	Wave speed in the fluid
f	0.022		Friction coefficient

Table 6.2 Parameter for H_{in} sequence calculation for f and b estimation

Symbol	Value	Units	Symbol	Value	Units
u_{min}	17	m	T_{in}	1.5	s
u_{max}	20	m	α	0.018	
Δu_{min}	0.8	m	N	10	
Δu_{max}	1.2	m			

- The downstream pressure head was set to be $H_{out}(t) = 5.7\ (m)$
- The observer was set with $\lambda = 1$, $S(0) = I$ and the initial conditions $\hat{\xi}(0) = \left[0.01913,\ 0.01913,\ 0.07,\ 112871\right]^T$

In addition, the parameters employed for the input calculation are listed in Table 6.2.

MATLAB Tutorial: Implementation of the Optimization Algorithm for the Estimation of the Friction Coefficient and the Wave Speed

The objective of this tutorial is to demonstrate how to implement the presented algorithm (6.5).

1. Create a new file called *Opt_NL_ObservadorFriccionVelOnda.m.*
2. The program minimizes the quadratic function $\sum_{k=0}^{N-1}\left(u_{ref} - u_k\right)^2$ subject to nonlinear constraints and bound variables, by using a sequential quadratic programming algorithm.
3. Enter the following program into the file

```
global N Uref Ts alpha U_dot_max U_dot_min
Ts = 1.5; % Time period of the sequence
alpha = 0.018; % Alpha parameter
% Maximum and minimum value for the difference
% between two consecutive elements of the sequence
U_dot_max = 1.2/Ts;
U_dot_min = 0.8/Ts;
N = 10; % Time windows steps
% Minimum and maximum value for each element
% of the sequence
lb = 17;
ub = 20;
lb = lb*ones(1,N);
ub = ub*ones(1,N);
uref = 18;
Uref = uref*ones(1,N); % Reference signal
% Finds a constrained minimum of the function @objfun
u0 = rand(1,N);
options = optimoptions(@fmincon,'Algorithm','sqp');
[x,fval] = fmincon(@objfun,u0,[],[],[],[],lb,ub,...
    @confuneq,options)
```

4. Write a file *objfun.m* for the objective function $\sum_{k=0}^{N-1} \left(u_{ref} - u_k\right)^2$.

```
function f = objfun(x)
global Uref
U = x;
f = sum((Uref-U).^2); % Objective function
```

5. Write a file *confuneq.m* for the nonlinear constraints.

```
function [c,ceq] = confuneq(x)
global N Ts alpha U_dot_max U_dot_min
U = x;
% Model parameters
D = 0.1016; % Pipe inner diameter
Ar = pi*(D/2)^2; % Pipe cross-section area
m = 1/(2*D*Ar);
dz = 95.979000472368213; % Pipe length
% Calculation of the difference between
% two consecutive elements of the sequence
uu = nan(1,N);
for i=1:N-1
    uu(i) = abs(U(i+1)-U(i));
end
uu(N) = abs(U(N)- U(1));
% Observability Gramian calculation
for i=1:N
    x0 = [0.019130256698075 0.019 0.07 1284^2]';
    y = x0(1);
    U = circshift(U',-(i-1))';
    A = [0 1 -m*y^2 0;0 0 0 (U(1)-2*y+U(1))/dz^2;...
         0 0 0 0;0 0 0 0];
    B = [0 0 0 0]';
    C = [1 0 0 0];
    n = size(A,1);
    p = size(B,2);
    AA = [A B;zeros(p,n) zeros(p,p)];
    AA_exp = expm(Ts*AA);
    Ad = AA_exp(1:n,1:n);
    Cd = C;
    phy1 = eye(size(Ad));
    gram_o = phy1'*(Cd'*Cd)*phy1;
    for l=1:N
        A = [0 1 -m*y^2 0;0 0 0 (U(1)-2*y+U(1))/dz^2;...
             0 0 0 0;0 0 0 0];
        AA = [A B;zeros(p,n) zeros(p,p)];
        AA_exp = expm(Ts*AA);
        Ad = AA_exp(1:n,1:n);
        phy2 = Ad*phy1;
        phy1 = phy2;
        gram_o = gram_o + phy2'*(Cd'*Cd)*phy2;
    end
```

```
    M = alpha*eye(n,n) - gram_o;
    cc(n*i-(n-1):n*i) = eig(M);
end
% Equality and inequality restrictions
ceq = [];
c = [cc';(U_dot_min*Ts-uu)';(uu-U_dot_max*Ts)'];
```

Figure 6.1 illustrates the optimal persistent $H_{in}(t)$ sequence obtained and used in the simulation, which is repeated over time. Each step of this sequence corresponds to a time period of $T_{in} = 1.5\ (s)$.

Some estimation results are presented in Fig. 6.2. In this figure, the results obtained with the optimal persistent input calculated are compared with those of two additional sine-like input signals. The first, labeled as "sin" in the figure, corresponds to the input signal proposed by Torres et al. (2015), i.e., $H_{in}(t) = 18 + 0.9sin(t)\ (m)$. The second one, labeled as "sin*" in the figure, corre-

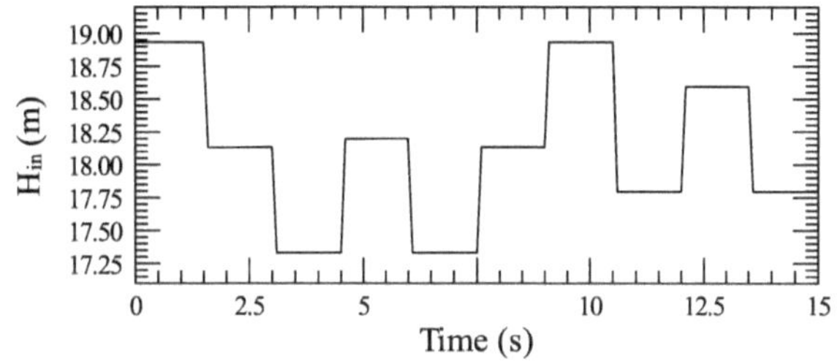

Fig. 6.1 Optimal regularly persistent $H_{in}(t)$ sequence for f and b estimation

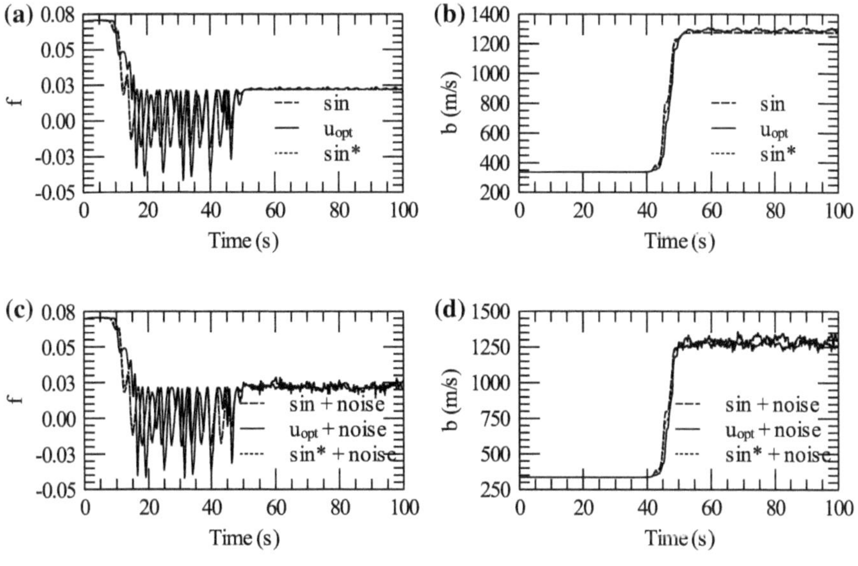

Fig. 6.2 Friction coefficient and wave speed estimation

sponds to a sine-like signal with the same average energy that the optimal persistent input obtained: $H_{in}(t) = 18 + 0.78sin(t)\ (m)$. Reported results in fact present the simultaneous estimation of the friction coefficient and wave speed, both with and without measurement noise.

From these figures, one can notice how the estimation is in effect well performed in all cases. Table 6.3 summarizes the average energy and IAE performance indexes obtained for each of the inputs considered. Higher energy inputs show better estimation accuracy; however, the performance of the optimal persistent input is sufficiently good (very close to those of the sine-like signals). The performance is even better if we consider the fact that a pulse signal, as shown in Fig. 6.1, is generated more easily in a control unit as a programmable logic controller, for example, than a sinusoidal one.

To study the effect of the frequency of the sine-like signal on the observer's performance, an input signal of the form $H_{in}(t) = 18 + 0.9sin(\omega t)\ (m)$ was considered where different values for ω were used, as illustrated in Fig. 6.3. The figures reveal as the frequency is increased the friction coefficient estimation is more affected by the measurement noise, while a reduction in the frequency results in a equivalent length estimation more affected by measurement noise. Additionally, no significant changes are observed in the energy of the signal; see Table 6.4.

Table 6.3 Input average energy and IAE performance index in f and b estimation

Signal	Avg. energy	f estimation IAE	b estimation IAE
u_{opt}	30.5	104.4	25.2
sin	40.68	97.69	24.77
sin*	30.5	99.11	24.93

Table 6.4 Frequency - Inputs average energy and IAE performance index in f and b estimation

Frequency	Avg. energy	f estimation IAE	b estimation IAE
4 Hz	40.45	120.7	25.29
3 Hz	40.5	108.1	25.04
2 Hz	40.59	98.65	24.8
1.75 Hz	40.61	96.8	24.71
1.5 Hz	40.63	94.68	24.66
1 Hz	40.68	90.79	24.7
0.5 Hz	40.71	88.96	25.46
0.25 Hz	40.71	95.42	26.81

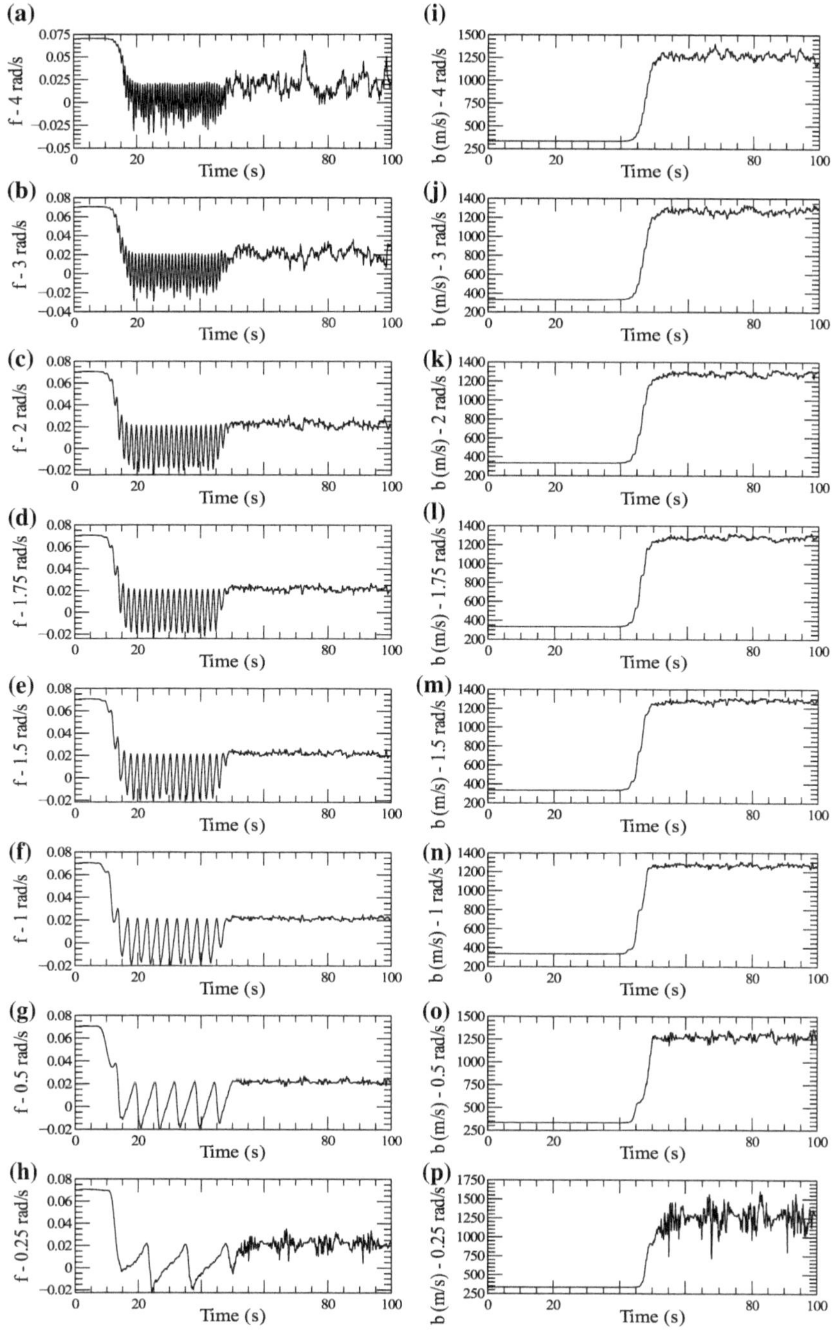

Fig. 6.3 Friction coefficient and equivalent length estimation with sine-like inputs

6.6.2 *Experimental Test: Estimation of the Friction Coefficient and The Equivalent Length*

In the observer designed in the previous section, three flow measurements along the pipeline are necessary: one measure at each boundary ($Q_{in}(t)$, $Q_{out}(t)$) and one in between. The observer works well enough according to simulation results, but in a real application high-velocity response flow sensors would be needed, and working with fewer sensors would be preferable.

In this section, the simultaneous estimation of the friction coefficient and the equivalent length of a pipeline is performed by using only boundary measurements ($H_{in}(t)$, $H_{out}(t)$, $Q_{in}(t)$) as proposed by Torres et al. (2016), but with an optimal regularly persistent input calculated through the algorithm (6.17). A prototype built in the Laboratorio de Hidrodinémica of the Instituto de Ingenieréa of the UNAM has been considered here. The prototype (Fig. 6.4) is equipped with:

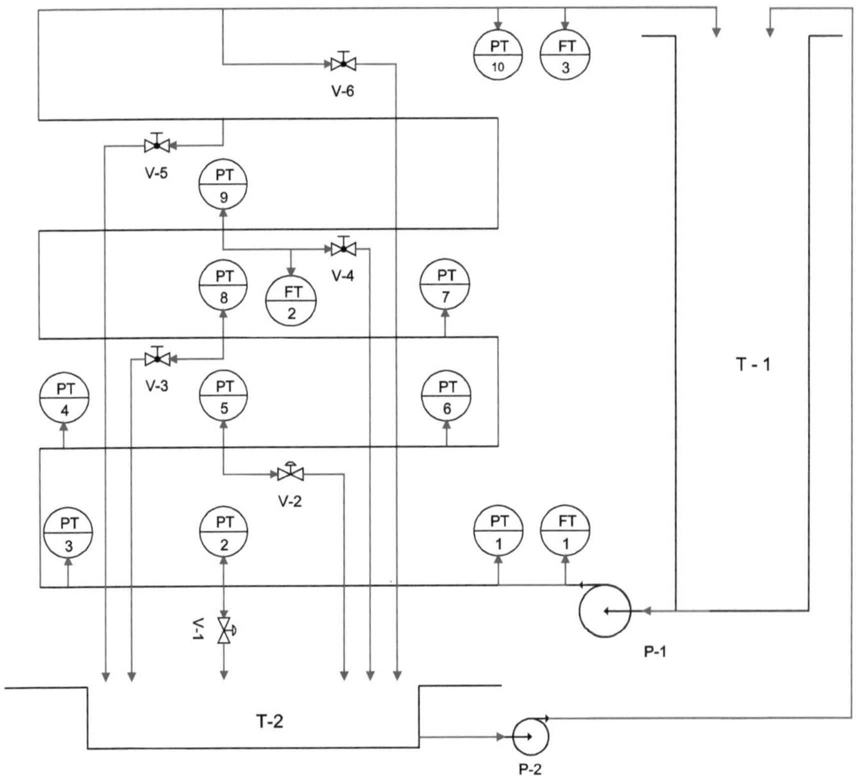

Fig. 6.4 Pipeline prototype scheme

- A 7.5 HP centrifugal pump (P-1), which provides the energy needed to recirculate the water from a reservoir (T-1) through a galvanized steel pipeline of 0.1016 m of diameter and 169.43 m of length.
- A Mitsubishi variable-frequency drive (VFD) which controls the rotational speed of the pump motor by a variation of the AC frequency in a range from 0 to 60 Hz.
- Six taps (V-1 to V-6) to simulate leaks.
- Eight pressure measurement intermediate points (PT-2 to PT-9).
- Flow and pressure sensors installed at both ends of the pipeline, as well as a flow sensor to measure the flow through tap No. 4 (FT-2).
- A recycling system for leak water composed by a tank (T-2) and a centrifugal pump (P-2).

If the parameters f, L are unknown, system (6.16) can be augmented (as done in 6.11) with the inclusion of f, L in the state vector of system (6.16), for designing the observer. The augmented system appears as follows:

$$\dot{\xi}(t) = \begin{bmatrix} 0 & 1 & -\frac{\xi_1(0)|\xi_1(0)|}{2\phi A_r} & 0 \\ 0 & 0 & 0 & \rho(t) \\ 0 & 0 & 0 & 0 \\ 0 & 0 & 0 & 0 \end{bmatrix} \xi(t), \tag{6.19}$$

where $\rho(t) = a_1(\dot{u}_1(t) - \dot{u}_2(t)), u_1(t) = H_{in}(t), u_2(t) = H_{out}(t), y(t) = Q^a_{in}(t)$ and $\xi(t) = \begin{bmatrix} Q^a_{in}(t) & Q^b_{in}(t) & f & \frac{1}{L} \end{bmatrix}^T$. Notice that the spatial step Δz in (6.16) has been replaced by equivalent length L. The observer was set with $\lambda = 0.25$, $S(0) = I$, and the initial conditions $\hat{\xi}(0) = \begin{bmatrix} 0.01913, & 0.01913, & 0.07, & \frac{1}{300} \end{bmatrix}^T$. The parameters used for the input calculation are listed in Table 6.5.

MATLAB Tutorial: Implementation of the Optimization Algorithm for the Estimation of the Friction Coefficient and the Equivalent Length

The objective of this tutorial is to demonstrate how to implement the presented algorithm (6.17).

1. Create a new file called *Opt_NL_ObservadorFriccionLeq.m*.
2. The program minimizes the quadratic function $\sum_{k=0}^{N-1}\left(u_{ref} - u_k\right)^2$ subject to nonlinear constraints and bounds, by using a sequential quadratic programming algorithm.

Table 6.5 Parameter for H_{in} sequence calculation for f and L estimation

Symbol	Value	Units	Symbol	Value	Units
u_{min}	17	m	T_{in}	6	s
u_{max}	20	m	α	0.00018	
Δu_{min}	1	m	N	6	
Δu_{max}	0.5	m	K^{21}	0.19426	

3. Enter the following program into the file

```
global N Uref Ts alpha U_dot_max U_dot_min
Ts = 6; % Time period of the sequence
alpha = 0.00018; % Alpha parameter
% Maximum and minimum value for the difference
% between two consecutive elements of the sequence
U_dot_max = 0.5;
U_dot_min = 0.1/Ts;
N = 6; % Time windows steps
% Minimum and maximum value for each element
% of the sequence
lb = 17;
ub = 20;
lb = lb*ones(1,N);
ub = ub*ones(1,N);
uref = 18.325;
Uref = uref*ones(1,N); % Reference signal
% Finds a constrained minimum of the function @objfun
u0 = rand(1,N);
options = optimoptions(@fmincon,'Algorithm','sqp');
[x,fval] = fmincon(@objfun,u0,[],[],[],[],lb,ub,...
    @confuneq,options)
% Pressure (m) to frequency (Hz) transformation
frq = 1.6667*x + 26.667
```

4. Write a file *objfun.m* for the objective function $\sum_{k=0}^{N-1} \left(u_{ref} - u_k\right)^2$.

```
function f = objfun(x)
global Uref
U = x;
f = sum((Uref-U).^2); % Objective function
```

5. Write a file *confuneq.m* for the nonlinear constraints.

```
global N Ts alpha U_dot_max U_dot_min
U = x;
% Model parameters
g = 9.81; % Gravitational acceleration
D = 0.1016; % Pipe inner diameter
Ar = pi*(D/2)^2; % Pipe cross-section area
a1 = g*Ar;
m = 1/(2*D*Ar);
% Calculation of the difference between
% two consecutive elements of the sequence
uu_d = nan(1,N);
for i=1:N-1
```

```
    uu_d(i) = (U(i+1)-U(i))/Ts;
end
% Calculation of the sequence first derivative
uu_d(N) = (U(N)-U(1))/Ts;
U = uu_d;
uu = nan(1,N);
for i=1:N-1
    uu(i) = abs(U(i+1)-U(i));
end
uu(N) = abs(U(N)- U(1));
% Observability Gramian calculation
for i=1:N
    x0 = [0.019130256698075 0.019 0.07 1/300]';
    y = x0(1);
    U = circshift(U',-(i-1))';
    A = [0 1 -m*y*abs(y) 0;...
         0 0 0 a1*(U(1)-0.19426*U(1));...
         0 0 0 0;0 0 0 0];
    B = [0 0 0 0]';
    C = [1 0 0 0];
    n = size(A,1);
    p = size(B,2);
    AA = [A B;zeros(p,n) zeros(p,p)];
    AA_exp = expm(Ts*AA);
    Ad = AA_exp(1:n,1:n);
    Bd = AA_exp(1:n,n+1:n+p);
    Cd = C;
    phy1 = eye(size(Ad));
    gram_o = phy1'*(Cd'*Cd)*phy1;
    for l=1:N
        A = [0 1 -m*y*abs(y) 0;...
             0 0 0 a1*(U(1)-0.19426*U(1));...
             0 0 0 0;0 0 0 0];
        AA = [A B;zeros(p,n) zeros(p,p)];
        AA_exp = expm(Ts*AA);
        Ad = AA_exp(1:n,1:n);
        phy2 = Ad*phy1;
        phy1 = phy2;
        gram_o = gram_o + phy2'*(Cd'*Cd)*phy2;
    end
    M = alpha*eye(n,n) - gram_o;
    cc(n*i-(n-1):n*i) = eig(M);
end
% Equality and inequality restriction
ceq = [];
c = [cc';(U_dot_min*Ts-uu)';(uu-U_dot_max*Ts)'];
```

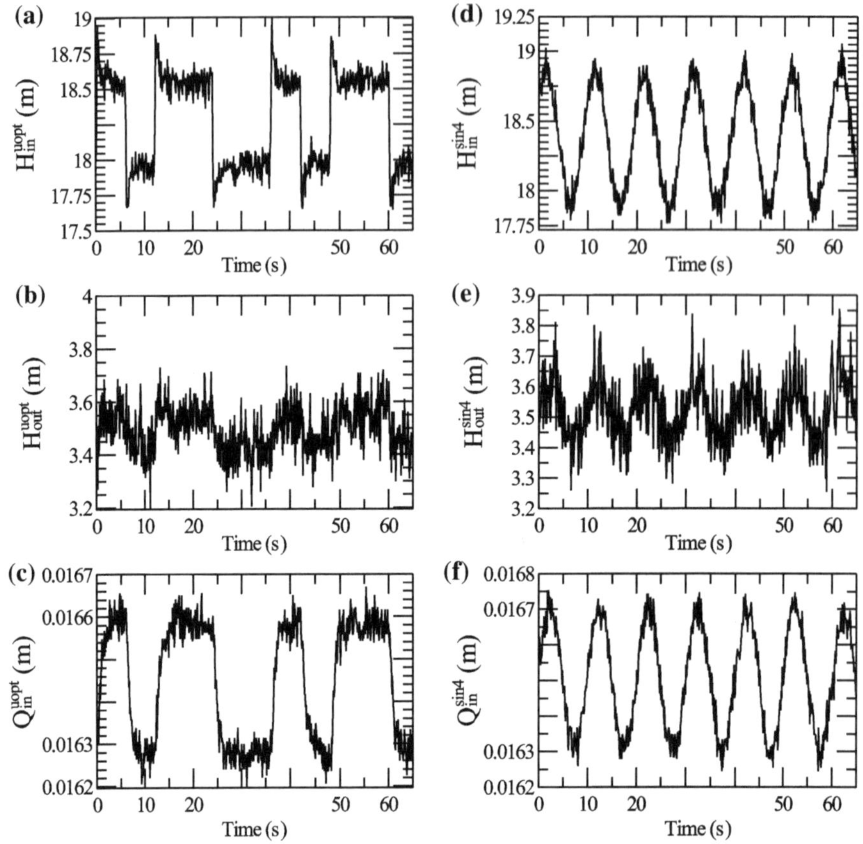

Fig. 6.5 Optimal regularly persistent $H_{in}(t)$, $H_{out}(t)$ for f and L estimation

The sequence obtained was

$$H_{in}^{u_{opt}}(k) = \left[\, 18.0309,\ 18.6251,\ 18.0157,\ 18.6233,\ 18.6309,\ 18.0251 \,\right]\ (m).$$

Each step of this sequence corresponds to a time period of $T_{in} = 6\ (s)$. To experimentally generate this pressures' sequence, the pump controller was set up with the sequence

$$VFD^{u_{opt}}(k) = \left[\, 56.7191,\ 57.7095,\ 56.6937,\ 57.7064,\ 57.7191,\ 56.7095 \,\right]\ (\mathrm{Hz}).$$

In Fig. 6.5(a)–(c), the pressure heads ($H_{in}^{u_{opt}}(t)$, $H_{out}^{u_{opt}}(t)$) and the input flow rate ($Q_{in}^{u_{opt}}(t)$) measurements are displayed, respectively.

In addition, simply to compare results, the estimation was also carried out by using several sine-like input signals. It was experimentally determined that sinusoids with

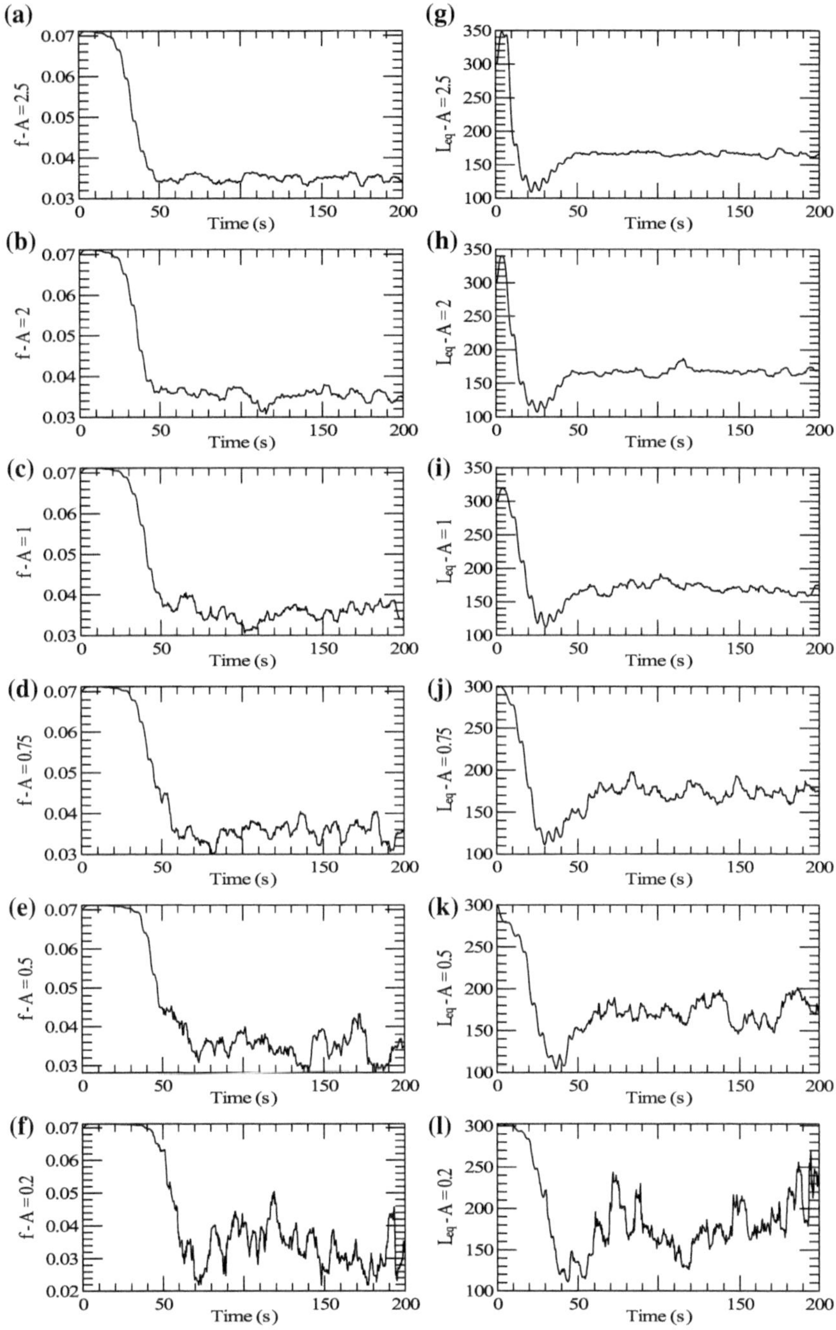

Fig. 6.6 Friction coefficient and equivalent length estimation with sine-like inputs

frequencies greater than 0.01 Hz cannot be generated by the pump. The use of very high or very slow frequencies could result in amplification of the measurement noise; see Fig. 6.3. Thus, signals of form $VFD(t) = 57.5 + A\sin(0.0628t)$ (Hz) were considered, where A was decreased from 2.5 to 0.2 as shown in Fig. 6.6. Table 6.6 gives the average energies and IAE performance indexes obtained for each of the inputs signal considered. Clearly, the lowest energy signal with which an acceptable estimation was performed corresponds to the amplitude $A = 0.75$. Therefore, for a given frequency of 0.01 Hz, the lowest energy signal with which the estimation performed well is the following:

$$VFD^{sin4}(t) = 57.5 + 0.75\sin(0.0628t) \text{ (Hz)}.$$

Table 6.6 Input average energy and IAE performance index in f and L estimation for sine-like inputs

Amplitude	Avg. energy	f estimation IAE	L estimation IAE
2.5	235	36.58	18.66
2	162.3	40.5	19.85
1	41.01	48.04	21.63
0.75	24	52.05	24.23
0.5	11.1	57.92	27.14
0.2	2.3	75.26	43.57

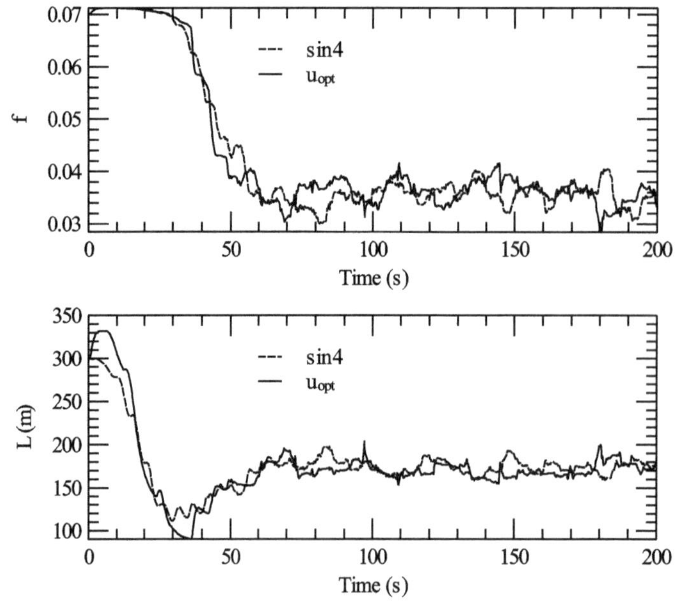

Fig. 6.7 Friction coefficient and equivalent length estimation

Table 6.7 Input average energy and IAE performance index in f and L estimation

Signal	Avg. energy	f estimation IAE	L estimation IAE
u_{opt}	21.64	52.27	27.92
sin	24.03	52.05	24.23
$u_{opt}^{R_\alpha=1.8}$	332.7	35.61	16.64

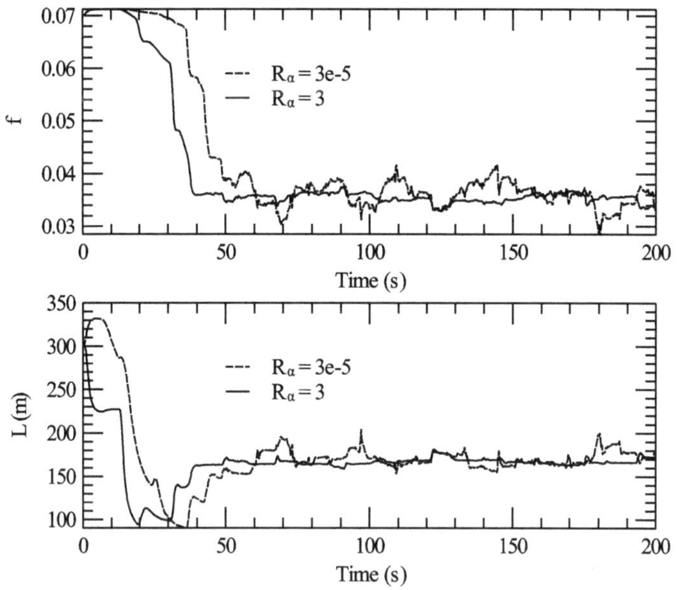

Fig. 6.8 Friction coefficient and equivalent length estimation for $R_\alpha = 3$ and $R_\alpha = 3 \times 10^{-5}$

In addition, the corresponding upstream and downstream pressure heads ($H_{in}^{sin4}(t)$, $H_{out}^{sin4}(t)$) and the input flow rates ($Q_{in}^{sin4}(t)$) are illustrated in Fig. 6.5d–f respectively.

Figure 6.7 shows the friction coefficient and equivalent length estimation results for both, the optimal regularly persistent input and the sine-like input. The convergence in this figure in both cases can be confirm; however, it is important to state that the energy of the sine-like input is higher than the energy of the optimal regularly persistent input calculated, see Table 6.7.

On the other hand, to analyze the effect of the algorithm parameters, which are the observability degree α and Gramian window N, on the observer performance, the ratio $R_\alpha = \frac{\alpha}{N}$ was increased from 3×10^{-5} to 3, as in Fig. 6.8 (and in a detailed manner in Fig. 6.9). The estimation results show clearly that by increasing R_α the estimation accuracy is improved, but there is a price to be paid for this improvement: an increment in the input energy, See Table 6.8.

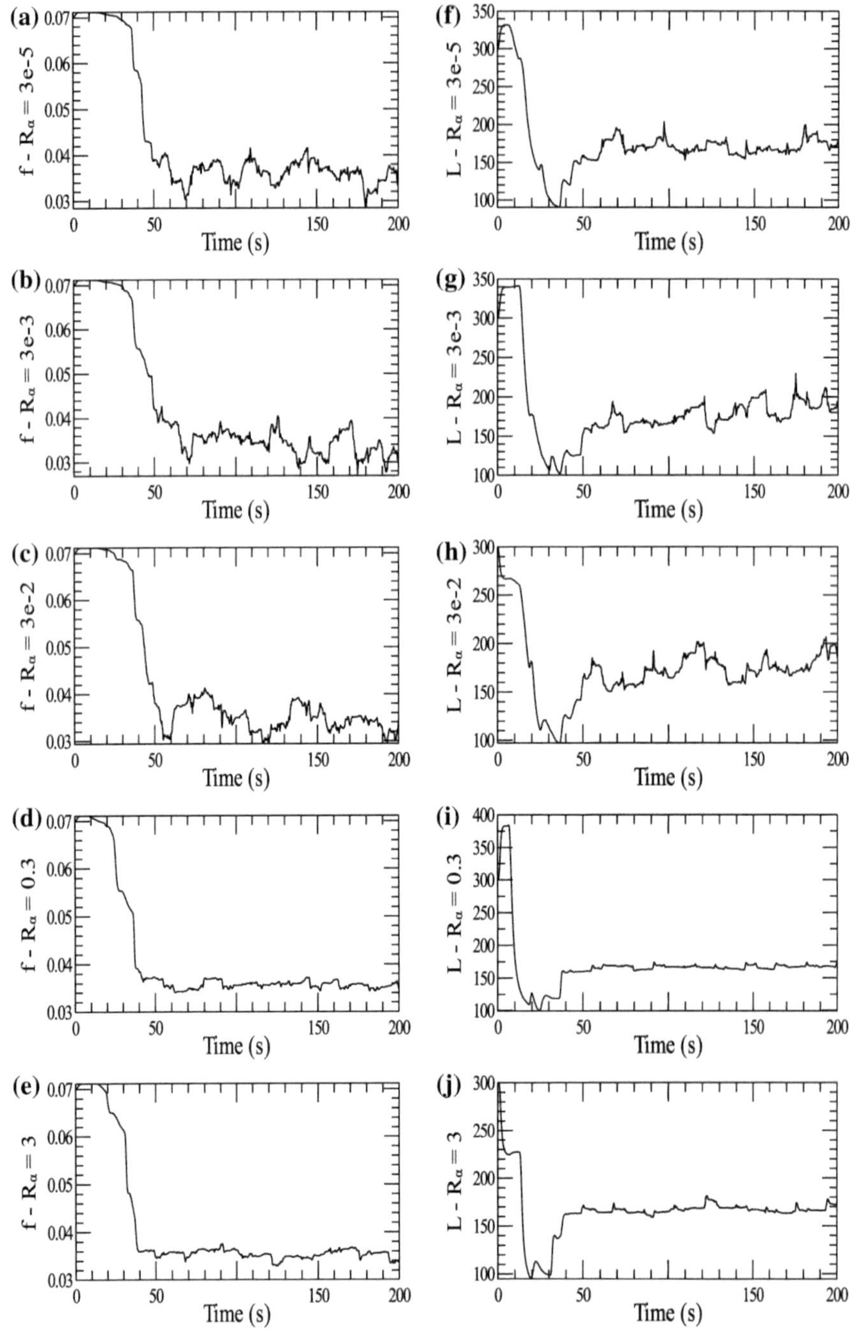

Fig. 6.9 R_α - Friction coefficient and equivalent length estimations

Table 6.8 R_α - Inputs average energy and IAE performance index in f and L estimations

R_α	Avg. energy	f estimation IAE	L estimation IAE
3×10^{-5}	21.64	52.27	27.92
3×10^{-3}	20.31	27.92	33.96
3×10^{-2}	30.23	52.23	26.69
0.3	354.5	35.9	21.12
3	332.7	35.61	16.64

6.7 Conclusions

In this chapter, a new methodology for calculating *regularly persistent inputs* for state affine systems has been presented. The proposed methodology is based on a optimization algorithm which has been described and validated through simulations and experimental test. Two representations of the transmission line as Liénard-like system have been employed to formulate exponential boundary observers. Such observers have been used to estimate the friction, the wave speed, and the equivalent length of a pipeline. Provided simulations and, even more importantly, experimental tests illustrated the good estimation results obtained with the *regularly persistent inputs* calculated through the proposed methodology. The use of the proposed methodology in problems of leak detection in pipeline, as well as its possible use for control purposes, will be part of future developments.

References

Besançon, G. (2007). *Nonlinear observers and applications*. Heidelberg: Springer.

Besançon, G., Bornard, G., & Hammouri, H. (1996). Observer synthesis for a class of nonlinear control systems. *European Journal of Control*, *2*(3), 176–192.

Besançon, G., Scola, I. R., & Georges, D. (2013). Input selection in observer design for non-uniformly observable systems. *IFAC Proceedings Volumes*, *46*(23), 664–669.

Besançon, G., & Ţiclea, A. (2007). An immersion-based observer design for rank-observable nonlinear systems. *IEEE Transactions on Automatic Control*, *52*(1), 83–88.

Besançon, G., Voda, A., & Jouffroy, G. (2010). A note on state and parameter estimation in a Van der Pol oscillator. *Automatica*, *46*(10), 1735–1738.

Campbell, S. L. & Nikoukhah, R. (2015). *Auxiliary signal design for failure detection*. Princeton: Princeton University Press.

Ţiclea, A., & Besançon, G. (2009). State and parameter estimation via discrete-time exponential forgetting factor observer. *IFAC Proceedings Volumes*, *42*(10), 1370–1374.

Isermann, R. & Münchhof, M. (2010). *Identification of dynamic systems: an introduction with applications*. Heidelberg: Springer.

Jauberthie, C., Bournonville, F., Coton, P., & Rendell, F. (2006). Optimal input design for aircraft parameter estimation. *Aerospace Science and Technology*, *10*(4), 331–337.

Kalaba, R., & Spingarn, K. (1977). Optimal input system identification for nonlinear dynamic systems. *Journal of Optimization Theory Application*, *21*(1), 91–102.

Levadi, V. (1966). Design of input signals for parameter estimation. *IEEE Transactions on Automatic Control*, *11*(2), 205–211.
Litman, S., & Huggins, W. (1963). Growing exponentials as a probing signal for system identification. *Proceedings of the IEEE*, *51*(6), 917–923.
Ljung, L. (1979). Asymptotic behavior of the extended Kalman filter as a parameter estimator for linear systems. *IEEE Transactions on Automatic Control*, *24*(1), 36–50.
Lu, L. (2010). Optimal inputs and sensitivities for parameter estimation in bioreactors. *Journal of Mathematical Chemistry*, *47*(3), 1154–1176.
Mehra, R. (1974a). Optimal input signals for parameter estimation in dynamic systems-survey and new results. *IEEE Transactions on Automatic Control*, *19*(6), 753–768.
Mehra, R. (1974b). Optimal inputs for linear system identification. *IEEE Transactions on Automatic Control*, *19*(3), 192–200.
Rafajłowicz, E. (1989). Time-domain optimization of input signals for distributed-parameter systems identification. *Journal of Optimization Theory and Applications*, *60*(1), 67–79.
Scola, I. R., Besançon, G., & Georges, D. (2013). Input optimization for observability of state affine systems, *46*(2), 737–742.
Shafer, M. F. (1984). Flight investigation of various control inputs intended for parameter estimation.
Sinha, N. K. & Kuszta, B. (1983). *Modelling and identification of dynamic systems*. New York: Springer.
Torres, L., Aguiñaga, J. A. D., Besançon, G., Verde, C., & Begovich, O. (2016). Equivalent Liénard-type models for a fluid transmission line. *Comptes Rendus Mécanique*, *344*(8), 582–595.
Torres, L., Besançon, G., Navarro, A., Begovich, O., & Georges, D. (2011). Examples of pipeline monitoring with nonlinear observers and real-data validation. In: *8th IEEE international multi-conference on signals systems and devices* (pp. 1–6).
Torres, L., Besançon, G., & Verde, C. (2015). Liénard-type model of fluid flow in pipelines: application to estimation. In *12th International conference on electrical engineering, computing science and automatic control (CCE)* (pp. 1–6). IEEE.
Van Loan, C. (1978). Computing integrals involving the matrix exponential. *IEEE Transactions on Automatic Control*, *23*(3), 395–404.
Yang, J., Wen, Y., & Li, P. (2008). Leak location using blind system identification in water distribution pipelines. *Journal of Sound and Vibration*, *310*(1), 134–148.
Zecchin, A. C., White, L. B., Lambert, M. F., & Simpson, A. R. (2013). Parameter identification of fluid line networks by frequency-domain maximum likelihood estimation. *Mechanical Systems and Signal Processing*, *37*(1), 370–387.

Chapter 7
Recursive Scheme for Sequential Leaks' Identification

Cristina Verde and Jorge Rojas

Abstract This chapter deals with the problem of detection and identification of multi-leaks in a single, horizontal pipeline assuming that only flow and pressure sensors at the ends of the pipeline are available. A characteristic of this monitoring scenario with simultaneous leaks is that the events are undistinguishable in steady state. This means the leaks could only be identified during transient behaviors. A monitoring problem, close to the simultaneous leaks' issue, is the leaks' scenario appearing in sequence. It is shown here that the isolation task is feasible with this scenario in the framework of model-based methods. Thus, a general recursive scheme which is formatted with three coupled nonlinear input–output equivalent models in steady state is proposed. Since the scheme is based on the equivalence model condition between one and multiple leaks', previous to the presentation of the scheme, the static relation between equivalent models for one and multiple leaks is derived by considering the friction as a function of the flows. The interrelated three input-output models have the property to retain the data of the past leaks', which allows an on-line identification of the new event during a time window for any arbitrary number of leaks. In particular the identifiers are implemented by using extended Kalman filters. The algorithm is tested with synthetic data simulated with Pipeline Studio software for a sequential set of three leaks, and it shows successful results.

7.1 Introduction

Although pipelines' network installation includes a SCADA (supervisory control and data acquisition) system and is protected against damage, leaks and breaks caused by natural phenomena, a failure of mechanical integrity, excavation activities and pressure surges. Moreover, in the case of hazardous fluid transportation, leaks may have a severely damaging impact on the environment as well as considerable human and economic losses. As demonstrated throughout this book, a considerable effort

C. Verde (✉) · J. Rojas
Instituto de Ingeniería, UNAM, CP 04510 Mexico City, Mexico
e-mail: verde@unam.mx

C. Verde and L. Torres (eds.), *Modeling and Monitoring of Pipelines and Networks*, Applied Condition Monitoring 7,
DOI 10.1007/978-3-319-55944-5_7

has been made to develop computational methods for leaks' location which can improve the monitoring, supervision and diagnosis task of the SCADA on a higher level. If flow and pressure are assumed to be measurable only at the ends of the line, diverse online monitoring has been suggested (Billman and Isermann 1987; Scott and Barrufet 2003; Lee et al. 2005; Ferrante and Brunone 2003). In particular, Verde et al. (2011) reported that the algorithms generate false positions if multiple leaks scenarios exist with only sensors at the ends and steady flow is assumed. The false location occurs because these faults are weakly isolable, in the context of fault detection and identification (Chen and Patton 1999). Thus, simultaneous leaks are detectable, but the isolability problem is not feasible after the transient behavior disappears. Verde et al. (2014) suggest a quasi real-time procedure by using the family of dynamic models of parameterized transient models for all two leaks scenarios. This family is obtained by considering the equivalence in the steady state of a leak at position z_e with two leaks. This method, however, is developed considering constant the friction along all the pipeline which is not correct in the case of a pipeline with branches and leaks. Thus, the multiple simultaneous leaks' location problem, however, is still open.

A realistic case, related to the above mentioned issue, is the leaks' scenario appearing in sequence. This scenario could occur if an earthquake and its replica cause leaks very close to each other in time. Another real condition is an old line in which small leaks produced by corrosion appear with certain frequency but not at the same time. This sequential leaks' scenario is the central issue of this chapter.

The basic idea for solving the sequential leaks' identification by considering the equivalent models for multiple leaks' is published by Verde and Rojas (2015) that requires four coupled nonlinear observers. Disadvantages of the scheme are the slow time response of the estimations, the assumption of a constant friction f in the whole pipeline and the absence of the stability proof of the coupled estimators.

Recently Delgado et al. (2016) introduced an algorithm to tackle the sequential leaks' problem as well, in which the state model order for the estimation is incremented by two each time a new leak occurs, and the friction factor is assumed not constant. Since all the serial estimators remain active, the complexity of the algorithm increases with the numbers of leaks. The reported identified errors are considerable before the leaks' sequence occurrence, which cannot be explained.

Convinced that is impractical increase the number of estimators, a simple scheme for leaks' location is proposed here, since the equivalent models retain leaks' history with the same dimension and structure. The scheme operates recursively by adjusting the identifiers each time an event occurs with fixed-order models and assuming the friction as a function of the flow. The detection logic is derived by induction and can be then used for any number of leaks without knowing the number of leaks a priori. An implementation of the scheme by using MATLAB 2007 (2008) is proposed with three extended Kalman filters (EKF), which identify the equivalent models and estimate the parameters of the latest leak present. Similar to previous methods based on a balance of mass, it is assumed that only flow and pressure sensors at the ends of the pipeline are available. The scheme performance is discussed with the results for a set of three leaks simulated with Pipeline Studio.

7.2 Fluid Model

According to Chaudhry (2014), by assuming constant convective velocity and fluid density, the fluid dynamics in a one-dimensional horizontal pipeline of length L without branches is modeled by the partial differential equations (PDE)

$$\begin{aligned}\frac{\partial Q(z,t)}{\partial t}+a_1\frac{\partial H(z,t)}{\partial z}+\phi\big(Q(z,t)\big)&=0,\\ \frac{\partial H(t)}{\partial t}+a_2\frac{\partial Q(z,t)}{\partial z}&=0,\end{aligned} \tag{7.1}$$

where t is the time variable $[s]$, $z \in [0, L]$ is the relative space variable $[m]$, $H(z,t)$ is the pressure head variable $[m]$ along the line, $Q(z,t)$ is the flow $[\mathrm{m^3s^{-1}}]$, and the constants associated with the physical characteristics of the pipe are

$$a_1 = gA, \qquad \text{and} \qquad a_2 = \frac{b^2}{gA}. \tag{7.2}$$

Within these constants, g is the gravitational acceleration $[\mathrm{ms^{-2}}]$, A the cross-section area of the pipe $[m^2]$, b the wave velocity in the fluid $[\mathrm{ms^{-1}}]$, and D the inner diameter of the pipe $[m]$. Moreover, in (7.1) the friction term is given by

$$\phi(Q(z,t)) = \frac{f(Q(z,t))}{2DA}Q(z,t)|Q(z,t)|, \tag{7.3}$$

where the nonlinear friction function

$$f(Q(z,t)) = \left\{\left(\frac{64}{R_e}\right)^8 + 9.5\left[\ln\left(\frac{\varepsilon_0}{3.7D}+\frac{5.74}{R_e^{0.9}}\right)-\left(\frac{2500}{R_e}\right)^6\right]^{-16}\right\}^{0.125}, \tag{7.4}$$

depends on the roughness of pipe ε_0 and the Reynolds number $R_e = \frac{VD}{A\nu}$ with the flow velocity given by $V = \frac{Q}{A}$, and the kinematic viscosity of the fluid ν. Figure 7.1 shows the function f with respect to R_e for a set of relations ε_0/D reported by Prabhata (1993). The high sensitivity of the function with respect to the Reynolds number R_e in the interval $[10^{-3}, 10^{-4}]$ is identified from the graphics. Thus, if the flow is not complete turbulence or flow changes occur during the operation of the line, the assumption of constant f is mistaken. The following subsection introduces an analysis of the variables' deviations in a branched pipeline when f is assumed constant in (7.1), even if the fluid is turbulent around the operation point.

7.2.1 *Friction Sensitivity in a Branched Pipeline*

Consider a pipeline of length $L = \sum_{i=1}^{n} L_i$ divided into n sections with a branch between each of them and the variables' description shown in Fig. 7.2. Similarly to (7.1), the description of the pressure head and flow for each section i is given by

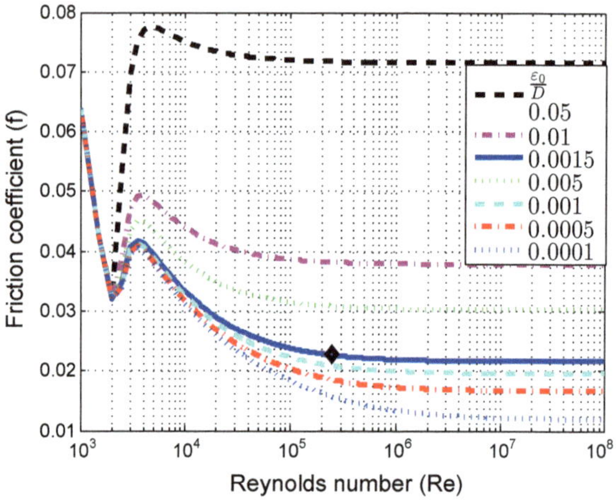

Fig. 7.1 Moody diagram for the friction function (7.4). The point marked with the symbol ◇ in the plotter corresponds to the normal operation point of the pipeline used as the case study

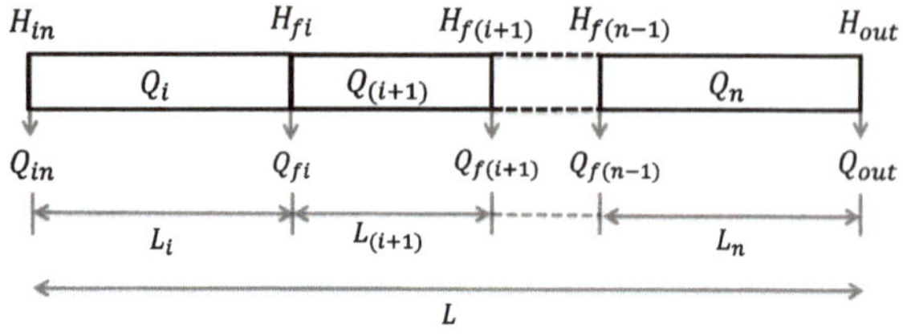

Fig. 7.2 Variables' description of a pipeline with n sections

$$\begin{aligned} \frac{\partial Q_i(z_i,t)}{\partial t} + a_1 \frac{\partial H_i(z_i,t)}{\partial z_i} + \phi_i\big(Q_i(z_i,t)\big) &= 0, \\ \frac{\partial H_i(z_i,t)}{\partial t} + a_2 \frac{\partial Q_i(z_i,t)}{\partial z_i} &= 0, \end{aligned} \tag{7.5}$$

where the spatial variable $z_i \in [0, L_i]$. According to Chaudhry (2014), at each branch between components the following boundary conditions satisfy:

$$Q_i(L_i,t) = Q_{i+1}(0,t) + Q_{bi}(t), \quad \text{and} \quad H_i(L_i,t) = H_{i+1}(0,t), \tag{7.6}$$

where $Q_{bi}(t)$ represents the outflow at interface i. The boundary conditions at the extremes of the pipeline are

$$H_1(0,t) = H_{in}(t), \quad \text{and} \quad H_n(L_n,t) = H_{out}(t). \tag{7.7}$$

As a consequence:

- The number of branches in a pipeline defines the cardinality of the set of PDE which describes the fluid.
- Branches in the line increase the space dimension of any approximated representation of the PDE and the order cannot be fixed, if the number of branches is unknown.

- Because of the boundary condition at each branch, the term $\phi_i\left(Q_i(z_i, t)\right)$ is distinct for each section even if the physical parameters of the line are the same.

The effect of the constant friction in the fluid model is analyzed by considering a pressure head $\overline{hp}(L_1, t)$ as a pattern at point L_1 and by calculating the normalized error

$$e(Q_{b1}) = \lim_{t\to\infty} \frac{\overline{hp}(L_1, t) - h_1(L_1, t)}{\overline{hp}(L_1, t)}, \tag{7.8}$$

as a deviation measure for a set of outflows $Q_{b1}(L_1, t)$. In this analysis, the numerical pattern $\overline{hp}(L_1, t)$ is generated by the simulator EPANET (EPA 2009) with the specific parameters of a water pilot plant reported by Cayetano (2016) and given in Table 7.1.

The first experiment consisted of generating a set of steady-state errors (7.8) for a different outflow Q_{b1} with a constant friction term, and the results are shown in Fig. 7.3. A linear increase of the error is identified as a function of the outflow. On the contrary, the errors by assuming a variable friction for each section remain bounded with a value of less than 0.05 %, as shown in Fig. 7.4.

Table 7.1 Pilot water pipeline with branches

Long $[L]$	169.43 [m]
Roughness $[\varepsilon]$	0.15 [mm]
Diameter $[D]$	0.1016 [m]
Wave velocity $[b]$	1330 [m/s]
Branch position L_1	84.715 [m]
Friction f_1	0.02
Friction f_2	0.02

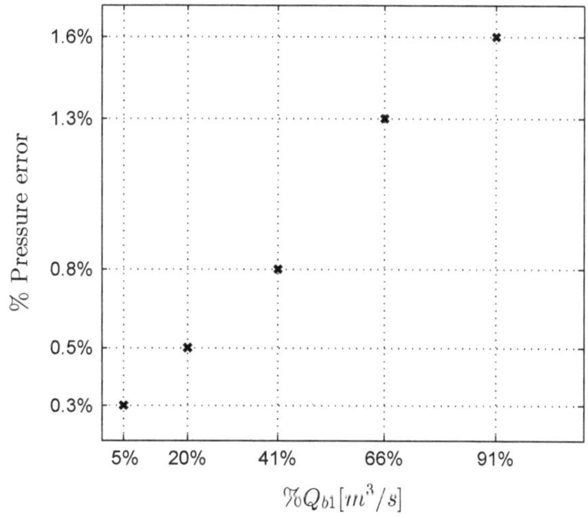

Fig. 7.3 Normalized errors assuming a constant friction for input pressure head $H_{in} = 18$ [m] and a set Q_{b1}

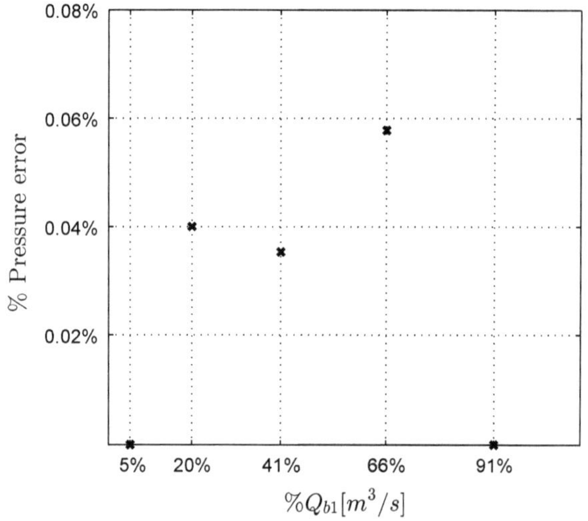

Fig. 7.4 Normalized errors assuming a variable friction as a function of the flow for input pressure head $H_{in} = 18$ [m] and a set Q_{b1}

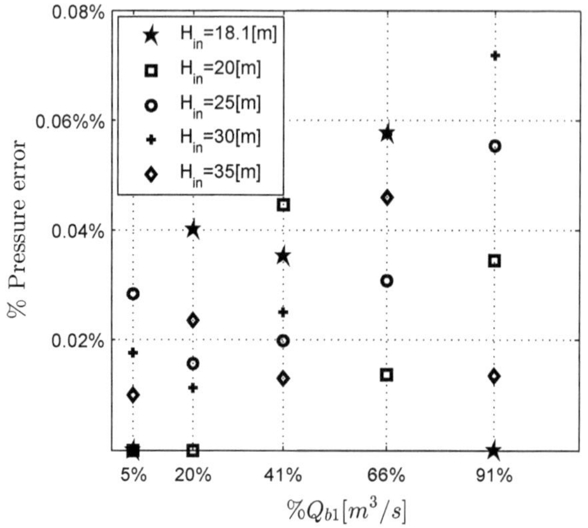

Fig. 7.5 Normalized errors assuming a variable friction as a function of the flow for different operation points

In the second experiment, the error (7.8) has been calculated with a finite dimension model of order ten and a variable friction for a set of operation points. Figure 7.5 shows the errors which are less than 0.08%, but they are not constant. The higher the input H_{in} and the outflow Q_{b1}, the higher is the error. These results confirm the numerical sensitivity of the variable at the branch point with respect to the friction f, even if the flow is in steady state. The boundary conditions together with the nonlinearity of the friction modify the properties of a single pipeline. Thus, to select feasible analytical nonlinear models for fault diagnosis in pipeline networks, a sensitivity study is indispensable.

Moreover, because the leak detection problem consists precisely of the identification of unknown branches positions and their outflows, any multiple leaks' location algorithm must consider the model (7.4) for the possible leaks' scenarios or design an online identification procedure for the friction term. Therefore, all computational leak detection methods developed by assuming a constant friction along the whole pipeline generate an error which only in the case of very small outflows could be neglected.

7.3 Input–Output Equivalent Models with Variant Friction

A characteristic of the system (7.5) is weakly distinguishable with respect to leaks if only the variable set $\{H_{in}, H_{out}, Q_{in}, Q_{out}\}$ is measurable (Verde 2003). Thus, some leaks' scenarios cannot be identified after their transient effects have been diminished. On the other hand, Korbicz et al. (2004) reported the steady-state relations which characterize all the two-leak events indistinguishable from the one- leak event. The reported relations assumed a constant friction f, however. Therefore, the equivalent relations considering the friction as a nonlinear function of the flows upstream and downstream from branches are here presented.

Consider Fig. 7.6 associated with structure of two pipelines interconnected by a branch and (7.5). The fluid behavior is then described by

$$\begin{aligned} \frac{\partial Q_{1e}(z,t)}{\partial t} + a_1 \frac{\partial H_{1e}(z,t)}{\partial z} + \phi\big(Q_{1e}(z,t)\big) &= 0, \\ \frac{\partial H_{1e}(z,t)}{\partial t} + a_2 \frac{\partial Q_{1e}(z,t)}{\partial z} &= 0 \end{aligned} \tag{7.9}$$

for $0 < z < L_e$ and by

$$\begin{aligned} \frac{\partial Q_{2e}(z,t)}{\partial t} + a_1 \frac{\partial H_{2e}(z,t)}{\partial z} + \phi\big(Q_{2e}(z,t)\big) &= 0, \\ \frac{\partial H_{2e}(z,t)}{\partial t} + a_2 \frac{\partial Q_{2e}(z,t)}{\partial z} &= 0, \end{aligned} \tag{7.10}$$

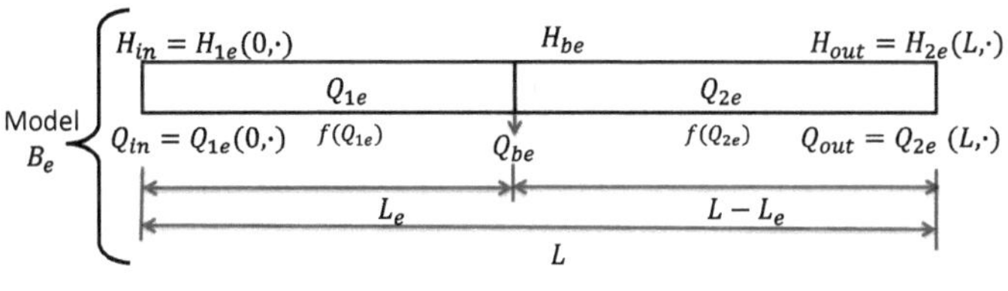

Fig. 7.6 Variables'description for model *Be* with one branch

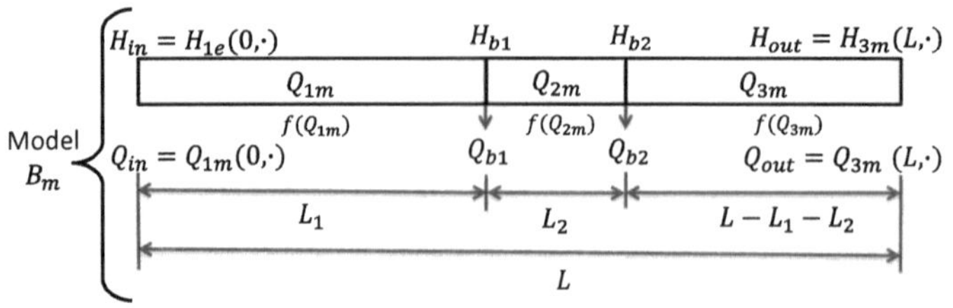

Fig. 7.7 Input–output variables' description for model B_m with two branches

for $L_e < z < L$, with the conditions at the interconnection point L_e

$$Q_{1e}(L_e, t) = Q_{2e}(0, t) + Q_{be}(t), \quad H_{1e}(L_e, t) = H_{2e}(0, t) = H_{fe}(t). \tag{7.11}$$

On the other hand, consider that Fig. 7.7 is associated with the structure of three pipelines interconnected by two branches and (7.5), then the fluid behavior is described by

$$\begin{aligned} \frac{\partial Q_{1m}(z,t)}{\partial t} + a_1 \frac{\partial H_{1m}(z,t)}{\partial z} + \phi\big(Q_{1m}(z,t)\big) &= 0, \\ \frac{\partial H_{1m}(z,t)}{\partial t} + a_2 \frac{\partial Q_{1m}(z,t)}{\partial z} &= 0 \end{aligned} \tag{7.12}$$

for $0 < z < L_1$ and by

$$\begin{aligned} \frac{\partial Q_{2m}(z,t)}{\partial t} + a_1 \frac{\partial H_{2m}(z,t)}{\partial z} + \phi\big(Q_{2m}(z,t)\big) &= 0, \\ \frac{\partial H_{2m}(z,t)}{\partial t} + a_2 \frac{\partial Q_{2m}(z,t)}{\partial z} &= 0, \end{aligned} \tag{7.13}$$

for $L_1 < z < L_2$ and by

$$\begin{aligned} \frac{\partial Q_{3m}(z,t)}{\partial t} + a_1 \frac{\partial H_{3m}(z,t)}{\partial z} + \phi\big(Q_{3m}(z,t)\big) &= 0 \\ \frac{\partial H_{3m}(z,t)}{\partial t} + a_2 \frac{\partial Q_{3m}(z,t)}{\partial z} &= 0, \end{aligned} \tag{7.14}$$

for $L_2 < z < L$-. These equations are related at the interconnection points L_1, L_2 by

$$\begin{aligned} Q_{1m}(L_1, t) &= Q_{2m}(0, t) + Q_{b1}(t), \quad H_{1m}(L_1, t) = H_{2m}(0, t) = H_{b1}(t), \\ Q_{2m}(L_2, t) &= Q_{3m}(0, t) + Q_{b2}(t), \quad H_{2m}(L_2, t) = H_{3m}(0, t) = H_{b2}(t). \end{aligned} \tag{7.15}$$

To obtain the equivalent relations between the steady-state variables of the scenarios with one and two branches, the fluid models given by the set (7.9), (7.10), and (7.11) and the set (7.12), (7.13), (7.14), (7.15) are considered. Moreover, it is assumed within in both models that they have the same physical parameters (L, A, b) and the same input–output behavior in steady state, but with their respective variant frictions. Since the relations are derived in steady state, in both sets of PDEs $\frac{\partial H_i(z)}{\partial t}$ and $\frac{\partial Q_i(z)}{\partial t}$ disappear, and the sets are linked by $Q_{be} = Q_{b1} + Q_{b2}$.

By integrating (7.9) and (7.10) with the respective limits, one gets for model B_e

$$H_e - H_{in} = -\frac{f(Q_{in})}{2DA^2g} Q_{in}^2 L_e, \tag{7.16}$$

$$H_{out} - H_e = -\frac{f(Q_{out})}{2DA^2g} Q_{out}^2 (L - L_e), \tag{7.17}$$

$$Q_{in} - Q_{out} = Q_{be}. \tag{7.18}$$

Similarly, by integrating (7.12), (7.13), (7.14) with the respective limits, one yields for model B_m

$$H_u - H_{in} = -\frac{f(Q_{in})}{2DA^2g} Q_{in}^2 L_u, \tag{7.19}$$

$$H_{b2} - H_{b1} = -\frac{f(Q_{2m})}{2DA^2g} Q_{2m}^2 (L_2), \tag{7.20}$$

$$H_{out} - H_{b2} = -\frac{f(Q_{out})}{2DA^2g} Q_{out}^2 (L - L_1 - L_2), \tag{7.21}$$

$$Q_{in} - Q_{out} = Q_{b1} + Q_{b2}. \tag{7.22}$$

By then adding (7.16) and (7.17), one attains

$$H_{out} - H_{in} = -\frac{f(Q_{in})}{2DA^2g} Q_{out}^2 \left(1 + \frac{Q_{be}}{Q_{out}}\right)^2 L_e - \frac{f(Q_{out})}{2DA^2g} Q_{out}^2 \left(L - L_e\right). \tag{7.23}$$

Similarly, by adding (7.19), (7.20) and (7.21) one gets

$$\begin{aligned} H_{out} - H_{in} = &-\frac{f(Q_{in})}{2DA^2g} Q_{out}^2 \left(1 + \frac{Q_{b1} + Q_{b2}}{Q_{out}}\right)^2 L_1 - \frac{f(Q_{2m})}{2DA^2g} Q_{out}^2 \left(1 + \frac{Q_{b2}}{Q_{out}}\right)^2 (L_2) \\ &- \frac{f(Q_{out})}{2DA^2g} Q_{out}^2 \left(L - (L_u + L_d).\right) \end{aligned} \tag{7.24}$$

Since (7.23) and (7.24) have the same left side term, one can write

$$\begin{aligned} L_e\left[f(Q_{in})\left(1 + \frac{Q_{be}}{Q_{out}}\right)^2 - f(Q_{out})\right] = L_1\left[f(Q_{in})\left(1 + \frac{Q_{b1} + Q_{b2}}{Q_{out}}\right)^2 - f(Q_{out})\right] \\ + L_2\left[f(Q_{2m})\left(1 + \frac{Q_{b2}}{Q_{out}}\right)^2 - f(Q_{out})\right], \end{aligned} \tag{7.25}$$

Table 7.2 Equivalent synthetic parameters

Model B_e	Model B_m
t= 10 [s]	t= 10 [s]
$Q_{be} = 2.35e^{-3}[\mathrm{m}^3/\mathrm{s}]$	$Q_{b1} = 1.2891e^{-3}[\mathrm{m}^3/\mathrm{s}]$, $Q_{b2} = 1.0649e^{-3}[\mathrm{m}^3/\mathrm{s}]$
$L_b = 73.17[\mathrm{m}]$	$L_1 = L_2 = 50.829[\mathrm{m}]$

in terms of the branched outflows, or

$$L_e\big(f(Q_{in})Q_{in}^2 - f(Q_{out})Q_{out}^2\big) = L_1\big(f(Q_{in})Q_{in}^2 - f(Q_{out})Q_{out}^2\big) + L_2\big(f(Q_{2m})Q_{2m}^2 - f(Q_{out})Q_{out}^2\big) \quad (7.26)$$

in terms of only the outflows at the pipeline ends. Relations (7.25) and (7.26) have not been reported before, and this equivalence between models allows the formulation of multiple leaks' diagnosis tasks with low-order models if the branches are considered as leaks' points.

An additional geometric relation between leaks' positions for the equivalent models is

$$L_1 < L_e < L_1 + L_2, \quad (7.27)$$

which was shown by Verde et al. (2007). This means the equivalent branch position in pipeline B_e is located between the branches' positions in pipeline B_m.

To determine the numerical difference generated by the equivalent relation, both models B_e and B_m are simulated during the transient response by using Pipeline Studio (2013) with the parameters of Table 7.1 together with the branched outflows and positions given in Table 7.2. Figure 7.8 describes the flow errors' evolution:

$$e_{in}(t) = Q_{in}(t)|_{B_e} - Q_{in}(t)|_{B_m}, \text{ and } e_{out}(t) = Q_{out}(t)|_{B_e} - Q_{out}(t)|_{B_m}. \quad (7.28)$$

As expected, the errors of the variables in both extremes are zero previous to the occurrence of the outflows in the branches. During the transient response, both differences deviate from zero, and when the transients disappear they tend to zero. The

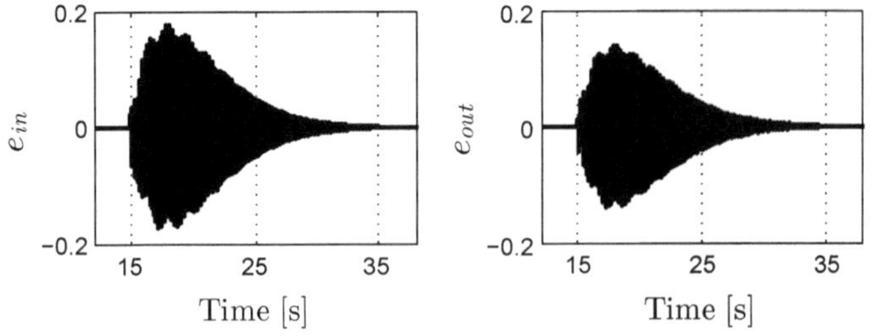

Fig. 7.8 Transient errors $e_{in}(t)$ and $e_{out}(t)$ with models B_e and B_m

parameters could be estimated only during approximately 15 [s] for this example. This equivalence between models can be generalized by induction for any number of branches.

7.4 Recursive Algorithm for Sequential Leaks' Location

Although models B_e and B_m are only equivalent in steady state, the PDEs (7.9) and (7.10) with (7.11) as well as the PDEs (7.12), (7.13), (7.14) with (7.15) allow the design of a recursive algorithm for the location of sequential leaks characterized by the dyad

$$(Q_f(j), L_f(j)),$$

with $Q_f(j)$ and $L_f(j)$ the outflow and position of the jth leak.

The key of the algorithm are three interrelated dynamic systems which describe the fluid and allow the parameters identification by stages. A property of the three systems with their respective parameters is their steady state equivalence by considering a branch outflow as a leak.

Thus, when the jth leak occurs the equivalent parameters

$$\theta_e(j) = \begin{pmatrix} \theta_{e1}(j) \\ \theta_{e2}(j) \end{pmatrix} = \begin{pmatrix} Q_{fe}(j) \\ L_e(j) \end{pmatrix},$$

within the dynamic model B_e are first identified. Later on in accordance with the events history the parameters of the two parallel dynamic models are identified. These models must have the steady-state equivalence property, and to simplify the logic of the algorithm, their variables are defined symmetrically with respect to spatial variable z. Figure 7.9 shows both proposed structures, one for upstream and other for downstream events, named B_u and B_d with respect to the equivalent position and to where the unknown parameter for each leak

$$\theta_u(j) = \begin{pmatrix} \theta_{u1}(j) \\ \theta_{u2}(j) \end{pmatrix} = \begin{pmatrix} Q_{uf1}(j) \\ L_{u1}(j) \\ Q_{uf2}(j) \\ L_{u2}(j) \end{pmatrix}, \quad \theta_d(j) = \begin{pmatrix} \theta_{d1}(j) \\ \theta_{d2}(j) \end{pmatrix} = \begin{pmatrix} Q_{df1}(j) \\ L_{d1}(j) \\ Q_{df2}(j) \\ L_{d2}(j) \end{pmatrix}, \tag{7.29}$$

has to be identified.

Figure 7.10 introduces the main stages and functions of the recursive algorithm and are summarized as follows.

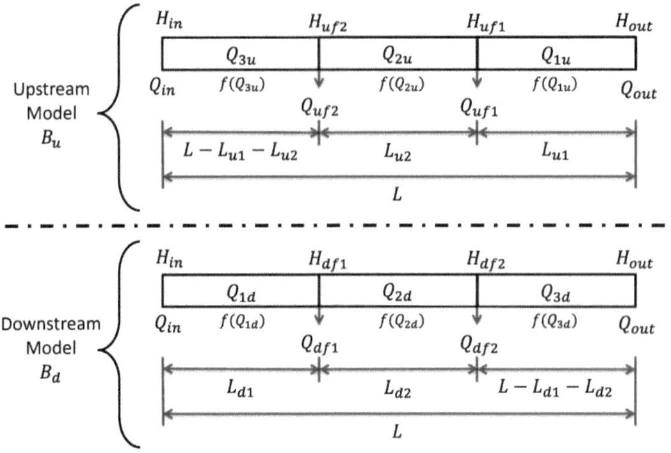

Fig. 7.9 Variables' description within models B_u and B_d with the positions L_{u2} and L_{d2} to be identified respectively

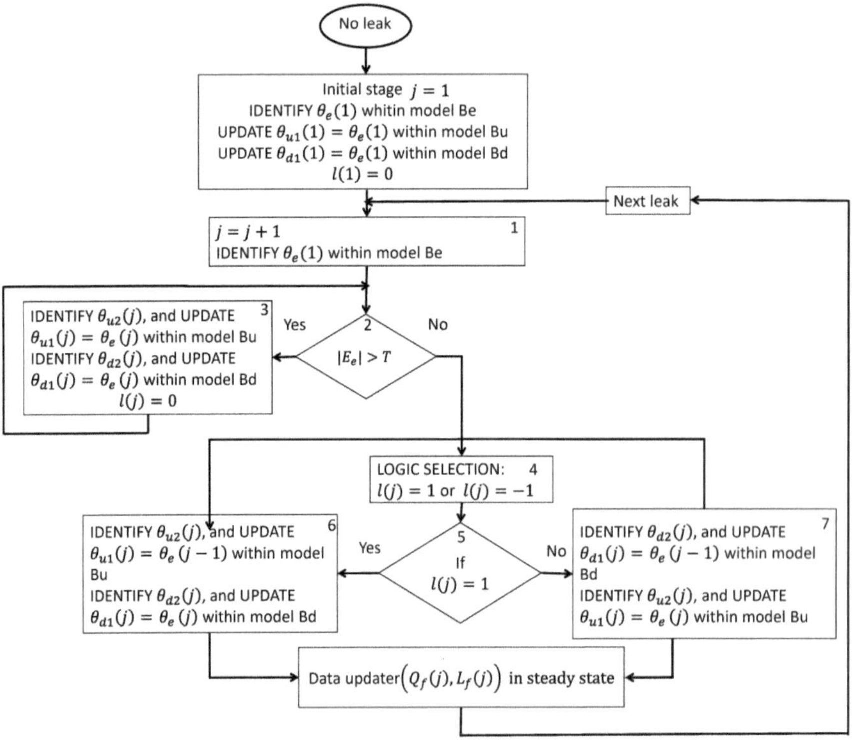

Fig. 7.10 Stages of the leaks' location algorithm

- **Initialize**. This stage estimates the parameter $\theta_e(1)$ within model B_e ((7.9), (7.10), (7.11)) for the first leak. Since the second leak could occur upstream or downstream from $L_e(1)$, both models B_{2u} and B_{2d} must be updated to be steady state equivalent to B_e by starting with the logic stage $\ell(1) = 0$. This is achieved by $\theta_{u1}(1) = \theta_e(1)$ and $\theta_{d1}(1) = \theta_e(1)$.
- **Increase** j **and estimate the parameters within** B_e. This stage specifically stores the number of leaks in the counter j, identifies $\theta_e(j)$ for each leak within the model given by (7.9), (7.10), and (7.11) and holds the logic state $\ell = 0$ during the convergence time.
- **Keep the equivalent leaks' condition**. This stage is formatted by the loop of functions 2 and 3 within the diagram. Its objective is the estimation of θ_u and θ_d during the transient response of the estimator of B_e for the last equivalent leak.
- **Select the orientation**. This stage determines the orientation of the jth leak by calculating the residual $sign(L_e(j-1) - L_e(j))$. This statement is based on the fact, shown by Verde et al. (2016), that the residual is positive if a leak occurs upstream of a previous one in the pipeline. The logic state $\ell = 1$ denotes that the leak is upstream referenced with respect to the previous position $L_e(j-1)$ within model B_e, and on the contrary $\ell = -1$ denotes a downstream leak.
- **Estimate**. This stage consists of two independent identifiers associated with models B_u and B_d and parameters $\theta_u(j)$ and $\theta_d(j)$ for the structures given in Fig. 7.9 and their respective set ((7.12), (7.13), (7.14), (7.15)). The conditions of how the parameters in both models should be identified and updated are derived by considering the equivalent (7.26), the inequality (7.27) and the logic state ℓ. Therefore, when the equivalent jth leak has been identified and ℓ is assigned, only the parameters of θ_{u2} and θ_{d2} must be identified, as well as the parameters of θ_{u1} and θ_{d1} are updated as functions of the estimation of θ_e. In particular, the procedures for each condition are the following.

  - For the upstream scenario, the unknown vector $\theta_{u2}(j)$ and $\theta_{d2}(j)$ are identified, and $\theta_{u1}(j)$ and $\theta_{d1}(j)$ are updated according to the previous and actual equivalent vector, respectively. This means $\theta_{u1}(j) = \theta_e(j-1)$ and $\theta_{d1}(j) = \theta_e(j)$ are updated in their respective models. This update is for holding the equivalence between B_u, B_d and B_e in steady state after the transient of any event.
  - For the downstream scenario, by symmetry the unknown vectors $\theta_{u2}(j)$ and $\theta_{d2}(j)$ within B_u and B_d are identified, with $\theta_{d1}(j)$ and $\theta_{u1}(j)$ updated to the previous and actual equivalent vector, respectively. Similar to the upstream scenario these conditions are required to hold the equivalent between the three models and to prepare the models for the next possible up-stream leak.

  Note that in both scenarios the updated parameters $\theta_{d1}(j)$ and $\theta_{u1}(j)$ can be considered as exogenous signals in B_d and B_u, which come from the output of the identifier of $\theta_e(j)$. This fact is described by the interconnecting lines between the blocks (2) and (6) and (7) in Fig. 7.10.
- **Update data**. This stage calculates the real parameters of the leaks L_{fi} and Q_{fi} by using (7.25) and the parameters of the history of the identified leaks θ_u and θ_d. Moreover, this stage prepares the variables for the occurrence of the next event.

A benefit of the proposed algorithm is that it can be implemented without additional instrumentation, and it holds fixed the order of the identifiers for models B_e, B_u, and B_d, even when more than two leaks have been estimated. As with all recursion, however an initialization step is required for $j = 1$ because of the absence of a reference position in a pipeline without leaks, Another advantage of the algorithm is its generality, because diverse identification procedures can be used for each of the nonlinear PDE.

7.5 Algorithm Implementation

This section describes a specific implementation of the leaks' location procedure given in Sect. 7.4. To simplify the parameters identification task, finite dimension models for the three structures B_e, B_u, and B_d are considered. For the three estimators, the pressure head vector is assumed to be known input:

$$u = (H_{in}, H_{out})',$$

and with the flows vector as known output

$$y = (Q_{in}, Q_{out})'.$$

For the identification task, extended Kalman filters (EKF) are designed. An advantage of the EKF is that its gain matrix $K(t)$ can be adjusted for all the identifiers each time a new event appears, and as a consequence the parameters converge locally for each event. Because the leak outflow Q_{fi} is modeled by

$$Q_{fi} = \lambda_{fi}\sqrt{H_{fi}}, \tag{7.30}$$

with the discharge coefficient λ_{fi} associated with the physical characteristics of the orifice. The identification of the constant λ_{fi} for each leak instead of Q_{fi} which changes with operation points and leaks' scenarios is recommended.

7.5.1 *Approximated Models and Estimators*

The finite dimension models of B_e, B_u, and B_d, can be obtained by using the spatial approximation

$$\frac{\partial v(z,t)}{\partial z} \backsimeq \frac{v_{z_{k+1}}(t) - v_{z_k}(t)}{\Delta z_k}, \tag{7.31}$$

where $v_{k+1}(t)$ and $v_k(t)$ correspond to the values of the variable $v(t)$ at two discrete contiguous spatial points and Δz_k is the discrete distance (Polyanin and Zaitsev 2014).

Thus, by substituting (7.31) in model B_e with one leak given by (7.9), (7.10) and (7.11) for two discrete sections, one gets

$$\begin{aligned}\dot{Q}_{in} &= \frac{a_1}{\theta_{e2}}(u_1 - H_e) - \phi(Q_{in}),\\ \dot{H}_{fe} &= \frac{a_2}{\theta_{e2}}(Q_{in} - Q_{out} - \theta_{e1}\sqrt{H_{fe}}),\\ \dot{Q}_{out} &= \frac{a_1}{L - \theta_{e2}}(H_e - u_2) - \phi(Q_{out}),\end{aligned} \tag{7.32}$$

where index j is omitted, θ_e is unknown and H_{fe} is the pressure at the unknown leak position. To simplify the notation, this model is written by

$$\begin{aligned}\dot{x}_e &= f_e(x_e, u, \theta_e),\\ y = C_e x_e &= \begin{pmatrix} 1\,0\,0 \\ 0\,0\,1 \end{pmatrix} x_e,\end{aligned} \tag{7.33}$$

with $x_e \in \mathbb{R}^3$. If θ_e is assumed constant and the augmented system

$$\begin{aligned}\dot{x}_{ea} &= \begin{pmatrix} \dot{x}_e \\ \dot{\theta}_e \end{pmatrix} = \begin{pmatrix} f_e(x_e, u, \theta_e) \\ 0 \end{pmatrix} = f_{ea}(x_{ea}, u),\\ y &= (C_e\ 0\ 0)\, x_{ea},\end{aligned} \tag{7.34}$$

is locally observable with output y, then an observer can be designed to identify θ_e (Isermann and Münchhof 2011). Thus, the estimated state $\hat{x}_{ea}$ is described by

$$\begin{aligned}\dot{\hat{x}}_{ea} &= f_{ea}(\hat{x}_{ea}, u) + K_e E_e,\\ E_e &= y - C_e \hat{x}_{ea},\end{aligned} \tag{7.35}$$

with the observer output error E_e. In addition, the gain matrix K_e is adjusted on line by applying the EKF described in Sect. 7.5.2 with the constant covariance matrices

$$Q_{ea} = diag(1,\ 1,\ 1,\ 10,\ 10^{10}), \quad R_{ea} = diag(1,\ 1).$$

Therefore, the matrix K_e is automatically adapted each time that a new event occurs. This could be a change in the operation point or a new leak. The equivalent parameters' vector $\hat{\theta}_e$ is formatted by the fourth and fifth variable of the state $\hat{x}_{ea}$. According to the algorithm, during the adaptation time the two systems B_d and B_u follow B_e, then $\ell = 0$ is imposed so long as $E_e > T$.

Similarly to the model B_e, by substituting (7.31) in the set (7.12), (7.13), (7.14), and (7.15) with three sections and the variables' description given in Fig. 7.9 for the models B_u and B_d respectively, one gets the approximated models:

$$\begin{aligned}
&\text{For model } B_d\\
\dot{Q}_{in} &= \frac{a_1}{\theta_{d12}}(H_{in} - H_{df1}) - \phi(Q_{in}),\\
\dot{H}_{df1} &= \frac{a_2}{\theta_{d12}}(Q_{in} - Q_{2d} - \theta_{d11}\sqrt{H_{df1}}),\\
\dot{Q}_{2d} &= \frac{a_1}{\theta_{d22}}(H_{df1} - H_{df2}) - \phi(Q_{2d}),\\
\dot{H}_{df2} &= \frac{a_2}{\theta_{d22}}(Q_{2d} - Q_{out} - \theta_{d21}\sqrt{H_{df2}}),\\
\dot{Q}_{out} &= \frac{a_1}{\theta_{d3}}(H_{df2} - H_{out}) - \phi(Q_{out}),\\
\theta_{d3} &= L - (\theta_{d12} + \theta_{d22}),
\end{aligned} \tag{7.36}$$

$$\begin{aligned}
&\text{For model } B_u\\
\dot{Q}_{in} &= \frac{a_1}{\theta_{u3}}(H_{in} - H_{uf2}) - \phi(Q_{in}),\\
\dot{H}_{uf2} &= \frac{a_2}{\theta_{u22}}(Q_{in} - Q_{2u} - \theta_{u11}\sqrt{H_{uf2}}),\\
\dot{Q}_{2u} &= \frac{a_1}{\theta_{u22}}(H_{uf2} - H_{uf1}) - \phi(Q_{2u}),\\
\dot{H}_{uf1} &= \frac{a_2}{\theta_{u12}}(Q_{2u} - Q_{out} - \theta_{u11}\sqrt{H_{uf1}}),\\
\dot{Q}_{out} &= \frac{a_1}{\theta_{u12}}(H_{uf1} - H_{out}) - \phi(Q_{out}),\\
\theta_{u3} &= L - (\theta_{u12} + \theta_{u22}).
\end{aligned} \tag{7.37}$$

In both models the parameters θ_u and θ_d are unknown and the pairs (H_{uf1}, H_{uf2}) and (H_{df1}, H_{df2}) are the pressures at the respective leaks' positions. Thus (7.36) and (7.37) can be written in the generic compact form

$$\begin{aligned}
\dot{x}_p &= f_p(x_p, u, \theta_{p1}, \theta_{p2}),\\
y &= C_p x_p = \begin{pmatrix} 1\,0\,0\,0\,0 \\ 0\,0\,0\,0\,1 \end{pmatrix} x_p,
\end{aligned} \tag{7.38}$$

for the label $p = \{u,\ d\}$. If the vector of parameters θ_{p2} is assumed constant and the augmented system

$$\begin{aligned} \dot{x}_{pa} &= \begin{pmatrix} \dot{x}_p \\ \dot{\theta}_{p2} \end{pmatrix} = \begin{pmatrix} f_p(x_p, u, \theta_{p1}, \theta_{p2}) \\ 0 \end{pmatrix} = f_{pa}(x_{pa}, u, \theta_{p1}), \\ y &= \begin{pmatrix} C_p \; 0 \; 0 \end{pmatrix} x_{pa}, \end{aligned} \tag{7.39}$$

is locally observable, then an observer can be designed to identify θ_{p2}. This condition is similar to that given for B_e.

According to the proposed algorithm, θ_{p1} is a exogenous input which depends on the logic state ℓ within (7.39). The description for the three possible scenarios can be simplified by defining an extended input vector where $*$ denotes previously identified vector.

- If the logic state $\ell = 1$, then $u_u = (u, \hat{\theta}_e^*)$ and $u_d = (u, \hat{\theta}_e)$,
- If the logic state $\ell = -1$, then $u_u = (u, \hat{\theta}_e)$ and $u_d = (u, \hat{\theta}_e^*)$, and
- If the logic state $\ell = 0$, then $u_u = u_d = (u, \hat{\theta}_e)$; this scenario appears so long as the output error E_e in (7.35) is outside of a specific convergence region defined by T.

Therefore, each observer is based on

$$\begin{aligned} \dot{x}_{p_{ext}} &= f_{p_{ext}}(x_{p_{ext}}, u_p), \\ y &= \begin{pmatrix} C_p \; 0 \; 0 \end{pmatrix} x_{p_{ext}}, \end{aligned} \tag{7.40}$$

where both state and input vectors are extended with the label $p = u$ for the upstream scenario and $p = d$ for the contrary.

Thus, the estimated state of (7.40) is described by

$$\dot{\hat{x}}_{p_{ext}} = f_{p_{ext}}(\hat{x}_{p_{ext}}, u_p) + K_{p_{ext}}(y - C_p \hat{x}_{p_{ext}}), \tag{7.41}$$

where $K_{p_{ext}}(t)$ is adjusted online by using an EKF with the constant covariance matrices

$$Q_{p_{ext}} = diag(1,\ 1,\ 1,\ 1,\ 1,\ 10,\ 10^{10}), \qquad R_{p_{ext}} = diag(1,\ 1).$$

The parameter's vector $\hat{\theta}_{p2}$ is formatted by the sixth and seventh variable of the estimated state $\hat{x}_{p_{ext}}$. Therefore, K_{pext} is automatically adapted each time that a new event occurs.

7.5.2 Extended Kalman Filter

Consider the nonlinear system (Gelb 1994)

$$\dot{x} = f(x, u) + w(t), \quad w(t) \sim \mathcal{N}(0, Q), \tag{7.42}$$

with a random noise $w(t)$, and the measurement model

$$y(t) = Cx + v(t), \quad v(t) \sim \mathcal{N}(0, R(t)),$$

includes a random noise $v(t)$ which is not correlated with $w(t)$. Moreover, assume that the initial condition $x(0) \sim \mathcal{N}(\hat{x}(0), P(0))$ and the expansion of $f(x, u)$ leads to

$$f(x, u) - f(\hat{x}, u) = F(x - \hat{x}) + \gamma(x, \hat{x}, u),$$

where the Jacobian matrix $F = \frac{\partial f(x,u)}{\partial x}$ is evaluated at $x = \hat{x}$ and $\gamma(x, \hat{x}, u)$ includes higher order terms.

Thus, the state estimation of the system around point $\hat{x}$ is given by

$$\dot{\hat{x}}(t) = f(\hat{x}, u, t) + K(t)(y - C\hat{x}), \tag{7.43}$$

where the gain matrix is calculated by $K(t) = P(t)C^T R^{-1}(t)$, and the error covariance matrix satisfies the dynamic Riccati equation:

$$\dot{P}(t) = (F(t) + I\eta)P(t) + P(t)(F(t) + I\eta)^T + Q(t) - P(t)CR^{-1}(t)CP(t), \tag{7.44}$$

with the Jacobian matrix $F(t)$ evaluated at each time. The $\eta > 0$ prescribes a degree of stability and increases the convergence domain of the estimation (Reif et al. 1998).

7.6 Simulation Results

The performance of the recursive scheme is assessed by simulating a water pilot pipeline with the sequence of three leaks given in Table 7.3. The synthetic data were generated by the industrial simulator Pipeline Studio by considering ten sections (Pipeline Studio 2013) and the whole location algorithm is implemented by Simulink from MATLAB (2008).

The time evolution of the pipeline flows together with the set of identified positions for the events' sequence described in Table 7.3 are summarized in Fig. 7.11. The scales of the plotters have been normalized to the variables' maximum values to clarify the presentation. The results of the specific estimated leaks sequence in steady state are summarized in Table 7.4. Errors of less then 2.5% are reported in all the scenarios. From the graphics of Fig. 7.11, the following statements are obtained.

Table 7.3 Parameters of the synthetic leak sequence

Inlet pressure H_{in}	18.65[m]	Outlet pressure H_{out}	4[m]
Leak orifice	1[cm]	Approximated outflow Q_{fi}	1^{-3}[m^3/s]
Leak position $L_f(1)$	118.60 [m]	Occurrence time t_{f1}	50[s]
Leak position $L_f(2)$	50.82 [m]	Occurrence time t_{f2}	100[s]
Leak position $L_f(3)$	84.71 [m]	Occurrence time t_{f3}	150[s]
$\hat{z}_e(0)$	80[m]	$\hat{z}_{u2}(0) = \hat{z}_{d2}(0)$	15[m]

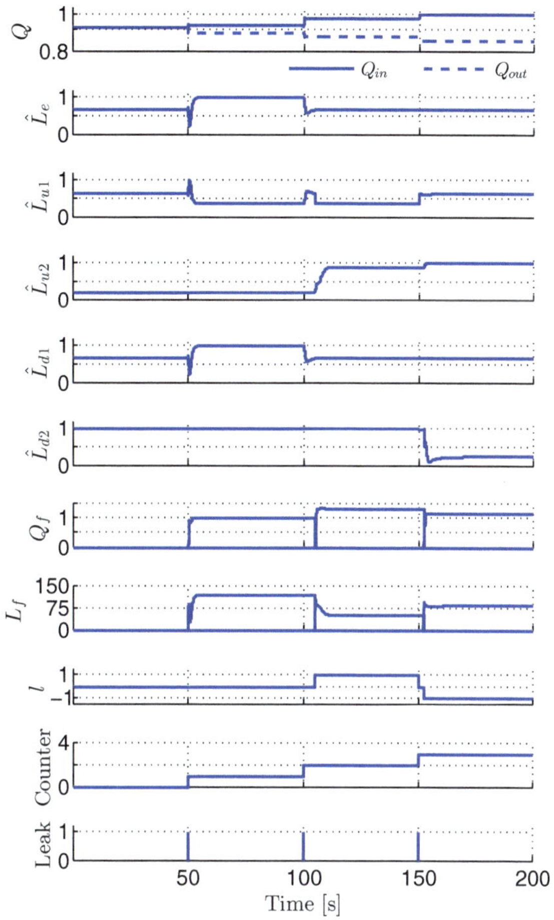

Fig. 7.11 Synthetic flows' evolution, estimated positions' sequence and control sequence where the values of $\hat{L}_e$, $\hat{L}_u$, $\hat{L}_d$, and Q_f are normalized

Table 7.4 Estimated leaks' parameters

$\hat{Q}_f(1) = 9.947e^{-4}[\mathrm{m}^3/\mathrm{s}]$	$\hat{L}_f(1) = 118.55[\mathrm{m}]$	$e_Q(1) = 2.2\%$	$e_L(1) = 0.03\%$
$\hat{Q}_f(2) = 9.862e^{-4}[\mathrm{m}^3/\mathrm{s}]$	$\hat{L}_f(2) = 50.76[\mathrm{m}]$	$e_Q(2) = 0.8\%$	$e_L(2) = 0.12\%$
$\hat{Q}_f(3) = 1.142e^{-3}[\mathrm{m}^3/\mathrm{s}]$	$\hat{L}_f(3) = 83[\mathrm{m}]$	$e_Q(3) = 1.0\%$	$e_L(3) = 2.02\%$

- The flows at the extremes of the pipeline are equal and remain constant so long as the pipeline is in normal condition. At the times $(50\,[s],\ 100\,[s],\ 150\,[s])$ the difference between inlet and outlet flows increases, indicating the occurrence of a leak.
- The estimation of the equivalent position $\hat{L}_e$ remains in the given initial condition and changes its value each time a leak appears. Note that the equivalent position is nonlinear with respect to the real leak position L_f of each leak. This effect highlights the fact that a multiple leaks' scenarios cannot be identified with a model with a single leak.

- When the second leak appears, the estimated positions given by $\hat{L}_{d2}$ and $\hat{L}_{u2}$ converge to the parameters, which produce three equivalent models in approximately four seconds. This can be seen from the time difference between the activation of the counter and the assignation of ℓ for each leak. Moreover, once state ℓ is assigned, the parameters $\hat{L}_{d1}$ and $\hat{L}_{u1}$ are updated and the parameters $\hat{L}_{d2}$ and $\hat{L}_{u2}$ are automatic estimated.
- The values of $Q_f(j)$ and $L_f(j)$ calculated with the data of $\hat{\theta}_{u1}$ and $\hat{\theta}_{d1}$ involve the transitory response of the identifiers. Low-pass filters could smooth the transients.

Thus, these results shown the feasibility of a simple recursive scheme for detecting and isolating a set of sequential leaks in a pipeline without an increase in the order and complexity of the model. A future challenge would be the minimization of the convergence of the equivalent leak position $\hat{L}_e$ for each leak.

7.7 Conclusions

This chapter discusses three problems associated with the monitoring of a horizontal pipeline in the particular case of multiple leaks'. The first problem is related with the sensibility analysis of the friction factor in a pipeline when branches and intermediate components are in the pipeline. Thus, any computational methods for pipeline monitoring introduce an error if a constant friction along the whole pipeline is assumed. This fact affects the leaks' location errors. Taking into account this conclusion, this chapter introduces the extension of the equivalent relation in steady state between one leak and multiple leaks. The derivation is achieved by considering the friction as a function of the flow downstream and upstream for each leak, and the static relation has not been published before. This result which is valid only in steady state of the fluid allows to design a sequential leaks' location scheme, which is the main contribution of this chapter. The recursive scheme includes three interrelated dynamic systems which describe the fluid by holding their structure for any number of leaks, and only the parameters are adapted each time a new leak occurs. In particular, the implementation of the algorithm with three extended Kalman filters as identifiers is applied to location a leaks' sequence in a water pipeline simulated with Pipeline Studio. The validation of the algorithm with a set of leaks shows successful results, and in future work practical scenarios including noise and uncertainties should assess its performance. In conclusion, a feasible simple recursive scheme has been developed to detect and isolate a set of sequential leaks in a pipeline without an increase in the order and complexity of the model. The future challenge for the algorithm is the convergence velocity improvement for the estimations.

Acknowledgements This work is supported by the Mexican Government Scholarship Program for International Students, DGAPA-UNAM IT100716, II-UNAM and CONACYT.

References

Billman, L., & Isermann, R. (1987). Leak detection methods for pipelines. *Automatica, 23*(3), 381–385.

Cayetano, S. R. (2016). *Identificación de parámetros de fricción en ductos con tomas laterales*. Master's thesis, Universidad Nacional Autónoma de México.

Chaudhry, M. C. (2014). *Applied hydraulic transients*. New York: Springer.

Chen, J., & Patton, R. J. (1999). *Robust model-based fault diagnosis for dynamic systems*. Boston, USA: Kluwer Academic Publishers.

Delgado, J., Besacon, G., Begovich, O., & Carvajal, J. (2016). Multi-leak diagnosis in pipelines based on extended Kalman filter. *Control Engineering Practice. 49*, 139–148.

EPA. (2009). Drinking water infrastructure needs survey and assessment. Technical report, U.S. Environmental Protection Agency (EPA).

Ferrante, M., & Brunone, B. (2003). Pipe system diagnosis and leak detection by unsteady-state tests: 1. *Harmonic Analysis. Advances in Water Resources, 26*, 95–105.

Gelb, A. (1994). *Applied optimal estimation*. MIT Press.

Isermann, R., & Münchhof, M. (2011). *Identification of dynamic systems*. Berlin: Springer.

Korbicz, J., Koscielny, J. M., Kowalczuk, Z., & Cholewa, W. (2004). *Fault diagnosis. models, artificial intelligence, applications*, Chapter detecting and locating leaks in transmission pipelines (pp. 821–864). Germany: Springer.

Lee, P., Vítkovsky, J., Lambert, M., Simpson, A., & Liggett, J. (2005). Leak location using the pattern of the frequency response diagram in pipelines: A numerical study. *Journal of Sound and Vibration, 284*, 1051–1075.

Pipeline Studio. (2013). Software. In *Energy Solutions International*. http://www.energy-solutions.com/.

Polyanin, A. D., & Zaitsev, V. F. (2014). *Handbook of nonlinear partial diffential equation*. Boca Raton: Chapman & Hall/CRC Press.

Prabhata, K. S. (1993). Design of a submarine oil pipeline. *Journal of Transportation Engineering, 119*, 159–170.

Reif, K., Sonnenmannn, F., & Unbehauen, R. (1998). An EKB-based nonlinear observer with a prescribed degree of stability. *Automatica, 34*(9), 1119–1998.

Scott, S., & Barrufet, M. (2003). Worldwide assessment of industry leak detection capabilities for single and multiphase pipelines. Internal Report 18133, Department of Petroleum Eng., Texas A&M University.

MATLAB. 2008. USA: MATLAB User Guide. The MathWorks Inc.

Verde, C. (2003). Accomodation of multi-leaks positions in a pipeline. *Safeprocess 03* (pp. 1041–1046). Washington DC: IFAC.

Verde, C., Molina, L., & Carrera, R. (2011). Practical issues of leaks diagnosis in pipelines. In *20th IFAC World Congress, August, Milan, Italy*.

Verde, C., Molina, L., & Torres, L. (2014). Parameterized transient model of a pipeline for multiple leaks location. *Journal of Loss Prevention in the Process Industries, 29*, 177–185. doi:10.1016/j.jlp.2014.02.013.

Verde, C., & Rojas, J. (2015). Iterative scheme for sequential leaks location. In *9th IFAC Symposium on Fault Detection, Supervision and Safety for Technical Processes SAFEPROCESS 2015*.

Verde, C., Visairo, N., & Gentil, S. (2007). Two leaks isolation in a pipeline by transient response. *Applied Water Resources, 30*, 1711–1721.

Verde, C., Torres, L., & González, O. (2016). Decentralized scheme for leaks' location in a branched pipeline. *Journal of Loss Prevention in the Process Industries*.

Chapter 8
Simulation of Gas Networks and Leak Detection Using Quadripole Models

Sara T. Baltazar, Paulo Lopes dos Santos and Teresa P. Azevedo Perdicoúlis

Abstract A cost-effective, accurate, and robust leak detection method is essential in gas network management in order to reduce inspection time and to increase reliability in the system. This work presents a model-based leakage detection method; the gas dynamics are described by a linearized system of partial differential equations that is further reduced to a one-dimensional spatial model. By using an electrical analogy, a pipeline can be represented by a two-port network, where mass flow behaves like current and pressure like voltage. Four transfer function quadripole models are then established to describe the gas pipeline dynamics, depending on the variables of interest at the pipeline boundaries. A leak detection method is devised by employing mass flow data at boundaries and pressure data at some point of the pipeline, as well as by assessing the effects of the leakage on the pressure and mass flow along the pipeline. A case study has been built from operational data supplied by REN Gasodutos (the Portuguese gas company) to show the advantages of the proposed models.

8.1 Introduction

Natural gas is mainly transported via pipelines (Adnan et al. 2014). The main goal in the management and operation of the pipeline network is to meet customer needs at the lowest price. It is also important to keep the network within safety levels, however, to avoid hazardous situations such as leakage, for instance Lin and Holbert (2006).

The leakage detection problem has been studied for more than 40 years (Luo et al. 2010); however there are still many open questions to be addressed in order to have simple and accurate leakage detectors with a small rate of false alarms (Turkowski et al. 2007). Distinct approaches exist to this problem: sensorial methods, hardware-based methods, and model-based methods (Geiger 2006; Murvay and Silea 2012).

S. T. Baltazar (✉) · T.P. Azevedo Perdicoúlis
UTAD, Vila Real, Portugal
e-mail: sarabaltazar@gmail.com

P. Lopes dos Santos
FEUP, Porto, Portugal

C. Verde and L. Torres (eds.), *Modeling and Monitoring of Pipelines and Networks*, Applied Condition Monitoring 7,
DOI 10.1007/978-3-319-55944-5_8

The latter type of these methods continuously monitor flow and pressure along the pipeline by using a mathematical model, and these are the simplest and most economical to implement. They can be combined with other types of methods to improve reliability or, for instance, to overcome the sparseness of mass flow measurements in the network (REN-Gasodutos 2014).

Trustworthy models are not easy to obtain because of the difficulty in determining the solution of the nonlinear partial differential equations (PDE) used to represent the gas dynamics in the pipelines. Regardless, non-horizontal models have been proposed (Geiger 2006; Verde 2005), and a linearization of the PDE system which can lead to simpler and more accurate models is proposed by Lopes dos Santos et al. (2010).

This chapter presents a new leakage detection technique based on a modular approach that considers a non-horizontal linearized model of the pipeline, discussed in Baltazar et al. (2014). By taking advantage of an analogy between gas pipelines and electrical circuits, where pressures are viewed as voltages and mass flows as electrical currents, different quadripole models are obtained; these describe the relation between pressure and mass flow at the pipeline ends, with the independent variables being chosen according to operation conditions.

Although quadripole models can only relate boundary values, intermediate values can also be obtained by partitioning the pipeline into two connected sections and then each section is modeled by a two-port model. A leakage is represented by a *current* source that extracts mass flow from the junction. The pipeline is instrumented with equally spaced pressure sensors and with two mass flow sensors at its boundaries, as is shown in Fig. 8.12. To mimic a real situation, noise is added to the data.

The quadripole models are used to overcome the scarcity of flow measurements in the network and allow the generation of the residuals if the pressure is measured at some intermediate points of the pipeline. Under normal conditions without leaks, the mass flow data at the inlet and outlet of the pipeline are used together with the models to estimate pressures at the ends and both pressures' and mass flows' variables at some intermediate points (L_i) of the line. On the other hand, the mass flows at points L_i are also calculated by using the measured pressures at the ends and in termediate points L_i, and mass flow at the inlet. Thus, the differences between the measured pressures at L_i and their estimated values generate leak residuals. Thus, a leakage is detected when both the residuals are *significantly* different from zero, and the differences between the mass flows in the normal conditions and the ones calculated from the pressures and inlet mass flow data deviate significantly. As the leakage effect propagates at speed sound, it is perceived first by the sensors closest to the point where it occurs. So the proposed method locates it in between these points.

We now describe the sections. In Sect. 8.2, the nonlinear second-order hyperbolic PDE system is linearized, and the transfer function (TF) models are then derived in Sect. 8.3. The four different quadripole models are derived in Sect. 8.4, and validated with real data in Sect. 8.5. By involving the quadripole models, intermediate pressure and mass flow descriptions for normal operation conditions are obtained in Sect. 8.6. The leakage is modeled in Sect. 8.7. In Sects. 8.8 and 8.9, the leakage detection method is presented. Some conclusions and directions for future work are presented in Sect. 8.10.

8.2 Hyperbolic Partial Differential Equations' System

In this work, the gas dynamics within the pipeline are represented by a nonlinear second-order hyperbolic system of PDE. This is a simplified model where certain assumptions have been imposed; for instance, we consider unidirectional flow, constant temperature and no viscous and no turbulent effects (Králik et al. 1988; Osiadacz 1987; Reddy et al. 2006). Hence,

$$\begin{cases} \frac{\partial q(\ell,t)}{\partial t} = -\mathscr{A}\frac{\partial p(\ell,t)}{\partial \ell} - \frac{fc^2}{2\mathscr{D}\mathscr{A}}\frac{q^2(\ell,t)}{p(\ell,t)} - \mathscr{A}g\sin\theta\frac{p(\ell,t)}{c^2} \\ \frac{\partial p(\ell,t)}{\partial t} = -\frac{c^2}{\mathscr{A}}\frac{\partial q(\ell,t)}{\partial \ell}. \end{cases} \tag{8.1}$$

where q is the node mass flow, ℓ is space, t is time, $\mathscr{A}$ is the cross-section area, p is the edge pressure drop, f is the friction factor, c is the isothermal speed of sound, $\mathscr{D}$ is the pipe diameter, g is the acceleration caused by gravity, and θ is the pipeline inclination angle. We write $p(\ell,t)$ and $q(\ell,t)$ because both pressure, p, and mass flow, q, depend on space and time.

This is a very complex model and, further simplifications are required in order to obtain a solution. Thus, we set

$$p(\ell,t) = \bar{p}(\ell) + \Delta p(\ell,t) \quad \text{and} \quad q(\ell,t) = \bar{q} + \Delta q(\ell,t),$$

where $\bar{p}(\ell)$ and $\bar{q}$ are the pressure and gas mass flow operational levels and $\Delta p(\ell,t)$ and $\Delta q(\ell,t)$ are the respective deviations from these values along the pipeline. By substituting pressure and mass flow in terms of their operational values in the nonlinear term of the first equation of (8.1), after ignoring terms of order equal or greater than 2, we have the following approximation:

$$\frac{q^2(\ell,t)}{p(\ell,t)} = \frac{(\bar{q}+\Delta q(\ell,t))^2}{\bar{p}(\ell)+\Delta p(\ell,t)} \approx \frac{\bar{q}^2}{\bar{p}(\ell)} + 2\frac{\bar{q}}{\bar{p}(\ell)}\Delta q(\ell,t) - \frac{\bar{q}^2}{\bar{p}^2(\ell)}\Delta p(\ell,t). \tag{8.2}$$

Moreover, the third term of the right side of (8.2) may also be ignored since gas distribution networks operate at very high pressure, approximately 80 *bar*. Thus, (8.2) is substituted in the first equation of (8.1)

$$\frac{\partial q(\ell,t)}{\partial t} = -\mathscr{A}\frac{\partial p(\ell,t)}{\partial \ell} - \frac{fc^2}{2\mathscr{D}\mathscr{A}}\frac{\bar{q}}{\bar{p}(\ell)}(\bar{q}+2\Delta q(\ell,t)) - \mathscr{A}g\sin\theta\frac{p(\ell,t)}{c^2}, \tag{8.3}$$

By assuming small flow oscillations, it can also be considered that

$$\bar{q} + 2\Delta q(\ell, t) \approx \bar{q} + \Delta q(\ell, t) = q(\ell, t).$$

Hence, the linearized model can be written by

$$\begin{cases} \dfrac{\partial q(\ell, t)}{\partial t} = -\mathscr{A}\dfrac{\partial p(\ell, t)}{\partial \ell} - 2\alpha q(\ell, t) - \beta \sin\theta p(\ell, t) \\ \dfrac{\partial p(\ell, t)}{\partial t} = -\dfrac{c^2}{\mathscr{A}}\dfrac{\partial q(\ell, t)}{\partial \ell}, \end{cases} \tag{8.4}$$

where $\alpha = \dfrac{fc^2}{4\mathscr{D}\mathscr{A}}\dfrac{\bar{q}}{\bar{p}(\ell)}, \qquad \beta = \dfrac{\mathscr{A}g}{c^2}.$

8.3 Transfer Function Model

The Laplace transform is applied to the linearized version of (8.1), to obtain the spatial system

$$\begin{bmatrix} \dfrac{\partial P(\ell, s)}{\partial \ell} \\ \dfrac{\partial Q(\ell, s)}{\partial \ell} \end{bmatrix} = \begin{bmatrix} -\dfrac{\beta}{\mathscr{A}}\sin\theta & -\dfrac{2\alpha + s}{\mathscr{A}} \\ -s\dfrac{\mathscr{A}}{c^2} & 0 \end{bmatrix} \begin{bmatrix} P(\ell, s) \\ Q(\ell, s) \end{bmatrix}, \tag{8.5}$$

where $P(\ell, s)$ and $Q(\ell, s)$ are the Laplace transform of the pressure and the mass flow, respectively. $P(0, s)$ and $Q(0, s)$ are, respectively, pressure and mass flow at the inlet. The solution of (8.5) can then be written as

$$\begin{bmatrix} P(\ell, s) \\ Q(\ell, s) \end{bmatrix} = \begin{bmatrix} E_{11}(\ell, s) & E_{12}(\ell, s) \\ E_{21}(\ell, s) & E_{22}(\ell, s) \end{bmatrix} \begin{bmatrix} P(0, s) \\ Q(0, s) \end{bmatrix}, \tag{8.6}$$

$$E_{11}(\ell, s) = \frac{e^{-\delta(s)T_d(\ell)}}{-2\gamma(s)}\left[\frac{g}{2c}\sin\theta\left(1 - e^{-2\gamma(s)T_d(\ell)}\right) - \gamma(s)\left(1 + e^{-2\gamma(s)T_d(\ell)}\right)\right], \tag{8.7}$$

$$E_{12}(\ell, s) = \frac{ce^{-\delta(s)T_d(\ell)}\left(1 - e^{-2\gamma(s)T_d(\ell)}\right)}{-2\gamma(s)}\frac{c\,(s + 2\alpha)}{\mathscr{A}}, \tag{8.8}$$

$$E_{21}(\ell, s) = \frac{e^{-\delta(s)T_d(\ell)}\left(1 - e^{-2\gamma(s)T_d(\ell)}\right)}{-2\gamma(s)}\frac{s\mathscr{A}}{c}, \tag{8.9}$$

$$E_{22}(\ell, s) = \frac{e^{-\delta(s)T_d(\ell)}}{2\gamma(s)}\cdot\left[\frac{g}{2c}\sin\theta\left(1 - e^{-2\gamma(s)T_d(\ell)}\right) + \gamma(s)\left(1 + e^{-2\gamma(s)T_d(\ell)}\right)\right], \tag{8.10}$$

with $E_{12}(\ell, s) = \frac{c^2(s+2\alpha)}{\mathscr{A}^2 s} E_{21}(\ell, s)$ the relation between the two off-diagonal transfer functions (TF) and

$$T_d(\ell) := \frac{\ell}{c}, \quad \gamma(s) := \sqrt{(s+\alpha)^2 + \frac{g^2}{4c^2}\sin^2\theta - \alpha^2}, \quad \delta(s) := \frac{g}{2c}\sin\theta - \gamma(s).$$

If we fix $\ell = L$ in (8.6), where L is the pipe length, $P(L, s)$ and $Q(L, s)$ are then associated with the Laplace transform of the outlet pressure and mass flow, respectively. Therefore, the following model relates inlet and outlet pressures and mass flows:

$$\begin{cases} P(L, s) = E_{11}(L, s)P(0, s) + E_{12}(L, s)Q(0, s) \\ Q(L, s) = E_{21}(L, s)P(0, s) + E_{22}(L, s)Q(0, s)\ . \end{cases} \tag{8.11}$$

Notice that $E_{ij}(\ell, s)$, $i, j = 1, 2$, are the TFs

$$E_{ij}(\ell, s) = \left.\frac{\mathscr{Y}_i(s)}{\mathscr{U}_j(s)}\right|_{\mathscr{U}_k=0}, \quad k = i \neq j\ , \tag{8.12}$$

where $\mathscr{Y}$ and $\mathscr{U}$ are outlet and inlet variables, respectively. In the sequel, for the sake of simplicity, we drop the argument L in $T_d(L)$ and $E_{ij}(L, s)$ for $i, j = 1, 2$, as well as in all TFs that may occur, since a pipeline of fixed length is always considered.

8.4 Quadripole Models

To take advantage of the aforementioned electrical analogy, we obtain different TF models for different choices of the independent variables (model input signals). In addition to the model (8.11), there are four two-port network (quadripole) model options. In this section, these models are derived, simplified, and approximated by simpler and more intuitive models.

8.4.1 Impedance Model

In the impedance model depicted in Fig. 8.1 and described in (8.13), the boundary pressures, $P(0, s)$ and $P(L, s)$ are written as functions of the mass flows, $Q(0, s)$ and $Q(L, s)$

$$\begin{cases} P(0, s) = Z_{11}(s)Q(0, s) - Z_{12}(s)Q(L, s) \\ P(L, s) = Z_{21}(s)Q(0, s) - Z_{22}(s)Q(L, s). \end{cases} \tag{8.13}$$

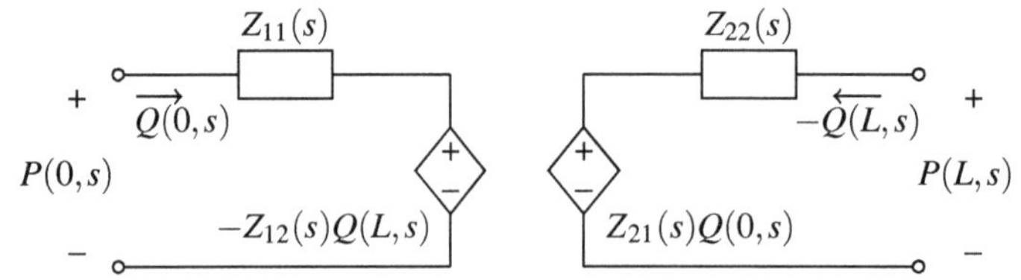

Fig. 8.1 Impedance quadripole

Impedances $Z_{ij}(s)$, $i, j = 1, 2$ are obtained from (8.7)–(8.10) and given by Baltazar et al. (2016b) as

$$Z_{11}(s) = \frac{c\gamma(s)}{\mathscr{A}s}\frac{1+e^{-2T_d\gamma(s)}}{1-e^{-2T_d\gamma(s)}} + \frac{g}{2\mathscr{A}s}\sin\theta\ , \tag{8.14}$$

$$Z_{12}(s) = \frac{2ce^{T_d\frac{g}{2c}\sin\theta}\gamma(s)}{\mathscr{A}s}\frac{e^{-T_d\gamma(s)}}{1-e^{-2T_d\gamma(s)}}\ , \tag{8.15}$$

$$Z_{21}(s) = \frac{2ce^{-T_d\frac{g}{2c}\sin\theta}\gamma(s)}{\mathscr{A}s}\frac{e^{-T_d\gamma(s)}}{1-e^{-2T_d\gamma(s)}}\ , \tag{8.16}$$

$$Z_{22}(s) = \frac{c\gamma(s)}{\mathscr{A}s}\frac{1+e^{-2T_d\gamma(s)}}{1-e^{-2T_d\gamma(s)}} - \frac{g}{2\mathscr{A}s}\sin\theta\ . \tag{8.17}$$

In Baltazar et al. (2014), we see that these TFs can be approximated by

$$\hat{Z}_{11}(s) = K_{Z_{11}}\left(\frac{s+\alpha}{s}\right) + \frac{g}{2\mathscr{A}s}\sin\theta\ , \tag{8.18}$$

$$\hat{Z}_{12}(s) = K_{Z_{12}}\left(\frac{s+\alpha}{s}\right)e^{-T_d s}\ , \tag{8.19}$$

$$\hat{Z}_{21}(s) = K_{Z_{21}}\left(\frac{s+\alpha}{s}\right)e^{-T_d s}\ , \tag{8.20}$$

$$\hat{Z}_{22}(s) = K_{Z_{22}}\left(\frac{s+\alpha}{s}\right) - \frac{g}{2\mathscr{A}s}\sin\theta\ , \tag{8.21}$$

where $K_{Z_{ij}}$, $i, j = 1, 2$, are the gains (Baltazar et al. 2016b).

8.4.2 Admittance Model

When considering the mass flows as a function of the pressures, the equivalent circuit shown in Fig. 8.2 describes the flow process given by

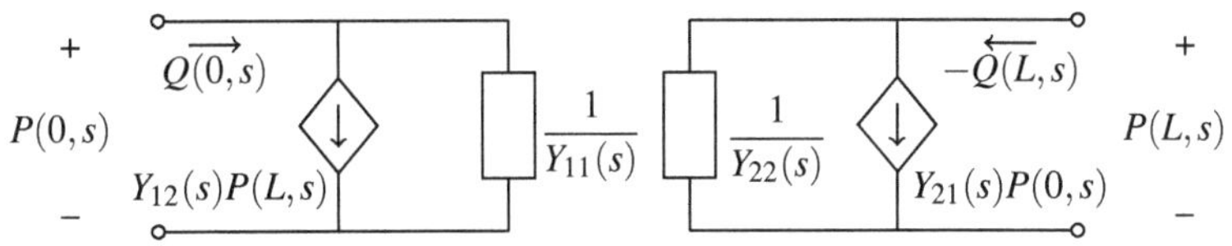

Fig. 8.2 Admittance quadripole

$$\begin{cases} Q(0,s) = Y_{11}(s)P(0,s) + Y_{12}(s)P(L,s) \\ Q(L,s) = -Y_{21}(s)P(0,s) - Y_{22}(s)P(L,s) \end{cases} \tag{8.22}$$

with

$$Y_{11}(s) = \frac{c\gamma(s)}{\mathscr{A}s}\frac{1+e^{-2T_d\gamma(s)}}{1-e^{-2T_d\gamma(s)}} + \frac{g}{2\mathscr{A}s}\sin\theta\,, \tag{8.23}$$

$$Y_{12}(s) = \frac{2ce^{T_d\frac{g}{2c}\sin\theta}\gamma(s)}{\mathscr{A}s}\frac{e^{-T_d\gamma(s)}}{1-e^{-2T_d\gamma(s)}}\,, \tag{8.24}$$

$$Y_{21}(s) = \frac{2ce^{-T_d\frac{g}{2c}\sin\theta}\gamma(s)}{\mathscr{A}s}\frac{e^{-T_d\gamma(s)}}{1-e^{-2T_d\gamma(s)}}\,, \tag{8.25}$$

$$Y_{22}(s) = \frac{c\gamma(s)}{\mathscr{A}s}\frac{1+e^{-2T_d\gamma(s)}}{1-e^{-2T_d\gamma(s)}} - \frac{g}{2\mathscr{A}s}\sin\theta\,. \tag{8.26}$$

Baltazar et al. (2014) revealed that this set of admittances can be approximated by

$$\hat{Y}_{11}(s) = K_{Y_{11}}\left(\frac{s+\alpha}{s+2\alpha}\right) - \frac{g\mathscr{A}}{2c^2(s+2\alpha)}\sin\theta\,, \tag{8.27}$$

$$\hat{Y}_{12}(s) = -K_{Y_{12}}\left(\frac{s+\alpha}{s+2\alpha}\right)e^{-T_d s}\,, \tag{8.28}$$

$$\hat{Y}_{21}(s) = -K_{Y_{21}}\left(\frac{s+\alpha}{s+2\alpha}\right)e^{-T_d s}\,, \tag{8.29}$$

$$\hat{Y}_{22}(s) = K_{Y_{22}}\left(\frac{s+\alpha}{s+2\alpha}\right) + \frac{g\mathscr{A}}{2c^2(s+2\alpha)}\sin\theta\,, \tag{8.30}$$

where $K_{Y_{ij}}$, $i, j = 1, 2$, are the gains (Baltazar et al. 2016b).

8.4.3 Hybrid Model

In the hybrid model, the system is considered to be driven by the inlet mass flow and outlet pressure with the inlet pressure and the outlet mass flow as outputs. It is depicted in Fig. 8.3 and described by

$$\begin{cases} P(0,s) = H_{11}(s)Q(0,s) + H_{12}(s)P(L,s) \\ Q(L,s) = -H_{21}(s)Q(0,s) - H_{22}(s)P(L,s) \end{cases} \tag{8.31}$$

where (Baltazar et al. 2016b)

$$H_{11}(s) = \frac{c}{\mathscr{A}}\frac{s+2\alpha}{\gamma(s)}\frac{1-e^{-2T_d\gamma(s)}}{1+e^{-2T_d\gamma(s)} - \frac{g}{2c}\sin\theta\left(1-e^{-2T_d\gamma(s)}\right)}, \tag{8.32}$$

$$H_{12}(s) = \frac{2e^{T_d\frac{g}{2c}\sin\theta}\gamma(s)e^{-T_d\gamma(s)}}{\gamma(s)\left(1+e^{-2T_d\gamma(s)}\right) - \frac{g}{2c}\sin\theta\left(1-e^{-2T_d\gamma(s)}\right)}, \tag{8.33}$$

$$H_{21}(s) = -\frac{2e^{-T_d\frac{g}{2c}\sin\theta}\gamma(s)e^{-T_d\gamma(s)}}{\gamma(s)\left(1+e^{-2T_d\gamma(s)}\right) - \frac{g}{2c}\sin\theta\left(1-e^{-2T_d\gamma(s)}\right)}, \tag{8.34}$$

$$H_{22}(s) = \frac{\mathscr{A}}{c}\frac{s\left(1-e^{-2T_d\gamma(s)}\right)}{s\left(1+e^{-2T_d\gamma(s)}\right) - \frac{g}{2c}\sin\theta\left(1-e^{-2T_d\gamma(s)}\right)}. \tag{8.35}$$

The H_{ij}, $i, j = 1, 2$ parameters can be approximated by Baltazar et al. (2016a)

$$\hat{H}_{11}(s) = K_{H_{11}}\left(\frac{s+2\alpha}{s+\alpha-\frac{g}{2c}\sin\theta}\right), \tag{8.36}$$

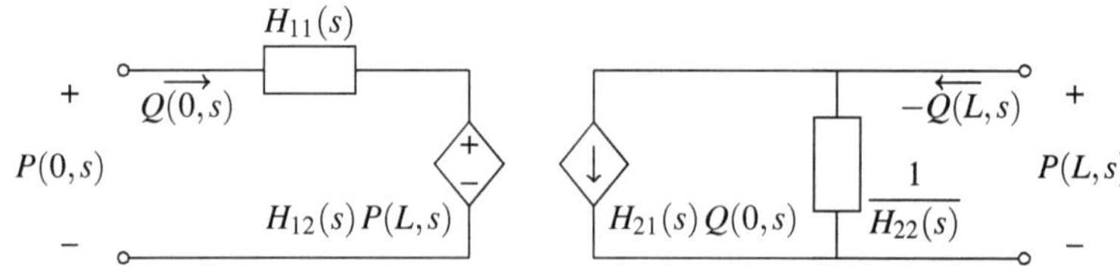

Fig. 8.3 Hybrid quadripole

$$\hat{H}_{12}(s) = K_{H_{12}} \left(\frac{(s+\alpha)\, e^{-T_d s}}{s + \alpha - \dfrac{g}{2c} \sin\theta} \right), \tag{8.37}$$

$$\hat{H}_{21}(s) = K_{H_{21}} \left(\frac{(s+\alpha)\, e^{-T_d s}}{s + \alpha - \dfrac{g}{2c} \sin\theta} \right), \tag{8.38}$$

$$\hat{H}_{22}(s) = K_{H_{22}} \left(\frac{s}{s + \alpha - \dfrac{g}{2c} \sin\theta} \right), \tag{8.39}$$

where $K_{H_{ij}}$, $i, j = 1, 2$, are the gains (Baltazar et al. 2016b).

8.4.4 Inverse Hybrid Model

For the inverse hybrid circuit, shown in Fig. 8.4, the sources are the inlet pressure and the outlet mass flow. Furthermore, the model is described by

$$\begin{cases} Q(0,s) = G_{11}(s)P(0,s) - G_{12}(s)Q(L,s) \\ P(L,s) = G_{21}(s)P(0,s) - G_{22}(s)Q(L,s)\,. \end{cases} \tag{8.40}$$

where according to Baltazar et al. (2016a, b)

$$G_{11}(s) = \frac{\mathscr{A}}{c}\frac{s+2\alpha}{\gamma(s)} \frac{s\left(1 - e^{-2T_d\gamma(s)}\right)}{\gamma(s)\left(1 + e^{-2T_d\gamma(s)}\right) + \dfrac{g}{2c}\sin\theta\left(1 - e^{-2T_d\gamma(s)}\right)}, \tag{8.41}$$

$$G_{12}(s) = -\frac{2e^{T_d \frac{g}{2c}\sin\theta}\gamma(s)e^{-T_d\gamma(s)}}{\gamma(s)\left(1 + e^{-2T_d\gamma(s)}\right) + \dfrac{g}{2c}\sin\theta\left(1 - e^{-2T_d\gamma(s)}\right)}, \tag{8.42}$$

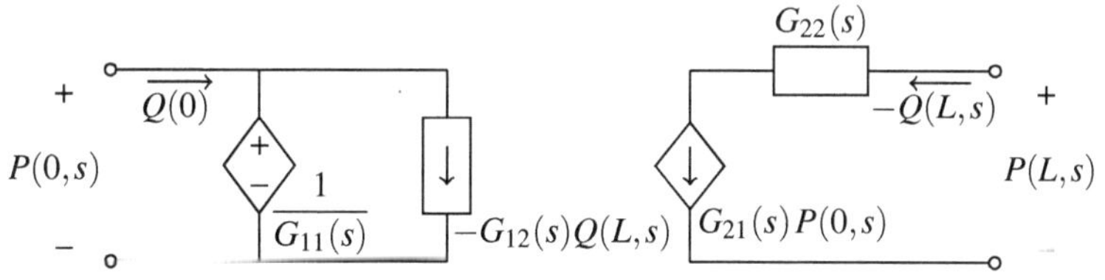

Fig. 8.4 Inverse hybrid quadripole

$$G_{21}(s) = \frac{2e^{-T_d \frac{g}{2c}\sin\theta}\gamma(s)e^{-T_d\gamma(s)}}{\gamma(s)\left(1+e^{-2T_d\gamma(s)}\right)+\frac{g}{2c}\sin\theta\left(1-e^{-2T_d\gamma(s)}\right)}, \tag{8.43}$$

$$G_{22}(s) = \frac{c}{\mathscr{A}}\frac{(s+2\alpha)\left(1-e^{-2T_d\gamma(s)}\right)}{\gamma(s)\left(1+e^{-2T_d\gamma(s)}\right)+\frac{g}{2c}\sin\theta\left(1-e^{-2T_d\gamma(s)}\right)}. \tag{8.44}$$

These functions can be approximated by

$$\hat{G}_{11}(s) = K_{G_{11}}\left(\frac{s}{s+\alpha+\frac{g}{2c}\sin\theta}\right), \tag{8.45}$$

$$\hat{G}_{12}(s) = -\hat{K}_{G_{12}}\left(\frac{(s+\alpha)\,e^{-T_d s}}{s+\alpha+\frac{g}{2c}\sin\theta}\right), \tag{8.46}$$

$$\hat{G}_{21}(s) = K_{G_{21}}\left(\frac{(s+\alpha)\,e^{-T_d s}}{s+\alpha+\frac{g}{2c}\sin\theta}\right), \tag{8.47}$$

$$\hat{G}_{22}(s) = K_{G_{22}}\left(\frac{s+2\alpha}{s+\alpha+\frac{g}{2c}\sin\theta}\right), \tag{8.48}$$

where $K_{G_{ij}}$ for $i, j = 1, 2$ are the gains.

8.5 Model Validation

To validate the TF models, a horizontal pipeline was simulated by using the data of one day of operation, supplied by the REN Gasodutos. These data are depicted in Fig. 8.5, where inlet and outlet pressures are represented by P_{iREN}, P_{oREN}, and inlet and outlet mass flows are Q_{iREN}, Q_{oREN}, respectively.

The data were collected with a sampling period of 2 min, producing a data set of cardinality $N_{data} = 720$, and the following parameters were considered in this case study:

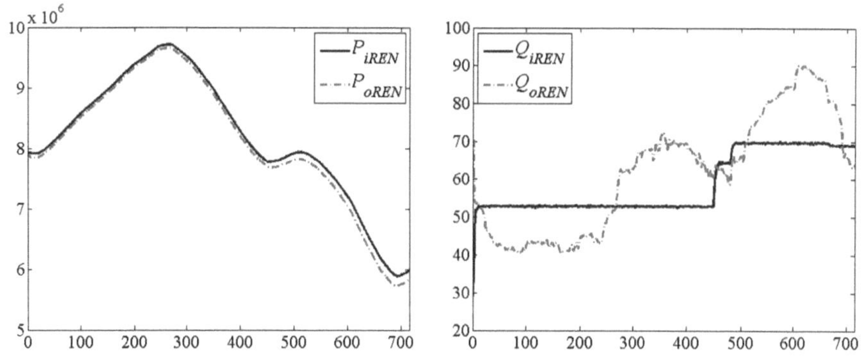

(a) Pressures (Pa) (b) Mass flows (kg/s)

Fig. 8.5 Inlet and outlet pressures and mass flows values from REN Gasodutos

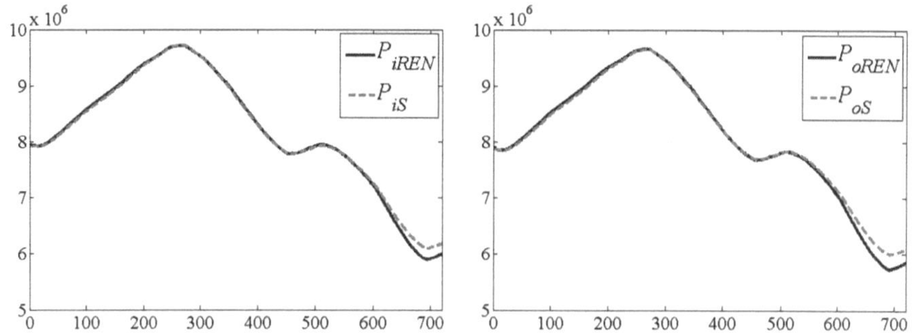

(a) Inlet pressures (b) Outlet pressures

Fig. 8.6 Measured (P_{iREN}, P_{oREN}) and simulated (P_{iS}, P_{oS}) (*in dashed*) pressures (Pa)

$$\begin{array}{ll} \text{Operational flow } \bar{q} = 90\,\text{kg/s} & \text{Operational pressure } \bar{p} = 8 \times 10^6\,\text{bar} \\ \text{Pipeline length } L_{REN} = 36\,\text{km} & \text{Cross-section area } \mathscr{A} = 0.49\,\text{m}^2 \\ \text{Friction factor } f = 7.9 \times 10^{-3} & \text{Speed of sound } c = 340\,\text{m/s} \\ \text{Diameter } \mathscr{D} = 0.793\,\text{m} & \text{Gravity } g = 9.8\,\text{m/s}^2. \end{array} \tag{8.49}$$

Figure 8.6 allows a comparison between the measured pressures and the simulated variables with the impedance model by using the inlet and outlet mass flows as independent variables. From the time evolution of the pressures, one can say that simpler linear quadripole models describe the gas behavior without significant loss of precision. Please refer to Baltazar et al. (2016a) for a detailed discussion of these models.

Note that if the pipeline is horizontal the following TFs of the models are symmetric:

$$
\begin{aligned}
&\hat{Z}_{11}(s) = \hat{Z}_{22}(s), && \hat{Y}_{11}(s) = \hat{Y}_{22}(s), && \hat{G}_{11}(s) = \hat{H}_{22}(s),\\
&\hat{G}_{21}(s) = \hat{H}_{12}(s), && \hat{Z}_{12}(s) = \hat{Z}_{21}(s), && \hat{Y}_{12}(s) = \hat{Y}_{21}(s),\\
&\hat{G}_{12}(s) = -\hat{H}_{12}(s) && \hat{G}_{22}(s) = \hat{H}_{11}(s), && \hat{H}_{12}(s) = \hat{H}_{21}(s),\\
&\hat{H}_{11}(s) = \frac{c^2\,(s+2\alpha)}{\mathscr{A}^2 s}\hat{H}_{22}(s).
\end{aligned}
$$

For the sake of simplicity, we omit the complex variable s in the rest of this chapter.

8.6 Junction Point

A junction in a pipeline is modeled by considering two sections, as shown in Fig. 8.7, and the whole model is given by two connected quadripoles as in Fig. 8.8.

By considering the simplified impedance model, we have for the left quadripole

$$
\begin{cases}
P(0) = \hat{Z}_{11}(x)Q(0) - \hat{Z}_{12}(x)Q(x)\\
\\
P(x) = \hat{Z}_{21}(x)Q(0) - \hat{Z}_{22}(x)Q(x)
\end{cases}
\tag{8.50}
$$

and for the right quadripole

$$
\begin{cases}
P(x) = \hat{Z}_{11}(L-x)Q(x) - \hat{Z}_{12}(L-x)Q(L)\\
\\
P(L) = \hat{Z}_{21}(L-x)Q(x) - \hat{Z}_{22}(L-x)Q(L)\ .
\end{cases}
\tag{8.51}
$$

By using the second equation of (8.50) and the first one of (8.51), the mass flow, $Q(x)$ for any x of the pipeline is given by

$$
Q(x) = \frac{L-x}{L}e^{-\frac{x}{c}s}Q(0) + \frac{x}{L}e^{-\frac{L-x}{c}s}Q(L)\ , \tag{8.52}
$$

and the pressure, $P(x)$, is given by

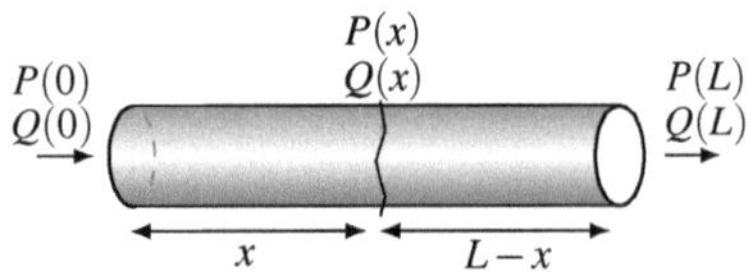

Fig. 8.7 Junction of two pipelines

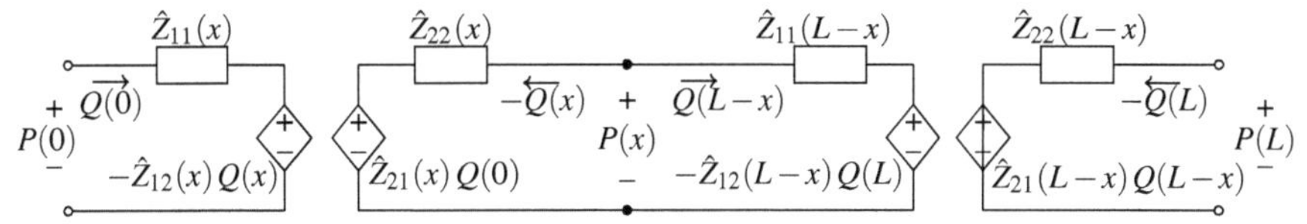

Fig. 8.8 Impedance quadripole model for a junction

$$P(x) = \frac{c}{\mathscr{A}\alpha T_d}\frac{s+\alpha}{s}\left[e^{-\frac{x}{c}s}Q(0) - e^{-\frac{L-x}{c}s}Q(L)\right] . \tag{8.53}$$

Similarly, by adopting the admittance model, the circuit depicted in Fig. 8.9 is obtained with the respective equations for the left quadripole

$$\begin{cases} Q(0) = \hat{Y}_{11}(x)P(0) + \hat{Y}_{12}(x)P(x) \\ Q(x) = -\hat{Y}_{21}(x)P(0) - \hat{Y}_{22}(x)P(x) , \end{cases} \tag{8.54}$$

and for the right quadripole:

$$\begin{cases} Q(x) = \hat{Y}_{11}(L-x)P(x) + \hat{Y}_{12}(L-x)P(L) \\ Q(L) = -\hat{Y}_{21}(L-x)P(x) - \hat{Y}_{22}(L-x)P(L) . \end{cases} \tag{8.55}$$

Therefore, by using the second equation of (8.9), the first one of (8.55) and the set of FT (8.27)–(8.30), we have

$$P(x) = \frac{L-x}{L}e^{-\frac{x}{c}s}P(0) + \frac{x}{L}e^{-\frac{L-x}{c}s}P(L) , \tag{8.56}$$

and

$$Q(x) = \frac{\mathscr{A}}{c\alpha T_d}\frac{s+\alpha}{s+2\alpha}\left[e^{-\frac{x}{c}s}P(0) - e^{-\frac{L-x}{c}s}P(L)\right] . \tag{8.57}$$

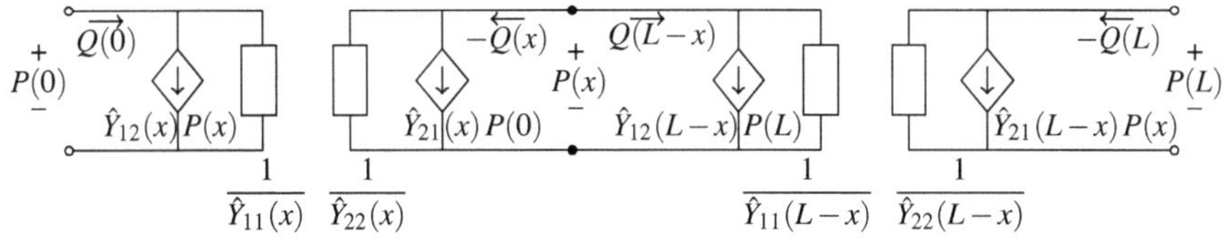

Fig. 8.9 Admittance quadripole model for a junction

By applying either the hybrid or the inverse hybrid models, intermediate pressure/mass flow values can also be represented in terms of either the inlet pressure and outlet mass flow or inlet mass flow and outlet pressure under normal conditions. As a consequence

$$Q(x) = -\frac{1}{1+\frac{\alpha^2 x(L-x)}{c^2}\frac{s(s+2\alpha)}{(s+\alpha)^2}}\left[e^{-\frac{x}{c}s}Q(0) + \frac{\mathscr{A}\alpha x}{c^2}\frac{s}{s+\alpha}e^{-\frac{L-x}{c}s}P(L)\right], \quad (8.58)$$

$$P(x) = -\frac{1}{1+\frac{\alpha^2 x(L-x)}{c^2}\frac{s(s+2\alpha)}{(s+\alpha)^2}}\left[\frac{\alpha\,(L-x]}{\mathscr{A}}\frac{s+2\alpha}{s+\alpha}e^{-\frac{x}{c}s}Q(0) - e^{-\frac{L-x}{c}s}P(L)\right], \quad (8.59)$$

$$P(x) = \frac{1}{1+\frac{\alpha^2 x(L-x)}{c^2}\frac{s(s+2\alpha)}{(s+\alpha)^2}}\left[e^{-\frac{x}{c}s}P(0) - \frac{\alpha x}{\mathscr{A}}\frac{s+2\alpha}{s+\alpha}e^{-\frac{L-x}{c}s}Q(L)\right], \quad (8.60)$$

$$Q(x) = \frac{1}{1+\frac{\alpha^2 x(L-x)}{c^2}\frac{s(s+2\alpha)}{(s+\alpha)^2}}\left[\frac{\mathscr{A}\alpha\,(L-x)}{c^2}\frac{s}{s+\alpha}e^{-\frac{x}{c}s}P(0) + e^{-\frac{L-x}{c}s}Q(L)\right]. \quad (8.61)$$

8.7 Leakage Model

Consider Fig. 8.10 which represents a pipeline with a leak at point x.

By using the quadripole models described in Sect. 8.4 and the junction model presented in Sect. 8.6, the gas pipeline with the leakage at point x_{Leak} is equivalent to the quadripole represented in Fig. 8.11. The leakage is represented by an ideal current source, because it is considered approximately constant and independent of the pressure variations during the detection time.

A leakage occurrence is associated with a mass flow loss at the leak point x_{Leak}, denoted by Q_{Leak}, which propagates to the left of x_{Leak} by $Q_{L_{Lk}}(x_{Leak})$ and to the right by $Q_{R_{Lk}}(x_{Leak})$; also a leakage produces an impact on mass flow/pressure along the pipeline.

Since the pipeline is described by a linear system, the superposition theorem can aid in calculating $P(L_i)$ and $Q(L_i)$, $i = 0, 1, 2, 3, 4$. By denoting

- variables left of the leak point with the subscript L and, vice versa variables to the right with the subscript R and,

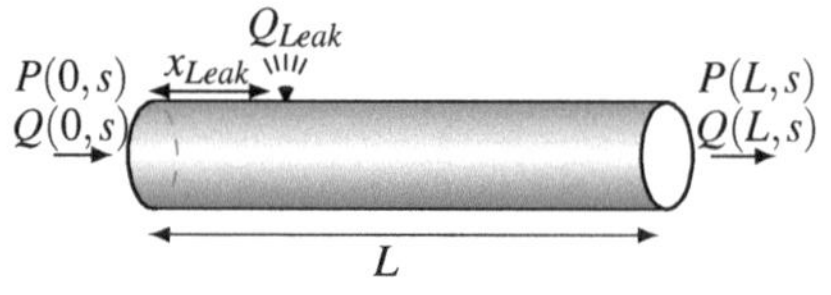

Fig. 8.10 Pipeline variable description with a leakage at point x_{Leak}

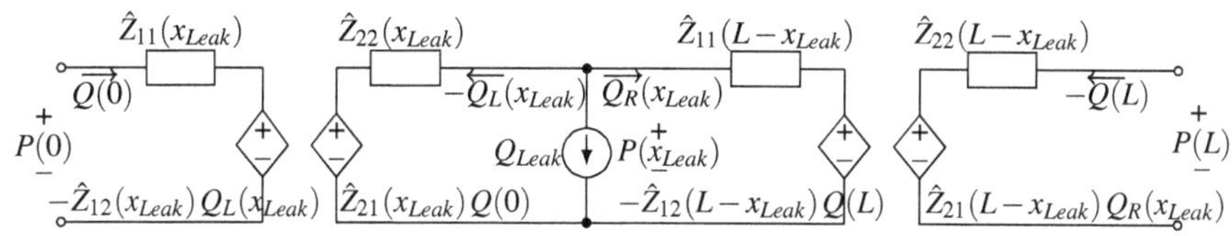

Fig. 8.11 Quadripole for a pipeline with a leak at point x_{Leak}

- subscript Lk for variables in leakage condition and nLk for the variables without leakage,

one obtains for every point of the pipeline the following relations:

$$Q_L(x_{Leak}) = Q_{L_{nLk}}(x_{Leak}) + Q_{L_{Lk}}(x_{Leak}), \tag{8.62}$$
$$Q_R(x_{Leak}) = Q_{R_{nLk}}(x_{Leak}) + Q_{R_{Lk}}(x_{Leak}), \tag{8.63}$$
$$P(L_i) = P_{nLk}(L_i) + P_{Lk}(L_i), \tag{8.64}$$
$$Q(L_i) = Q_{nLk}(L_i) + Q_{Lk}(L_i), \tag{8.65}$$

where $P_{nLk(L_i)}$ is calculated with $Q_{Leak} = 0$ and $P_{Lk(L_i)}$ is calculated with the nominal mass flows $Q_{nLk}(L0) = Q_{nlk}(L_4) = 0$. By substituting (8.64)–(8.65) in (8.62)–(8.63) and by using the superposition principle together with the Kirchhoff current law (KCL), we obtain expressions for the mass flows $Q_{L_{LK}}$ and $Q_{L_{LK}}$ to the left and right of the leak respectively given by

$$Q_{L_{Lk}}(x_{Leak}) = \frac{x}{L_4} Q_{Leak}, \tag{8.66}$$

$$Q_{R_{Lk}}(x_{Leak}) = -\frac{L_4 - x}{L_4} Q_{Leak}. \tag{8.67}$$

By using (8.66)–(8.67) after a few simple calculations, the influence of the leakage on the pressure at the pipeline endpoints can be expressed as

$$P_{Lk}(L_0) = -\hat{Z}_{11}(L_4)\, e^{-\frac{x_{Leak}}{c} s} Q_{Leak}, \tag{8.68}$$
$$P_{Lk}(L_4) = -\hat{Z}_{11}(L_4)\, e^{-\frac{L_4 - x_{Leak}}{c} s} Q_{Leak}. \tag{8.69}$$

Moreover, the influence of the leakage on the pressure at the points, L_i, for $i = 1, 2, 3$, from the point of view of inlet L_0 or outlet L_4 is reduced to

$$Q_{Lk}(L_i) = \frac{L_i - L_{0,4}}{L_4} e^{-\frac{|L_i - x_{Leak}|}{c} s} Q_{Leak}, \tag{8.70}$$

$$P_{Lk}(L_i) = -Z_{11}(L_4) e^{-\frac{|L_i - x_{Leak}|}{c} s} Q_{Leak}. \tag{8.71}$$

8.8 Quadripole Models for Leakage Detection and Location

In this section, the quadripole models described in Sect. 8.4 are used to detect and locate the time of a possible leakage. Section 8.8.1 describes the detection and location procedure by using only the data at the ends of the pipeline. Finally, an estimation of the leakage mass outflow is proposed in Sect. 8.8.2.

8.8.1 Leakage Detection and Location

According to the principle of model-based fault detection, a residual must be obtained from the model and system data. Here the residual is generated by considering the difference between pressure/mass flow pipeline profiles obtained from data and the profiles obtained with the model under normal conditions. Thus, the nominal profile corresponds to the intermediate pressures and mass flows calculated by (8.53) while using the measured mass flows at the pipeline boundaries. The operational profile comprises measured pressures along the pipeline and intermediate mass flows calculated with (8.50) and (8.51) by assuming the measured pressures and mass flow at one of the boundaries as known. The differences between the two profiles of both pressures and mass flows are then calculated for monitoring the leak condition. Furthermore, the differences of the mass flows are integrated along time since the residuals are very small in the leak condition. Moreover, the integral filters the high frequency noise. A leakage is perceived whenever the pressure differences and the integrated mass flow differences surpass predefined thresholds.

By considering the pressure representation at a certain observation point of the pipeline L_i, given by (8.64), and by substituting the terms respectively under normal condition (8.53) and in leakage situation (8.71), we obtain

$$P(L_i) = \frac{c}{\mathscr{A}\alpha T_d}\frac{s+\alpha}{s}\left[e^{-\frac{L_i}{c}s}Q(0) - e^{-\frac{L_4-L_i}{c}s}Q(L_4) - e^{-\frac{|L_i-x_{Leak}|}{c}s}Q_{Leak}\right]. \tag{8.72}$$

From this equation and (8.53) with $x = x_{Leak}$, one gets

$$\Delta P(L_i) = P_{nLk} - P_{(L_i)} = \frac{c}{\mathscr{A}\alpha T_d}\frac{s+\alpha}{s}e^{-\frac{|L_i-x_{Leak}|}{c}s}Q_{Leak}. \tag{8.73}$$

On the other hand, by defining the mass flow from the outlet point of view, one has

$$Q_c(L_4) = Q(L_4) + e^{-\frac{|L_i-x_{Leak}|-L_4+L_i}{c}s}Q_{Leak}, \tag{8.74}$$

then by using (8.74) in (8.72)

$$P(L_i) = \frac{c}{\mathscr{A}\alpha T_d}\frac{s+\alpha}{s}\left[e^{-\frac{L_i}{c}s}Q(0) - e^{-\frac{L_4-L_i}{c}s}Q_c(L_4)\right], \tag{8.75}$$

and finally from (8.74) the mass flow difference at the outlet is

$$\Delta Q(L_4) = Q_c(L_4) - Q(L_4) = e^{-\frac{|L_i - x_{Leak}| - L_4 + L_i}{c} s} Q_{Leak}. \tag{8.76}$$

Similarly, by defining the mass flow from the inlet point of view,

$$Q_c(L_0) = Q(L_0) + e^{-\frac{|L_i - x_{Leak}| - L_i}{c} s} Q_{Leak}, \tag{8.77}$$

and by using (8.74) and (8.77), (8.72) can be written as

$$P(L_i) = \frac{c}{\mathscr{A} \alpha T_d} \frac{s + \alpha}{s} \left[e^{-\frac{L_i}{c} s} Q_c(0) - e^{-\frac{L_4 - L_i}{c} s} Q(L_4) \right]. \tag{8.78}$$

This means the mass flow difference at the inlet is

$$\Delta Q(L_0) = Q_c(L_0) - Q(L_0) = e^{-\frac{|L_i - x_{Leak}| - L_i}{c} s} Q_{Leak}. \tag{8.79}$$

From this equation we conclude that a leakage can be detected, by using the differences between the calculated (assuming no leaks) and the measured pressures along the pipeline and mass flows at the boundaries. Since $\Delta Q(L_0)$ and $\Delta Q(L_4)$ are equal to Q_{Leak} delayed by a certain amount of time dependence of the leak location, these values can also be used to estimate the leakage outflow and its location. The differences between the intermediate mass flows and pressures are also dependent on the leak location and can help to increase the robustness of the detector.

8.8.2 Leakage Outflow

The leakage outflow can be estimated from the expected value of the mass flow inlet/outlet variation (8.76) and (8.79), after the transient response produced by the leakage can be neglected, as

$$Q_{Leak} = \frac{\mathscr{E}\{\Delta Q(L_4)\} + \mathscr{E}\{\Delta Q(L_0)\}}{2}, \tag{8.80}$$

where $\mathscr{E}\{v\}$ denotes the expected value of v.

8.9 Case Study

In this section, the described procedure for leak diagnosis in a pipeline with a set of intermediate observation points is applied to a case study. Since we did not have real data from a long pipeline, we built a case study. We consider a pipeline with a length

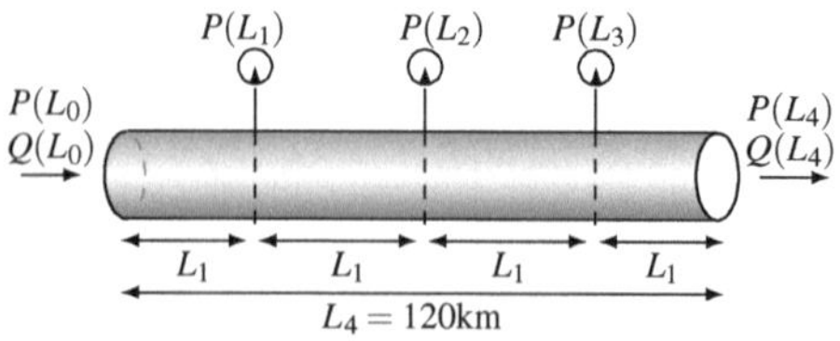

Fig. 8.12 Gas pipeline instrumented with five equally spaced pressure sensors and two mass flow sensors at the boundaries

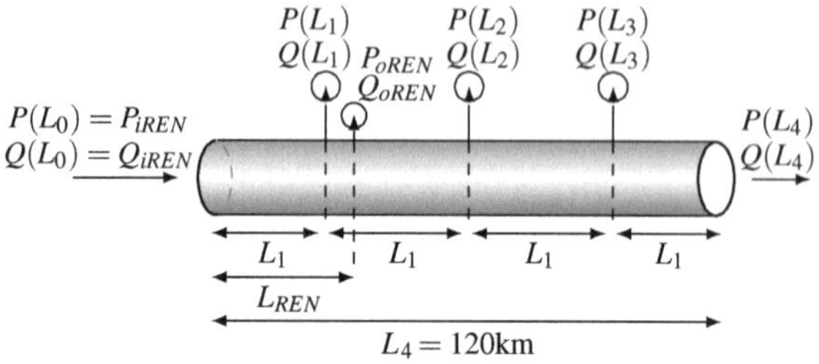

Fig. 8.13 Pipeline with pressures' and mass flows' observation points

of 120 km, four observation points 30 km apart, and leaks of different sizes simulated at different locations of the pipeline; see Fig. 8.12. The physical parameters are the same as the ones of the operational data (8.49). The pipeline was simulated by using inlet pressure P_{iREN} and mass flow Q_{iREN} data of one day of operation, provided by the Portuguese gas company, REN Gasodutos, and is depicted in Fig. 8.5 of Sect. 8.5. We start by considering the leakage located at 5 km from the inlet point $P(L_0)$.

In the Fig. 8.13, L_{REN} corresponds to the length of the REN pipeline ($L_{REN} = 36$ km), and P_{oREN} and Q_{oREN} are REN outlet pressure and mass flows, respectively. P_{oREN} and Q_{oREN} were used to validate the generated data.

Section 8.9.1 describes the simulation results without leaks. In Sect. 8.9.2 the behavior of the pipeline with a leakage of 5% is analyzed. The results of the quadripole leakage detector are presented and discussed in Sect. 8.9.3 for a leakage of 1% of the nominal mass flow.

8.9.1 Pipeline Profiles Without a Leak

By assuming the data to be known at the inlet ($P(L_0) = P_{iREN}$, $Q(L_0) = Q_{iREN}$), pressures $P(L_i)$ and mass flows $Q(L_i)$, $i = 1, 2, 3, 4$ were calculated by using these data and the model (8.50), without leaks. The calculated pressure and mass flow profiles are depicted in Fig. 8.14. To validate these graphics, pressure $P(L_{REN})$ and the mass flow $Q(L_{REN})$, computed with quadripole models and the calculated outlet data ($P(L_4)$ and $Q(L_4)$), were compared to the P_{oREN} and Q_{oREN} data provided by REN; both quantities showed good agreement.

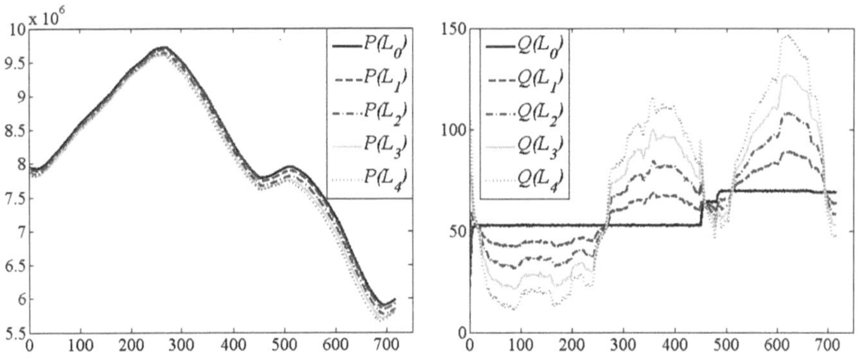

(a) Pressures (Pa) (b) Mass flows (kg/s)

Fig. 8.14 Case study pressure and mass flow values at stations L_i for $i = 0, 1, 2, 3, 4$

In a real-world situation, the pressures and the intermediate mass flows of the normal operation profile are calculated from the boundary mass flows $Q(L_0)$ and $Q(L_4)$. Therefore, in order to mimic a real situation, we also calculated the normal profile from the boundary mass flows; in the absence of noise, the differences were negligible.

8.9.2 Pipeline Profiles with Leak

By considering the pipeline with the same parameters as above, a leakage of 5% of the nominal mass flow was simulated at 5 km from the inlet point, between positions L_0 and L_1, with starting time $N = 300$ (corresponding to instant $t = 600$ min $= 10$ h, since the data sample period is 2 min).

Figure 8.15 allows a comparison between the pressures at the section boundaries in normal profile and operational profile with a leak where a pressure loss is visible that starts when the abnormal condition is present.

The differences between the normal and operational mass flow profiles are also small, as shown in Fig. 8.16. Notice that the mass flow differences can be easily masked by measurement noise. To increase detector's robustness, these differences are integrated, which induces a drift that sooner or later supersedes the noise.

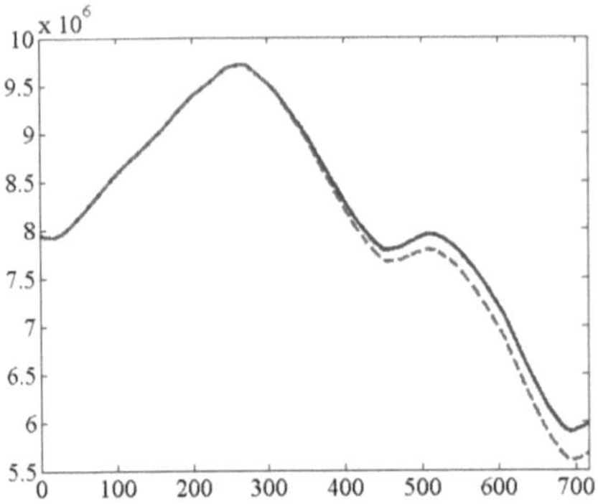

(a) Pressure values at L_0.

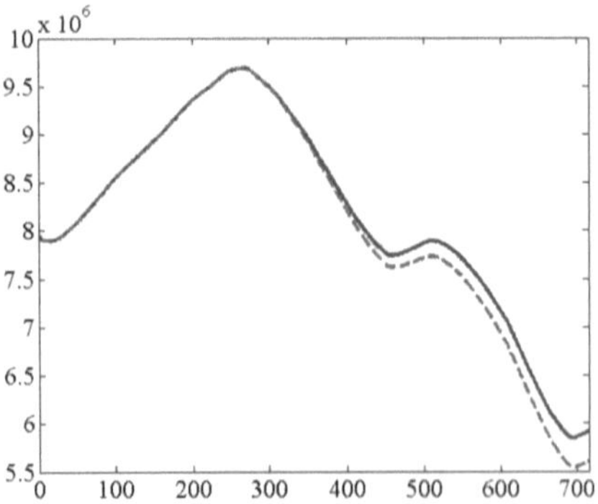

(b) Pressure values at L_1.

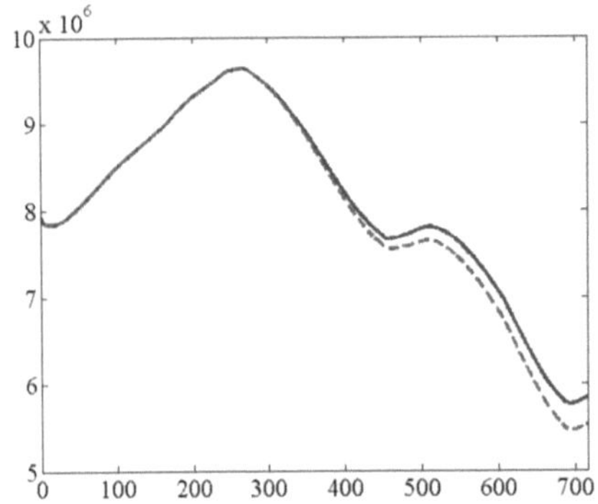

(c) Pressure values at L_2.

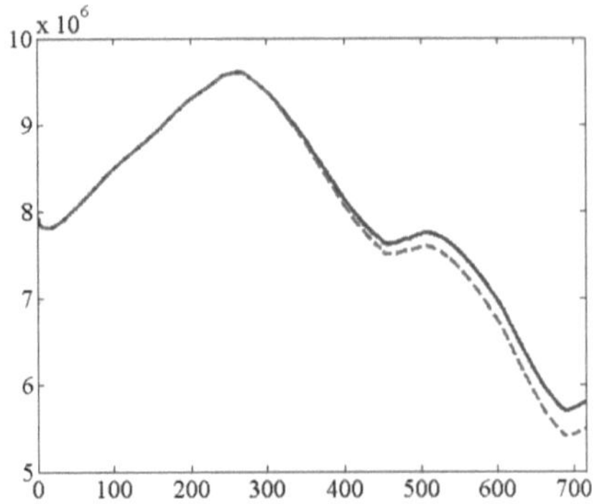

(d) Pressure values at L_3.

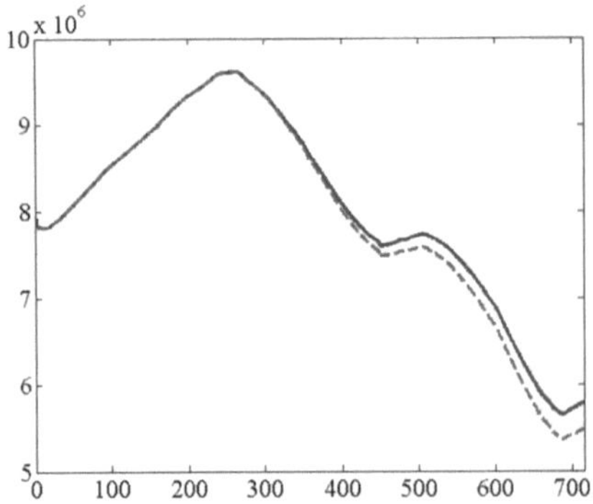

(e) Pressure values at L_4.

Fig. 8.15 Normal profile (*continuous line*) and operational profile (*dashed line*) pressures (Pa) at points L_i for $i = 0, 1, 2, 3, 4$

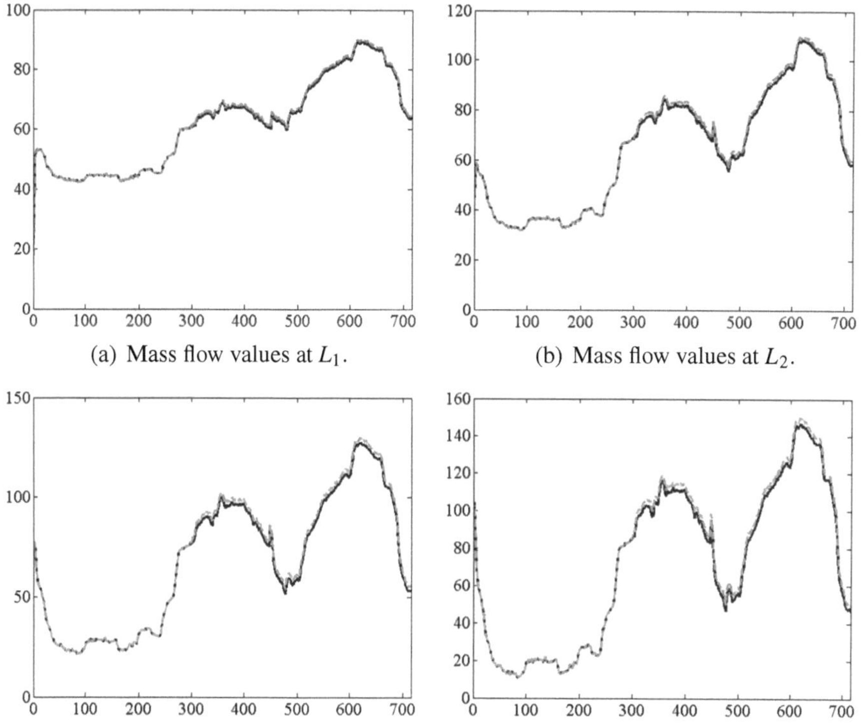

(a) Mass flow values at L_1.

(b) Mass flow values at L_2.

(c) Mass flow values at L_3.

(d) Mass flow values at L_4.

Fig. 8.16 Normal profile (*continuous line*) and operational profile (*dashed line*) mass flows (kg/s) at stations L_i for $i = 1, 2, 3, 4$

8.9.3 Leakage Detection Results

To mimic a real situation, we consider the measured boundary mass flows $Q(L_0)$ and $Q(L_4)$, and the pressures at the observation points $P(L_i)$ for $i = 0, 1, 2, 3, 4$ disturbed with Gaussian noise. Three different noise levels are applied to the data: $SNR = 20\,\text{dB}$, $10\,\text{dB}$ and $0\,\text{dB}$, where SNR stands for *signal to noise ratio* defined as $SNR = 20 \log \left| \frac{X_{rms}}{N_{rms}} \right|$, with X_{rms} and N_{rms} being the square root of the mean square value of the undisturbed signal, $x(t)$, and noise, $n(t)$, respectively. The undisturbed values were simulated from the pressure and mass flow values at the pipeline inlet, provided by REN Gasodutos (P_{iREN} and P_{oREN}), by using the quadripole depicted in Fig. 8.11; measured values are denoted by $Q_m(L_0)$, $Q_m(L_4)$ and $P_m(L_i)$ for $i = 0, 1, 2, 3, 4$. Although the study has been performed for the three noises levels, here we only report the most extraneous situation, i.e., a $SNR = 0\,\text{dB}$.

By applying the quadripole detectors' method to a leakage case (namely of 1%), the normal operation profile was generated by calculating the pressures

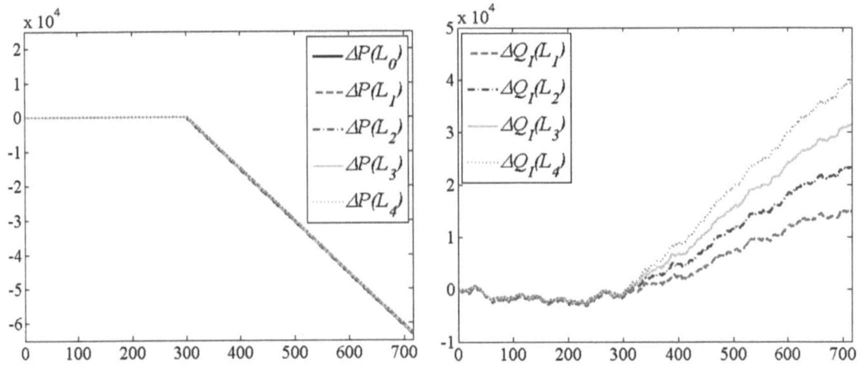

(a) Pressures difference (b) Integrated mass flow difference

Fig. 8.17 Pressure difference $\Delta P(L_i)$ for $i = 0, 1, 2, 3, 4$ and integrated mass flows difference $\Delta Q_I(L_i)$ for $i = 1, 2, 3, 4$, with SNR= 0 dB noise of a 1% leakage

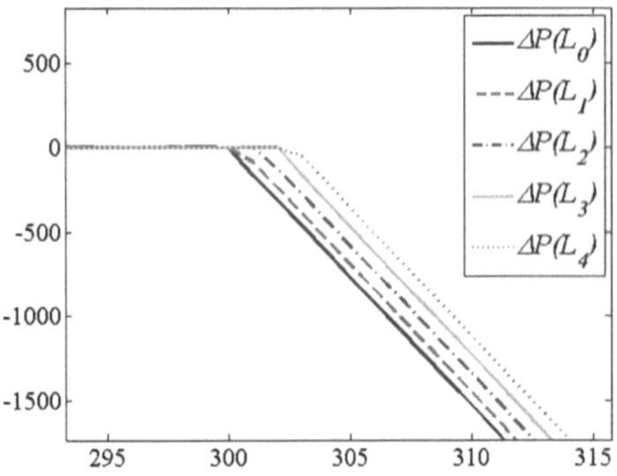

Fig. 8.18 Pressures' difference $P_{nLK}(L_i) - P_m(L_i)$ for $i = 0, 1, 2, 3, 4$, zoom around the instant at which the leakage starts: $N = 300 = 10\,\text{h}$

along the pipeline $P(L_i)$ for $i = 0, 1, 2, 3, 4$ and the intermediate flows $Q(L_i)$, $i = 1, 2, 3$, from the measured boundary mass flows $Q_m(L_0)$ and $Q_m(L_4)$. These are denoted as $P_{nLK}(L_i)$ for $i = 0, 1, 2, 3, 4$ and $Q_{nLK}(L_i)$ for $i = 1, 2, 3$. Next, an operational profile was generated by calculating the mass flows $Q(L_i)$ for $i = 1, 2, 3, 4$, from the inlet measured mass flow $Q_m(L_0)$ and the measured pressures $P_m(L_i)$ for $i = 0, 1, 2, 3, 4$. These calculated mass flows are denoted as $Q_c(L_i)$ for $i = 1, 2, 3, 4$. Both $\Delta P(L_i) = P_{nLK}(L_i) - P_m(L_i)$ for $i = 0, 1, 2, 3, 4$, and the integrated $\Delta Q(L_i) = Q_c(L_i) - Q_{nLK}(L_i)$ for $i = 1, 2, 3$ and $\Delta Q(L_4) = Q_c(L_4) - Q_m(L_4)$ are depicted in Fig. 8.17, which clearly shows the occurrence of a leak at instant $N_{data} = 300$. These figures show how a 1% leakage can be easily perceived in the control room even with a high level of noise.

Further analysis of the pressure difference shown in Fig. 8.18 leads to the conclusion that the leakage was first perceived by sensors L_0 and L_1, and that it took 2 min to be perceived by sensor L_1 (at 25 km from the leak point) and 7 min by sensor L_4 (outlet).

Table 8.1 Leakage outflow evaluation, in percentage

Leakage (%)	Estimated outflow (%)	Relative error (%)
1	$0.9924 \simeq 1$	2.7
5	$4.9969 \simeq 5$	0.5
10	$9.9987 \simeq 10$	0.3

To increase the sensitivity of the method, we impose that the leakage must be perceived by at least two sensors; in this example, it took 2 min to be detected, and 65 kg of fluid were lost.

To estimate the leakage outflow, we use (8.80). In Table 8.1 we consider three different leakage outflows with their respective outflow estimation in 50 Monte Carlo simulations which were performed in every situation.

8.10 Conclusion and Future Work

A hyperbolic PDE system that represents the gas dynamics in the pipelines is linearized and discretized in space. By employing an electrical analogy, the pipeline is next described by TF quadripole models. Pressure and mass flow values were simulated with these quadripole models and were approximated to the ones simulated by SIMONE®. From this we concluded that the model is able to capture the system dynamics. This TF quadripole approach allows the modeling of connection of two pipelines, making it possible to assess the values of pressure and mass flow at intermediate points. A leakage detection method based on the TF quadripole models was proposed. A case study was built that included the simulation of a leakage. For this case study, the leakage was detected and located, and its outflow was estimated by the proposed method. Since the influence of the pipelines' inclination on the network gas dynamics should not be overlooked, in future work these models also must be tested with operational data that consider different inclination angles and that improve the leakage location.

References

Adnan, N. F., Ghazali, M. F., Amin, M., Malik, A., & Ariffin, A. (2014). Leak detection in MDPE gas pipeline using dual-tree complex wavelet transform.

Baltazar, S. T., Azevedo Perdicoúlis, T., & Lopes dos Santos, P. (2014). Quadripole models for gas networks. In *Proceedings of XVI congreso latino Americano de control automático (CLCA)*. Cancún, México.

Baltazar, S. T., Azevedo Perdicoúlis, T., & Lopes dos Santos, P. (2016a). Quadripole Models for simulation and leak detection on gas pipelines. In *47th PSIG annual meeting*. Vancouver, BC, Canada.

Baltazar, S. T., Azevedo Perdicoúlis, T., & Lopes dos Santos, P. (2016b). Quadripole models on gas pipelines. https://drive.google.com/open?id=0B6Tb93qVJr18Wk10TzltZTdIbkE.
Geiger, G. (2006). State-of-the-art in leak detection and localisation. In *Proceedings of Pipeline Technology 2006 Conference*. Hannover, Germany: Hannover Messe.
Králik, J., Stiegler, P., Vostrý, Z., & Závorka, J. (1988). *Dynamic modelling of large-scale networks with application to gas distribution*. Prague: Elsevier.
Lin, K., & Holbert, K. E. (2006). Extension of power line fault location techniques to pressurized line diagnostics. In *38th North American power symposium*. Carbondale, IL, USA.
Lopes dos Santos, P., Azevedo Perdicoúlis, T. P., Jank, G., & Martins de Carvalho, J. (2010). A lumped transfer function model for high pressure gas pipelines. In *49th IEEE conference on decision and control (CDC)*. Atlanta, GA, USA.
Luo, H., Su, H., & Zhan, S. (2010). An intelligent monitoring system for the anticorrosive layer on underground oil pipeline. In *2010 International conference on measuring technology and mechatronics automation* (Vol 1, pp. 513–516). IEEE.
Murvay, P.-S., & Silea, I. (2012). As survey on gas leak detection and localization techniques. *Journal of Loss Prevention in the Process Industries*, *25*(6), 966–973.
Osiadacz, A. J. (1987). *Simulation and analysis of gas networks*. Spon, London: E. & F.N.
Reddy, H. P., Narasimhan, S., & Bhallamudi, S. M. (2006). Simulation and state estimation of transient flow in gas pipeline networks using a transfer function model. *Industrial & Engineering Chemistry Research*, *45*, 3853–3863.
REN-Gasodutos (2014). REN-Gasodutos: Personal Communication.
Turkowski, M., Bratek, A., & Slowikowski, M. (2007). Methods and systems of leak detection in long range pipelines. *Journal of Automation, Mobile Robotics & Intelligent Systems*, *8*, 39–46.
Verde, C. (2005). Accommodation of multi-leak location in a pipeline. *Control Engineering Practice*, *13*, 1071–1078.

Chapter 9
Features of Demand Patterns for Leak Detection in Water Distribution Networks

Marcos Quiñones-Grueiro, Cristina Verde and Orestes Llanes-Santiago

Abstract This chapter presents a data-driven based approach for detection of leaks in water distribution networks in which the demand is formed by a known periodic pattern plus a stochastic variable. The leak detection method is based on an adaptation of the dynamic principal component analysis (DPCA), and it is assumed that only pressures at selected consumption nodes are measured. Since the variables of water distribution networks (WDNs), even in normal conditions, are nonstationary and time-correlated the data are preprocessed with a periodic transformation previous to the application of DPCA. The proposed approach is validated with the Hanoi network model. The performance is evaluated with three indexes: the leak detection rate, the false alarm rate, and the delay of the detection with respect to the leak's occurrence time. All of them are satisfactory for diverse leaks' scenarios, and the proposed approach presents an improvement in the leak detection rate of approximately 70% as compared with the traditional PCA and DPCA methods.

9.1 Notation

Parameter	Description
a	Number of principal components
b_n	Number of branches connected to node n
$d_n(t)$	Demand at node n
$d_{n_\psi}(t)$	Periodic function of the demand at node n
$d_{n_\xi}(t)$	Random term of the demand at node n
D	Mahalanobis norm
D_{th}	Threshold for the Mahalanobis norm

M. Quiñones-Grueiro (✉) · O. Llanes-Santiago
Universidad Tecnológica de la Habana José Antonio Echeverría (CUJAE), Calle 114, #11901 e/ Ciclovìa y Rotonda, Marianao, Havana, Cuba
e-mail: marcosqg88@gmail.com

C. Verde
Instituto de Ingeniería-UNAM Coyoacan, 04510 México D.F., Mexico

C. Verde and L. Torres (eds.), *Modeling and Monitoring of Pipelines and Networks*, Applied Condition Monitoring 7,
DOI 10.1007/978-3-319-55944-5_9

$f_n(t)$	Outflow of a leak at node n
G	Number of water sources in a loop
$h_g(t)$	Pressure head associated with a water source
$h_l(t)$	Pressure head associated with a water drop
J	Number of demand cycles
L	Number of water drops in a loop
m	Number of measured variables
P	Transformation matrix of data X
$q_i(t)$	Flow of the branch i
R	Correlation matrix of data X
S	Covariance matrix of data X
SPE	Squared prediction error of the transformed data
t	Sampling time
T^2	Statistic measure of PCA
X	Historical data matrix for the training process
$x_i(t)$	component i of vector $x(t)$
$x(t_{new})$	Data vector at new time
$x(t)$	Data vector
$\tilde{x}_a(t)$	Vector of principal components
$x^*(t)$	Vector of transformed variables
α	Flow exponent
$\varepsilon(t)$	Random variable for the demand model $d_{n_\xi}(t)$
$\mathscr{E}\{x_i\}$	Expected value of variable x_i
τ	Number of sampling times of a period
Γ	Time period for the demand
θ	Parameter associated with the pressure head

9.2 Introduction

Water distribution networks (WDNs) are complex nonlinear dynamic systems continuously delivering drinking water to different types of consumers with an admissible pressure at each node of the network. One important concern of the distribution service system is ensuring the nominal pressure in all nodes at minimal cost.

Leaks produce a significant loss in the fluid transportation system and additional effects, such as low pressure, to final consumers. Moreover, leaks damage the network infrastructure and can produce environmental issues with serious economic and ecological consequences (Pérez et al. 2014). It is estimated that leakages can represent up to 30% of the total amount of extracted water in some WDNs (Farley and Trow 2003).

Leakage monitoring may be performed through maintenance routines or under suspected leak conditions (Lambert 1994). Some new technologies for locating leaks require high-cost specific instrumentation, such as optical fiber, and sometimes the isolation task entails a partial shutdown of the network. On the other hand, automatic techniques based on measurements of flow and pressure by using sensors installed in some nodes of the network are the most common.

Many leak detection and isolation techniques have been proposed where the main specifications are a quick response and a low false alarm rate. Some of these are based on a mathematical model which represents the physical laws describing the pipeline. The procedures presented by Brunone and Ferrante (2001) and Verde et al. (2016), however, assume a single demand node and cannot be applied to leak detection in a WDN. The number of variables and constraints involved in describing a network makes the use of analytical models for the leak detection infeasible.

To overcome this shortcoming, and taking into account the great amount of information generated by instrumentation and SCADA (supervisory control and data acquisition systems), historical data-based methods have been widely used for fault detection and isolation in both academic research and industrial applications (Chiang et al. 2001; Kruger and Xie 2012; Macgregor and Cinar 2012).

Pattern recognition and statistical monitoring techniques have also been adapted to solve fault detection problems in the last few years. Gabrys and Bargiela (1999) and Arsene et al. (2012) proposed artificial neural networks and confidence limit analysis to solve the leak detection and location problem in WDNs by considering uncertainties in the measurements. These papers assumed flow measurements in multiple nodes; however, only input flow is usually available in WDNs owing to the high costs of the instrumentation (Ferrandez-gamot et al. 2015; Misiunas et al. 2005).

Principal component analysis (PCA) tools (Duzinkiewicz et al. 2008; Gertler et al. 2010; Nowicki et al. 2012) have also been applied for detecting and locating leaks in WDNs. Duzinkiewicz et al. (2008) considered the flow measurement for a set of nodes. Gertler et al. (2010) assumed the pressure to be measured in all nodes. This assumption is not feasible because of the high sensors prices. Moreover, the uncertainties of measurements and demand have not been considered in these papers. Other works only consider uncertainties in the measurements (Arsene et al. 2012; Gabrys and Bargiela 1999).

Recently, Cugueró-Escofet et al. (2015) and Ferrandez-gamot et al. (2015), considered a demand disturbance at each node for leak detection and location issues. The leak monitoring is, however, recommended when the demand is almost constant—in the early morning. The variability of the demand throughout the day is then not taken into account. A time-varying demand with uncertainty is modeled by Misiunas et al. (2005), but uncertainty is only considered as a constant white noise added to the demand throughout the day, which is not realistic.

In summary, the main limitations of the aforementioned approaches are as follows: (1) one measures flow in many nodes and pressure in almost all nodes, (2) uncertainties are only produced by the measured data, and (3) the demand is almost constant over the day. A novel monitoring approach is therefore proposed here which includes a preprocessing of the data to overcome the mentioned issues. A WDN model which integrates the effect of disturbance in the demand and the noise in the measurements is considered here. In addition, a periodic transformation for the data together with dynamic principal component analysis (DPCA) allows leak detection in WDN without an analytical model of the network. In particular, the data are periodically transformed to obtain weakly stationary variables, and DPCA is then incorporated for managing the time-correlated and redundant data. To identify the leakages, the statistic measure T^2 and the Square Prediction Error are applied, and as a case study the Hanoi network is used.

9.3 Water Distribution Network

A water distribution network is a complex system of interconnected pipes with many branches and at least one water source. The model under normal operation conditions and the leak scenarios are described in this section.

9.3.1 Normal Operation

According to the first principles, the behavior of the variables in a WDN with N nodes and $t \in \mathbb{Z}$ associated with the sampling time is described by the following laws:

- The net inflow must be equal to the net outflow for any node $n \in N$ of the network; i.e.,

$$\sum_{i=1}^{b_n} q_i(t) = d_n(t) \tag{9.1}$$

  where b_n is the number of branches connected to the node n, $q_i(t)$ denotes the flow of the branch i, and $d_n(t)$ is the respective demand.
- The sum of pressure heads around any loop of the network is equal to zero. Thus, a loop with G water sources and L water drops is modeled by

$$\sum_{g=1}^{G} h_g(t) + \sum_{l=1}^{L} h_l(t) = 0 \tag{9.2}$$

  where pressure heads $h_g(t)$ and $h_l(t)$ are associated with sources and drops respectively.
- The relation between flow $q(t)$ and pressure head $h(t)$ for any component of the network is modeled by

$$h(t) = \theta q^{\alpha}(t) \tag{9.3}$$

  where the parameter θ depends on the specific component and the exponent α could have a value close to 2 (Houghtalen and Hwang 2010).

9.3.2 Leaks and Consumer Demand

The main uncertainty in a WDN is produced by the consumer, even if the network is healthy. Thus, the variability of the demand during the day deteriorates the automatic monitoring if a deterministic approach is considered.

A leak with outflow f_n at node n modifies (9.1) and it becomes

$$\sum_{i=1}^{b_n} q_i(t) = d_n(t) + f_n(t) \tag{9.4}$$

One can see from (9.4) that the leak outflow $f_n(t)$ modifies the mass balance similarly as the demand $d_n(t)$. Therefore, to distinguish a leak from a demand deviation, some properties for the demand must be known. The periodicity of the demand for specific time intervals allows the discrimination between demand and leak, and this property has been used by Alvisi et al. (2007), Zhou et al. (2002) and Soyer and Roberson (2014).

The demand at each node $d_n(t)$ is considered a periodically stationary process modeled as

$$d_n(t) = d_{n_\psi}(t) + d_{n_\xi}(t) \tag{9.5}$$

where $d_{n_\psi}(t)$ is a periodic function with period Γ of τ sampling times, which describes the periodic behavior of the outflow at node i, and $d_{n_\xi}(t)$ represents the uncertainty around the periodic function. In particular, the uncertain demand is described by an autoregressive stationary process

$$d_{n_\xi}(t) = \phi_0 + \phi_1 d_{n_\xi}(t-1) + \phi_2 d_{n_\xi}(t-2) + \cdots + \phi_p d_{n_\xi}(t-p) + \varepsilon_n(t) \tag{9.6}$$

where $\varepsilon_n(t)$ is a Gaussian process with zero mean value and time dependent autocorrelation, because its variability changes over the period Γ (Papoulis 1991).

In a WDN, the number of measured variables and its sampling time play an important role for achieving satisfactory automatic leak monitoring systems. Here a data subset of cardinality m which form the measurements vector with flows and pressure heads is assumed

$$x(t) = [q_1(t), q_2(t), \ldots, h_1(t), h_2(t), \ldots]^T \in \Re^m \tag{9.7}$$

The selection of an appropriated set of measurable variables for the monitoring of a network is discussed in the references (Pérez et al. 2011; Casillas et al. 2015).

By considering the demand model (9.5) under normal conditions for each node of the network and a set of historical data $x(t)$ as a starting point, the following section introduces an adaptation of a well-known monitoring approach which allows the detection of a leak in the network without an analytical model.

9.4 A Novel Monitoring Approach: Periodic-Dynamic Principal Component Analysis (P-DPCA)

Multivariate statistical methods allow the monitoring of a system using the historical data which represent the process behavior under normal conditions (Chiang et al. 2001). In this framework, a data set from a process under normal operation

$$X = \begin{bmatrix} x_1(1) & x_2(1) & \cdots & x_m(1) \\ x_1(2) & x_2(2) & \cdots & x_m(2) \\ \vdots & \vdots & \vdots & \vdots \\ x_1(p) & x_2(p) & \cdots & x_m(p) \end{bmatrix} \in \Re^{p \times m} \tag{9.8}$$

is a called training matrix, and it is formed by p observations of vector $x(t)$. Thus, the covariance matrix can be estimated by $S = (X'X)/(p-1)$ with the mean value vector given by $\mu = \mathbb{E}\{x\}$.

A simple method for detecting faults consists of checking if a new observation $x(t_{new}) \in \Re^m$ is close to the nominal values with a certain tolerance. For random variables with Gaussian distributions, the Mahalanobis norm

$$D = (x(t_{new}) - \mu_x)' S^{-1} (x(t_{new}) - \mu_x) \tag{9.9}$$

estimates how much the variables deviate from the statistical properties of X. Thus, a system is in normal condition if the inequality

$$D < D_{th} \tag{9.10}$$

is satisfied, where D_{th} is a threshold which can be calculated according to the procedure given by Chiang et al. (2001).

The main difficulty for the estimation of (9.9) is the inversion of matrix S for a high dimension m or if the variables are correlated. Principal component analysis (PCA) is the most popular technique for reducing the dimension of X, and together with the statistics T^2 and SPE instead of Mahalanobis norm, can be used for a fault detection problem (Chiang et al. 2001; Kruger and Xie 2012).

9.4.1 Principal Component Analysis

According to Kruger and Xie (2012) for the fault detection task, the traditional PCA assumes the linearity of the process and

C1: Each variable x_i of matrix (9.8) has a Gaussian distribution function.

C2: Each variable x_i of matrix (9.8) is weakly stationary. Due to the different magnitudes of the components of vector $x(t)$, each variable $x_i(t)$ is scaled to zero mean and unit variance.

C3: The vector $x(t)$ is not time-correlated.

Under these conditions, PCA allows a dimensionality reduction of data space by preserving most of the process' information in terms of variability. A vector of uncorrelated variables $\tilde{x}_a(t) \in \Re^a$ with $a < m$, called principal components, is obtained for each data vector $x(t) \in \Re^m$ by using $\tilde{x}_a(t) = x(t)P$ where the matrix $P \in R^{m \times a}$ is given by the eigenvectors associated with the a most significant eigenvalues of the correlation matrix R of the historical data matrix X.

Since the measured variables (9.7) of the WDN do not satisfy conditions C2 and C3, the network data must be preprocessed before the application of PCA. The preprocessing procedure is the main contribution of this work. Thus, to overcome the nonstationarity of a variable a periodic transformation is applied, and later on, an extended matrix is formed by taking into account the time correlation of the variables.

9.4.1.1 Periodic Transformation

A periodically stationary variable of $x(t)$ with the structure and the features given by (9.5) can be transformed into a weakly stationary process by using its periodic expected value at each t. This statement is formalized as follows.

Fact: Let $x_i(t)$ be a periodically stationary process with the structure given by (9.5) and

$$t' = \begin{cases} t & 0 < t \leq \Gamma \\ t(mod(\Gamma)) & t > \Gamma \end{cases} \tag{9.11}$$

where $t' = t(mod(\Gamma))$ is the remainder of the division of t by Γ, then the process

$$x_i^*(t') = x_i(t') - \mathcal{E}_{t'}\{x_i\} \tag{9.12}$$

is weakly stationary, where the periodic expected value of x_l for t' can be estimated off-line by

$$\mathcal{E}_{t'}\{x_i\} \simeq \frac{1}{J+1} \sum_{j=0}^{J} x_i(t' + j\tau) , \tag{9.13}$$

with large enough $J + 1$ periods of the variable $x_i(t)$ and $J \in \mathbb{Z}$.◊

Note that since the proposed time-variant transformation depends on historical data for (9.8), the time-variant periodic mean $\mathcal{E}_{t'}\{x_i\}$ must be obtained a priori for each component of $x(t)$.

The fact can be straightforwardly proven as follows.

From the features of the variable $x_i(t')$, the set of equations for $J+1$ periods can be written

$$\begin{aligned} x_i(t') &= x_{i_\psi}(t') + x_{i_\xi}(t') \\ x_i(t'+\tau) &= x_{i_\psi}(t'+\tau) + x_{i_\xi}(t'+\tau) \\ \vdots \quad &= \quad \vdots \qquad\qquad\quad \vdots \\ x_i(t'+J\tau) &= x_{i_\psi}(t'+J\tau) + x_{i_\xi}(t'+J\tau) \end{aligned}$$

Hence, by adding the left and right sides of these equations, the relation

$$\begin{aligned} \sum_{j=0}^{J} x_i(t'+j\tau) &= \textstyle\sum_{j=0}^{J} \left(x_{i_\psi}(t'+j\tau) + x_{i_\xi}(t'+j\tau)\right) \\ &= (J+1)x_{i_\psi}(t') + \textstyle\sum_{j=0}^{J} x_{i_\xi}(t'+j\tau) \end{aligned} \tag{9.14}$$

can be written. By considering the time-variant mean of $x_i(t')$ given in (9.13) and the mean of $x_{i_\xi}(t')$ given by

$$\mathcal{E}_{t'}\{x_{i_\xi}\} \simeq \frac{1}{J+1} \sum_{j=0}^{J} x_{i_\xi}(t'+j\tau)$$

with J a large enough value, (9.13) is equivalent to

$$\mathcal{E}_{t'}\{x_i\} \simeq x_{i_\psi}(t') + \mathcal{E}_{t'}\{x_{i_\xi}\} \tag{9.15}$$

Thus, if the time-variant mean (9.15) for each t' is substituted in the transformation equation (9.12), this is reduced to

$$x_i^*(t') = x_{i_\xi}(t') - \mathcal{E}_{t'}\{x_{i_\xi}\} \tag{9.16}$$

Therefore, since $x_{i_\xi}(t')$ is weakly stationary, the variable $x_i^*(t')$ is as well a weakly stationary variable.

Thus, the respective transformed matrix $X^* \in \Re^{p\times m}$ with $p = (J+1)\tau$ is straightforwardly formed, and each variable satisfies the condition C2. This adaptation of PCA with a preprocessing is called P-PCA.

9.4.1.2 Dynamic Principal Component Analysis

To use the PCA method for time-correlated data, Ku et al. (1995) proposed a method called DPCA by modifying the matrix (9.8) to reflect static and dynamic relations in the singular values of the covariance matrix. Thus, the training data matrix takes the form

$$X_l = \begin{bmatrix} x_1(1) & \cdots & x_1(1-l) & \cdots & x_m(1) & \cdots & x_m(1-l) \\ x_1(2) & \cdots & x_1(2-l) & \cdots & x_m(2) & \cdots & x_m(2-l) \\ \vdots & \ddots & \vdots & \ddots & \vdots & \ddots & \vdots \\ x_1(p) & \cdots & x_1(p-l) & \cdots & x_m(p) & \cdots & x_m(p-l) \end{bmatrix} \in \Re^{p\times(lm)} , \qquad (9.17)$$

and the number of lags l is selected according to Rato and Reis (2013). Thus, the data of matrix X_l hold the condition C3 and can be directly used for a PCA model.

9.4.2 Fault Detection with P-DPCA

The P-PCA and DPCA methods described above can be combined to obtain suitable conditions for the application of traditional PCA. This new method is called P-DPCA. Thus, the preprocessing data together with the lags of $x^*(t)$ describe the historical training matrix of the network under normal conditions given by

$$X_l^* = \begin{bmatrix} x_1^*(1) & \cdots & x_1^*(1-l) & \cdots & x_m^*(1) & \cdots & x_m^*(1-l) \\ x_1^*(2) & \cdots & x_1^*(2-l) & \cdots & x_m^*(2) & \cdots & x_m^*(2-l) \\ \vdots & \ddots & \vdots & \ddots & \vdots & \ddots & \vdots \\ x_1^*(p) & \cdots & x_1^*(p-l) & \cdots & x_m^*(p) & \cdots & x_m^*(p-l) \end{bmatrix} \in \Re^{p\times(lm)} , \qquad (9.18)$$

and the respective transformation $P_l^* \in \mathbb{R}^{m\times a}$ is formed by the eigenvectors of the most significant eigenvalues of the data covariance matrix R_l^* calculated by using X_l^*.

To estimate deviations from the normal condition for a new datum $x(t_{new})$, the statistic measure T^2, and the Squared Prediction Error (SPE) are considered.

The norm T^2

$$T^2_{new} = x_l^*(t_{new})' P_l^* \Lambda_l^{*-1} P_l^{*\prime} x_l^*(t_{new}) \qquad (9.19)$$

measures deviations in the principal component directions where the diagonal matrix Λ_l^* is formed with the first a eigenvalues of the covariance matrix, and the norm SPE

$$SPE_{new} = r'r \ , \quad \text{with} \quad r = (I - P_l^* P_l^{*\prime}) x_l^*(t_{new}) \qquad (9.20)$$

measures deviations with respect to the PCA model where I is the identity matrix.

The threshold T^2_α and SPE_α for the respective statistics' measures (9.19) and (9.20) can be adjusted according to the sensitivity of the detector and a boundary false alarm rate, as explained by Chiang et al. (2001).

9.5 Case Study: Hanoi WDN

The case study is a benchmark which describes the WDN from Hanoi, Vietnam, presented by Fujiwara and Khang (1990). The configuration of the network is given in Fig. 9.1. The network is formed by 32 nodes and 34 pipes organized in three loops and two branches. The system is gravity fed by a single reservoir, and the parameters are given in Table 9.1. The pipe diameters were obtained from Sedki and Ouazar (2012). Three different demand patterns are considered, one for each loop with its respective branches. The demand pattern for each zone is shown in Fig. 9.2a. It is assumed that $m = 5$, and the sensors of pressure head are located at nodes (3, 10, 16, 23, 25).

9.5.1 *Model Simulator*

The EPANET package (Rossman 2000) is used to simulate this network together with MATLAB. The Hazen–Williams equation is used for the calculation of the friction factor with a roughness coefficient of 120 (Houghtalen and Hwang 2010). The node demands are simulated with daily periodicity ($\Gamma = 24$ hours) and a sampling period of 5 min. The demand $d_{n_\xi}(t)$ at each node is specifically modeled by

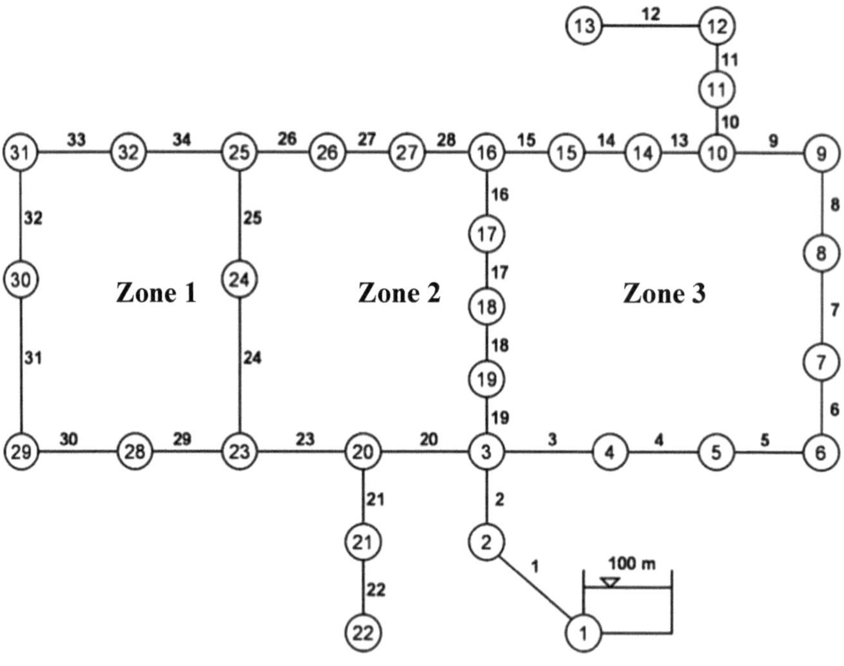

Fig. 9.1 Configuration of the Hanoi network

Table 9.1 Link and node parameters of the Hanoi network

Link	Length (m)	Diameter (mm)	Node	Maximum demand (m^3/h)
1	100	1016	1	0
2	1350	1016	2	890
3	900	1016	3	850
4	1150	1016	4	130
5	1450	1016	5	725
6	450	1016	6	1005
7	850	1016	7	1350
8	850	1016	8	550
9	800	1016	9	525
10	950	762	10	525
11	1200	609.6	11	500
12	3500	609.6	12	560
13	800	508	13	940
14	500	406.4	14	615
15	550	304.8	15	280
16	2730	304.8	16	310
17	1750	406.4	17	865
18	800	609.6	18	1345
19	400	508	19	60
20	2200	1016	20	1275
21	1500	508	21	930
22	500	304.8	22	485
23	2650	1016	23	1045
24	1230	762	24	820
25	1300	762	25	170
26	850	508	26	900
27	300	304.8	27	370
28	750	304.8	28	290
29	1500	406.4	29	360
30	2000	304.8	30	360
31	1600	304.8	31	105
32	150	406.4	32	805
33	860	406.4		
34	950	609.6		

$$d_{n_\xi}(t) = 0.5 d_{n_\xi}(t-1) + \varepsilon_n(t) \,, \tag{9.21}$$

and the variance of $\varepsilon_n(t)$ is assumed to be a function of the periodic demand pattern given by $\sigma^2_{\varepsilon_n}(t) = 0.03 d_{n_\psi}(t)$. Figure 9.2b shows the evolution of the uncertainty

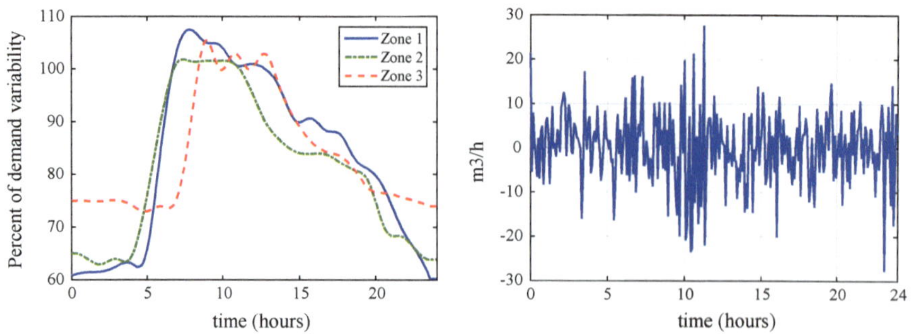

(a) Percent of variability for the three zones (b) Evolution of $\varepsilon_{12}(t)$ during a day

Fig. 9.2 Evolutions of the demand pattern and its deviation

$\varepsilon_{12}(t)$ throughout a day for node 12 located at zone 3. For all simulated scenarios, white noise is added to the measured variables, and only a single leak occurs.

9.5.2 P-DPCA Model and Its Validation

The training data set is obtained by simulating the network with the given features for the demands during 30 days. Thus, the transformation (9.12) with $J = 30$ is applied to the set of historical data with $m = 5$. For the generation of matrix (9.18), the lag parameter $l = 3$ is used because by adding more lags new linear relationships do not appear since a small key singular value and the lowest key singular value ratio are obtained according to method 1 presented by Rato and Reis (2013). To retain 99% of the variability, 10 principal components are chosen. The initial thresholds T_α^2 and SPE_α are estimated by considering a confidence interval of 0.05.

The performance of the P-DPCA model with respect to false alarms is tested with the false alarm rate (FAR). This rate is calculated by dividing the number of detected leaks for a set S of normal condition data by the cardinality of S during a day. Thus, the thresholds of the statistic measures have been adjusted based on a maximum FAR of 5%, and the test with a new data set produced a FAR of 3.52%.

9.6 Results

9.6.1 Leak Detection Test

The proposed approach is compared with PCA, DPCA, and P-PCA under the same leak scenarios to demonstrate its advantages, and only single leaks are emulated. Each of the 744 leak scenarios is characterized by its node and occurrence time. The performances are evaluated considering two indexes.

- The fault detection rate (FDR) is defined as the number of times that the leak is detected divided by the total number of samples where there is a leak present during a specific time window. This FDR allows a sensitivity analysis in terms of the leak's location and the occurrence time over the period of a day.
- The fault detection delay (FDD) is defined as the time interval between the leak's occurrence and its detection.

To easily visualize the behavior of the indexes, the information in the graphs is compacted, and box plots are drawn. The symbol $\odot$ denotes the median, the height of the box represents 75% of the results, the dashed line $--$ expresses the minimum and maximum rate, and the symbol $+$ describes the outlier of the distribution.

All the leaks were simulated with an emitter coefficient of 10 (Rossman 2000), and the leak size varied between 50 and 100 m^3/h.

Figure 9.3 illustrates the FDR results with P-DPCA. One can see that the highest boxes with the minimal median occur during the maximum demand time. Figures 9.4, 9.5 and 9.6 show the FDRs obtained with PCA, DPCA and P-PCA respectively. From these graphs one can conclude the following.

- The performance of PCA shown in Fig. 9.4 is very poor and in the time interval of [8–23] the detection is not feasible. Such infeasibility justifies the suggestion reported by Pérez et al. (2014) that the network should be tested in the very early morning when the variability is low.
- The performance of DPCA shown in Fig. 9.5 has similar results to PCA. This means the absence of the stationary condition of the variables makes the use of PCA infeasible even when the dynamic of the time-correlated variables is taken into account.

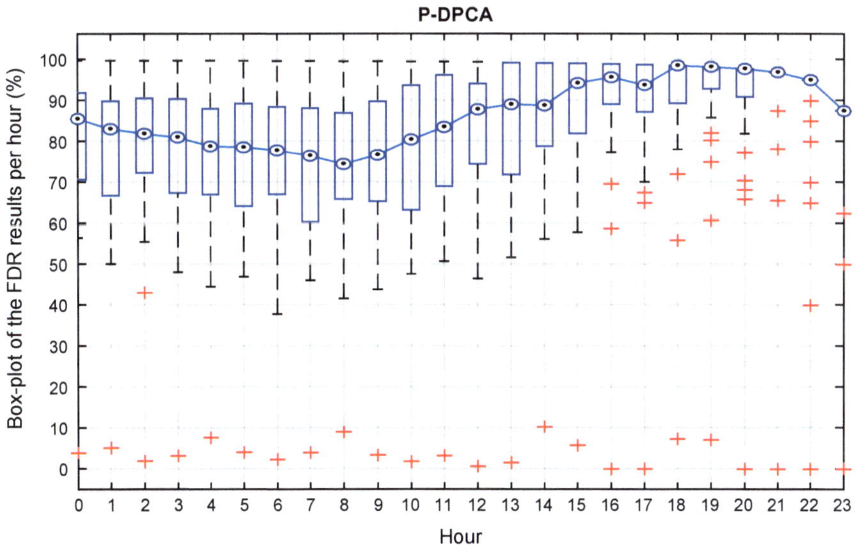

Fig. 9.3 FDR behavior by using P-DPCA for all leaks at a specific occurrence hour

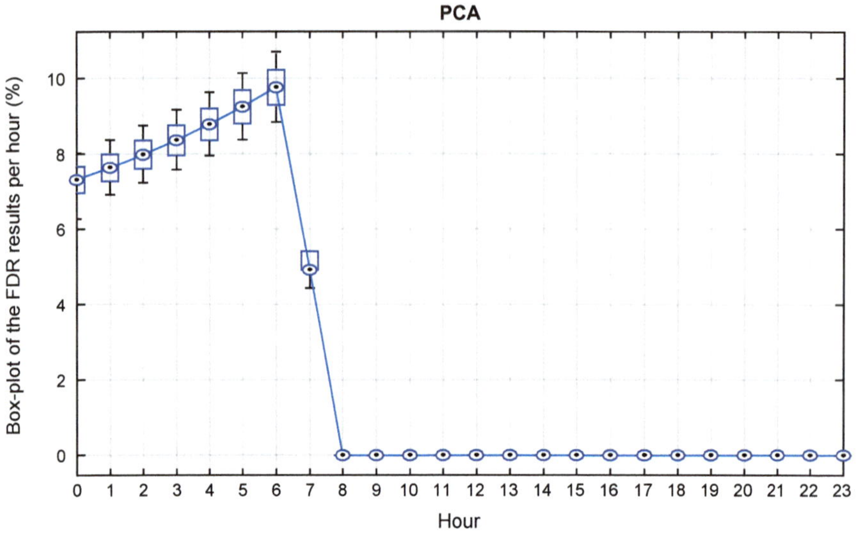

Fig. 9.4 FDR behavior by using PCA for all leaks at a specific occurrence hour

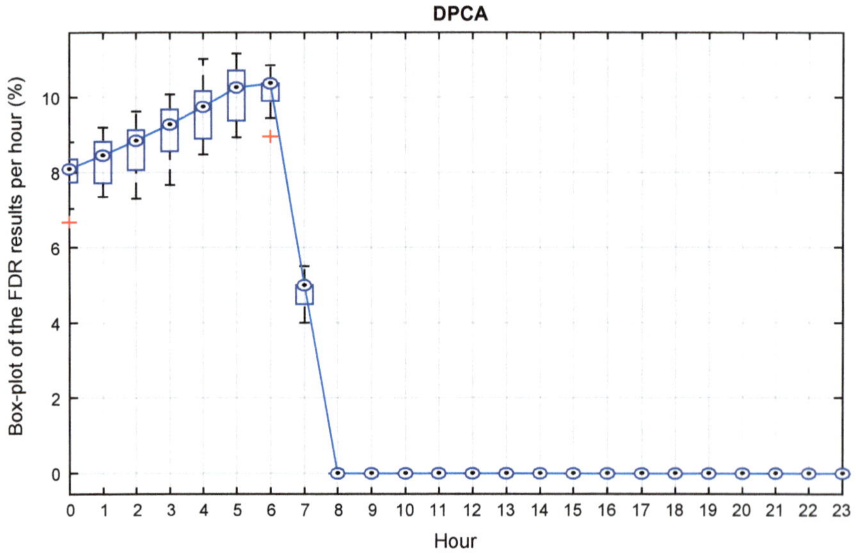

Fig. 9.5 FDR behavior by using DPCA for all leaks at a specific occurrence hour

- The FDR of P-PCA shown in Fig. 9.6 is higher than that obtained with PCA and DPCA but does not achieve the rate of P-DPCA. Thus, both the periodically stationary process and the dynamic of the pressures play a role in the monitoring of a WDN.

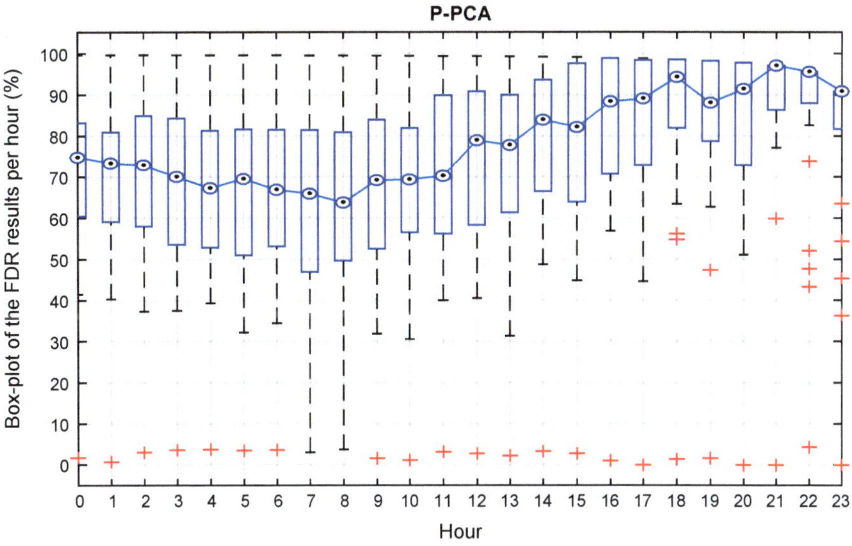

Fig. 9.6 FDR behavior by using P-PCA for all leaks at a specific occurrence hour

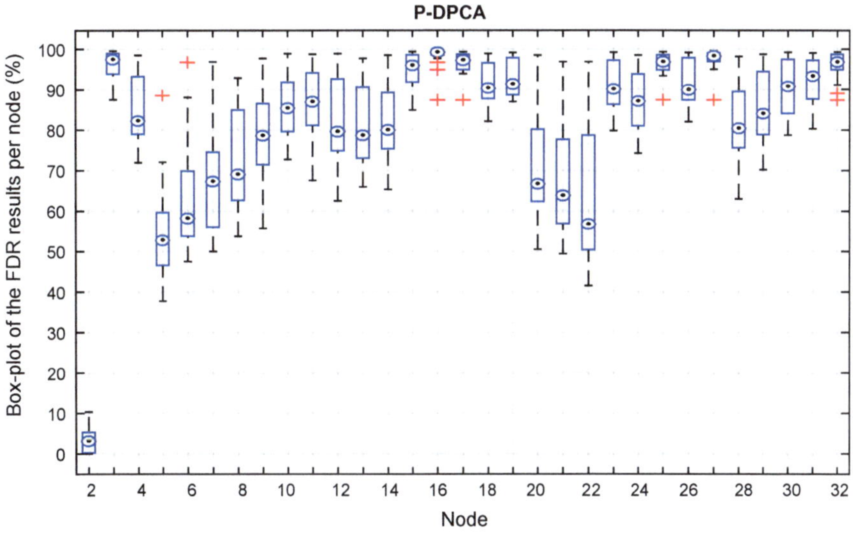

Fig. 9.7 FDR behavior with the P-DPCA model for each leak over a day

A second experiment is presented here. The FDR is evaluated by simulating a single leak over a period of 24 hours for the 31 nodes. The graphs for the P-DPCA and P-PCA models are shown in Figs. 9.7 and 9.8 respectively. One can conclude the following.

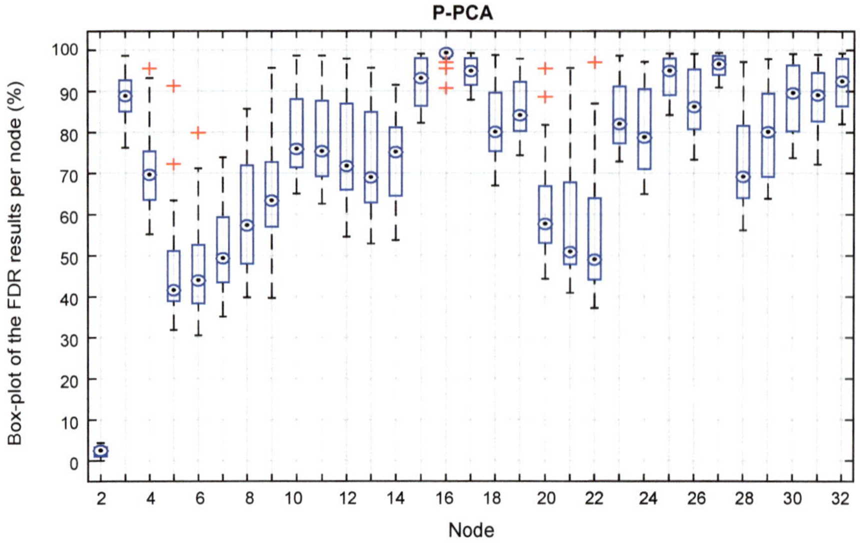

Fig. 9.8 FDR behavior with the P-PCA model for each leak over a day

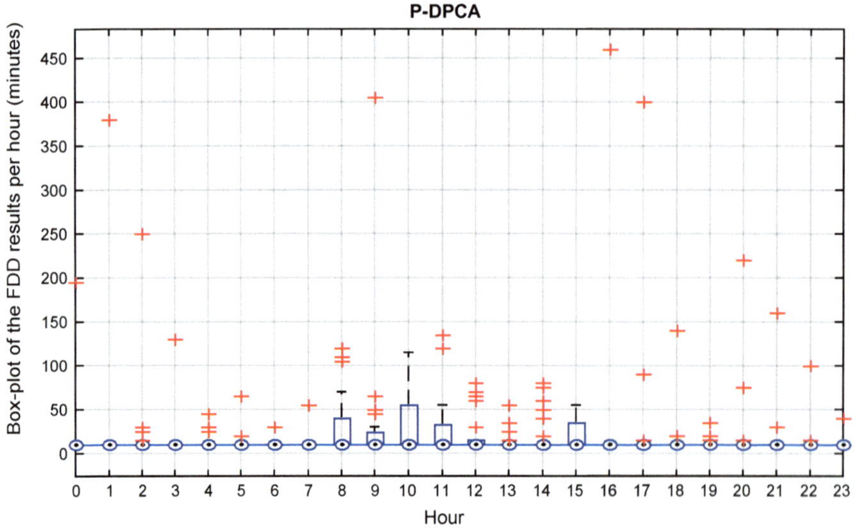

Fig. 9.9 FDD behavior with the P-DPCA for each hour

- In most of the scenarios, the FDRs exceed the value of 60% independently from the time occurrence and the node of the leak.
- The poorest performance is achieved for leaks at nodes $\{2, 5, 6, 7, 8, 20, 21, 22\}$. The insensitivity of the model in detecting a leak at node 2 is notable. This can be explained because the variables at node 3 are dominated by the pressure of the

Table 9.2 Performance of the compared models μ/σ

Model	FDR	FDD (minutes)
PCA	2.5/0.16	380/1.13
DPCA	4/0.19	350/1.25
P-PCA	74/21	20/30
P-DPCA	81/20.8	18/42

reservoir. Moreover, the insensitivity to a leak in the other nodes can be explained because of their long distance from the measurement points as verified in Fig. 9.1 and Table 9.1.

- The FDR graphs with the P-DPCA and P-PCA models are very similar. This means the time correlation of the pressures does not drastically alter the sensitivity of the detector.

To analyze the delay of the detector, the FDD is evaluated by simulating a leak in all the nodes for each hour of the period. Figure 9.9 shows the FDD, and the delay is around 10 minutes, except when the demand reaches its peak in the interval of [8, 11] hours.

To conclude, the mean values and standard deviations (μ/σ) of the FDR and FDD for each model considering all the data are given in the Table 9.2. The superiority of the P-DPCA model is demonstrated, because it presents the best FDR with an improvement of approximately 70% compared to the PCA and DPCA. Moreover, P-DPCA achieves the best FDD with a difference of approximately 300 min with respect to PCA or DPCA.

9.7 Conclusions

A data-based monitoring procedure is presented for leak detection in WDNs with demand patterns. The key of the procedure is the model assumed for the water demand in each node for a specific time interval. In particular, the demand is formed by a periodic plus a stochastic function, which allows the obtaining of weakly stationary signals and is the main contribution of this work. The Hanoi WDN is used as a case study with time-varying demand patterns. The comparison among the new approach and the standard ones with respect to the FDR and the FDD proves the superiority of the proposal with an improvement in the leak detection rate of approximately 70% compared to the PCA and DPCA methods. Note that the demand uncertainty at peak hours produces a reduction in the FDD.

Acknowledgements Project supported by the Mexican Government Scholarship Program for International Students. In addition, the authors acknowledge the support provided by DGAPA-UNAM IT100716, II-UNAM and Universidad Tecnológica de la Habana José Antonio Echeverría (CUJAE).

References

Alvisi, S., Franchini, M., & Marinelli, A. (2007). A short-term, pattern-based model for water-demand forecasting. *Journal of Hydroinformatics*, *9*(1), 39–50.

Arsene, C. T. C., Gabrys, B., & Al-dabass, D. (2012). Decision support system for water distribution systems based on neural networks and graphs theory for leakage detection. *Expert Systems with Applications*, *39*(18), 13214–13224.

Brunone, B., & Ferrante, M. (2001). Detecting leaks in pressurised pipes by means of transients. *Journal of Hydraulic Research*, *39*(5), 539–547.

Casillas, M. V., Garza-Castañón, L. E., & Puig, V. (2015). Sensor placement for leak location in water distribution networks using the leak signature space. *9th IFAC Symposium on Fault Detection, Supervision and Safety of Technical Processes* (pp. 214–219). Paris, France: IFAC.

Chiang, L. H., Rusell, E., & Braatz, R. D. (2001). *Fault detection and diagnosis in industrial systems*. London, England: Springer.

Cugueró-Escofet, P., Blesa, J., Pérez, R., Cugueró-Escofet, M. A., & Sanz, G. (2015). Assessment of a leak localization algorithm in water networks under demand uncertainty. *9th IFAC Symposium on Fault Detection, Supervision and Safety of Technical Processes* (pp. 226–231). Paris: France.

Duzinkiewicz, K., Borowa, A., Mazur, K., Grochowski, M., Brdys, M. A., & Jezior, K. (2008). Leakage detection and localisation in drinking water distribution networks by multiregional PCA. *Studies in Informatics and Control*, *17*(2), 135–152.

Farley, M., & Trow, S. (2003). *Losses in water distribution networks: A practitioners' guide to assessment. Monitoring and Control*. London, UK: IWA Publishing.

Ferrandez-gamot, L., Busson, P., Blesa, J., Tornil-sin, S., Puig, V., Duviella, E., et al. (2015). Leak localization in water distribution networks using pressure residuals and classifiers. *9th IFAC Symposium on Fault Detection, Supervision and Safety of Technical Processes* (pp. 2–7). Paris, France: IFAC.

Fujiwara, O., & Khang, D. B. (1990). A two-phase decomposition method for optimal design of looped water distribution networks. *Water Resources Research*, *26*(4), 539–549.

Gabrys, B., & Bargiela, A. (1999). Neural networks based decision support in presence of uncertainties. *Journal of Water Resources Planning and Management*, *125*(2), 272–280.

Gertler, J., Romera, J., Puig, V., & Quevedo, J. (2010). Leak detection and isolation in water distribution networks using principal component analysis and structured residuals. *Conference on Control and Fault Tolerant Systems* (pp. 1–6). Nice, France: IEEE.

Houghtalen, R., & Hwang, N. H. C. (2010). *Fundamentals of hydraulic engineering systems*. Prentice Hall.

Kruger, U., & Xie, L. (2012). *Statistical monitoring of complex multivariate processes*. West Sussex, UK: Wiley.

Ku, W., Storer, R. H., & Georgakis, C. (1995). Disturbance detection and isolation by dynamic principal component analysis. *Chemometrics and Intelligent Laboratory Systems*, *30*, 179–196.

Lambert, A. (1994). Accounting for losses: The bursts and background concept. *Water and Environment Journal*, *8*(2), 205–214.

Macgregor, J., & Cinar, A. (2012). Monitoring, fault diagnosis, fault tolerant control and optimization: Data driven methods. *Computers and Chemical Engineering*, *47*, 111–120.

Misiunas, D., Vitkovsky, J., Olsson, G., Lambert, M., & Simpson, A. R. (2005). Failure monitoring in water distribution networks. *Water Science and Technology*, *4*(4–5), 503–511.

Nowicki, A. D. A. M., Grochowski, M. I., & Duzinkiewicz, K. A. (2012). Data-driven models for fault detection using kernel PCA: A water distribution system case study. *International Journal of Applied Mathematical Computer Science*, *22*(4), 939–949.

Papoulis, A. (1991). *Probability, random variables, and stochastic processes* (3rd ed.). McGraw-Hill.

Pérez, R., Puig, V., Pascual, J., Quevedo, J., Landeros, E., & Peralta, A. (2011). Methodology for leakage isolation using pressure sensitivity analysis in water distribution networks. *Control Engineering Practice*, *19*, 1157–1167.

Pérez, R., Sanz, G., Puig, V., Quevedo, J., Cugueró-Escofet, M. A., Nejjari, F., et al. (2014). Leak Localization in Water Networks. *IEEE Control Systems Magazine*, *34*(4), 24–36.
Rato, T. J., & Reis, M. S. (2013). Defining the structure of DPCA models and its impact on process monitoring and prediction activities. *Chemometrics and Intelligent Laboratory Systems*, *125*, 74–86.
Rossman, L. A. (2000). *EPANET 2 Users Manual*. United States Envionmental Protection Agency.
Sedki, A., & Ouazar, D. (2012). Hybrid particle swarm optimization and differential evolution for optimal design of water distribution systems. *Advanced Engineering Informatics*, *26*(3), 582–591.
Soyer, R., & Roberson, J. A. (2014). Urban water demand forecasting a review of methods and models. *Journal of Water Resources Planning and Management*, *140*(2), 146–159.
Verde, C., Torres, L., & González, O. (2016). Decentralized scheme for leaks' location in a branched pipeline. *Journal of Loss Prevention in the Process Industries*, *43*, 18–28.
Zhou, S. L., McMahon, T. A., Walton, A., & Lewis, J. (2002). Forecasting operational demand for an urban water supply zone. *Journal of Hydrology*, *259*(1–4), 189–202.

Chapter 10
Leak Localization in Water Distribution Networks Using Pressure Models and Classifiers

Adrià Soldevila, Sebastian Tornil-Sin, Joaquim Blesa, Rosa M. Fernandez-Canti and Vicenç Puig

Abstract This chapter proposes an architecture and an associate methodology for leak localization in Water Distribution Networks (WDN) that are based on pressure models and classifiers. In a first stage of the proposed architecture, residuals are obtained by comparing available pressure measurements with the estimations provided by a WDN model. In a second stage, a classifier is applied to the residuals with the aim of determining the leak location. The classifier is trained with data generated by simulation of the WDN under different leak scenarios and uncertainty conditions. Several classification approaches are considered and compared. The proposed methodology is tested both using synthetic and experimental data with real WDNs of different sizes. The comparison with the current approaches shows a performance improvement.

10.1 Introduction

The traditional approach to leakage control is a passive one, whereby the leak is repaired only when it becomes visible. Recently developed acoustic instruments (Khulief et al. 2012) allow the localization of invisible leaks, but unfortunately, their application over a large-scale water network is very expensive and time-consuming.

A. Soldevila · S. Tornil-Sin · R.M. Fernandez-Canti · V. Puig (✉)
Research Center for Supervision, Safety and Automatic Control (CS2AC),
Rambla Sant Nebridi, S/n, 08022 Terrassa, Spain
e-mail: vpuig@iri.upc.edu

A. Soldevila
e-mail: adria.soldevila@upc.edu

S. Tornil-Sin
e-mail: stornil@iri.upc.edu

R.M. Fernandez-Canti
e-mail: rosa.mari.fernandez@upc.edu

S. Tornil-Sin · J. Blesa · V. Puig
Institut de Robòtica i Informàtica Industrial (CSIC-UPC), Carrer Llorens Artigas, 4-6,
08028 Barcelona, Spain
e-mail: jblesa@iri.upc.edu

C. Verde and L. Torres (eds.), *Modeling and Monitoring of Pipelines and Networks*, Applied Condition Monitoring 7,
DOI 10.1007/978-3-319-55944-5_10

A viable solution is to divide the network into district metered areas (DMA), where the *flow* and the *pressure* at the input are measured (Lambert et al. 2003; Puust et al. 2010), and to implement a permanent leakage monitoring since leakages increase the flow and decrease the pressure head at the DMA entrance. Various empirical studies (Lambert 2001; Thornton and Lambert 2005) propose mathematical models to describe the leakage flow with respect to the pressure at the leakage location. The best practice in the analysis of DMA flows consists of estimating the leakage when the demand is minimal. This typically occurs at night, when the customers' demand is low, and the leakage component is at its maximum percentage of the total flow consumption (Puust et al. 2010). Therefore, practitioners monitor the DMA or groups of DMAs for detecting, locating, and estimating the leakage level by analyzing the minimum night flow (Puust et al. 2010). Leakage detection may not be easy, however, because of unpredictable variations in consumer demands and measurement noise, as well as long-term consumption trends and seasonal effects.

Several works have been published addresing leak localization methods for WDNs. See Puust et al. (2010) and references therein. For example, in Colombo et al. (2009), a review of transient-based leak detection methods is offered. In Yang et al. (2008), a method is proposed for identifying leaks by using blind spots based on the analysis of acoustic and vibrations signals (Fuchs and Riehle 1991) together with models of buried pipelines which allow the prediction of wave velocities (Muggleton et al. 2002). In Wu and Sage (2006), Genetic Algorithms were proposed to solve an optimization problem allowing the quantification and location of water losses. More recently, Mashford et al. (2009) have developed a method for locating leaks and estimating their outflow by using support vector machines (SVM), which analyze data obtained by a set of pressure sensors in a pipeline network. Another set of methods is based on the inverse transient analysis (Covas and Ramos 2001; Kepler et al. 2011). The main idea is to analyze the pressure data collected over the occurrence of transitory events by means of the minimization of the difference between the observed and the calculated parameters. In Ferrante and Brunone (2003a, b), it is shown that unsteady-state tests can be used for pipe diagnosis and leak detection. The transient-test based methodologies use the equations for transient flow in pressurized pipes in frequency domain and then information about pressure waves is taken into account, too. More recently, the use of k-nearest neighbors (k-NN) and neuro-fuzzy classifiers for leak localization purposes has been proposed by Soldevila et al. (2016) and Wachla et al. (2015).

Model-based leak detection and isolation techniques have also been studied starting with the seminal paper of Pudar and Liggett (1992), which formulates the leak detection and localization problem as a least squares parameter estimation problem. Unfortunately, the parameter estimation of water network models is not an easy task (Savić et al. 2009). The problem of leak localization in WDNs can be addressed as a particular case of fault detection and isolation (FDI) in dynamic systems (Blanke et al. 2006). DMA hydraulic behavior is described by a nonlinear model expressed as a set of algebraic equations with no explicit solution that can only be solved by using numerical methods such as the one proposed by Todini and Pilati (1988). This fact limits the applicability of most model-based FDI approaches that require transforming or manipulating the model to generate a set of residuals with the desired FDI specifications. Thus, only primary (direct) residuals, which are sensitive to more than one leak, could be generated because DMAs typically are formed by a dense mesh of highly interconnected pipes. This fact added to the reduced number of sensors makes the isolation task difficult. For this reason, specific model fault diagnosis methods for leak localization should be developed. A first contribution in this line can be found in Pérez et al. (2011, 2014), where a model-based method that relies on pressure measurements and leak sensitivity analysis is proposed. This method consists of computing online residuals, that is, the differences between the measurements and their estimations obtained by using the hydraulic network model and defining respective thresholds that take into account the modeling uncertainty and the noise. When some of the residuals violate their threshold, the leak signature is matched to a leak sensitivity matrix to determine which of the possible leaks is present. Although this approach is efficient under ideal conditions, its performance decreases because of the nodal demand uncertainty and measurement noise. This method has been improved by Casillas et al. (2012) by taking into account an analysis along a time horizon. This work presents a comparison of several leak isolation methods. Note that in cases where flow measurements are available, leaks could be detected easily, since it is possible to establish simple mass balance in the pipes. See, for example, the work of Ragot and Maquin (2006), where a methodology to isolate leaks is proposed by using fuzzy analysis of the residuals. This method calculates the residuals between the flow measurements and their estimation by using a model without leaks. Although the use of flow measurements is feasible in large water networks, this does not occur when there is a dense mesh of pipes with only flow measurements at the entrance of each DMA. In this situation, water companies consider as a feasible approach the possibility of installing some pressure sensors inside the DMAs, because they are inexpensive and easy to install and maintain.

In this chapter, a new model-based approach for leak localization in WDNs using pressure models and classifiers is presented. This methodology is intended to be used after the leak has been detected by means of the analysis of the night DMA water demands (Puust et al. 2010) and after the application of the validation and reconstruction methodology described by Cugueró-Escofet et al. (2016) to the sensors used for leak localization. Following a model-based methodology successfully tested in Pérez et al. (2011, 2014), a pressure model of the considered WDN is used in a first stage to compute residuals that are indicative of leaks. In a second stage, a classifier is applied to the obtained residuals to determine the leak location. This online scheme relies on a previous offline work in which the network model is obtained, and the classifier is trained with data generated by extensive simulations of the network. These simulations consider three types of uncertainties: leaks with different magnitudes in all the nodes of the network, differences between the estimated and real consumer water demands and noise in pressure sensors. The underlying idea is to obtain a classifier that can distinguish the leak location independently of the unknown real leak magnitude and the presence of uncertainties associated with the water demands and the pressure measurements.

10.2 Background and Motivation

10.2.1 *Principle of Model-Based Leak Location Approaches*

Model-based approaches aim to locate leaks in a water distribution network by comparing pressure measurements with their estimations obtained by using a hydraulic network model. Usually, this methodology is used for locating leaks within a given leak size range defined by the water network management company. The minimum size is related to the sensor resolution and modeling/demand uncertainty, and the maximum size is defined as the value such that the leak behaves as a burst that can be seen in the street. Model-based leak localization methods are based on comparing the monitored pressure disturbances caused by leaks at certain inner nodes of the DMA network with the theoretical pressure disturbances caused by all potential leaks obtained by using their respective model (Pérez et al. 2014). This comparison employs the residual vector $r \in \mathbb{R}^{n_s}$, obtained from the difference between the measured pressure at DMA inner nodes $p \in \mathbb{R}^{n_s}$ and the pressure at these nodes calculated by using the network model considering a leak-free scenario $\hat{p}_o \in \mathbb{R}^{n_s}$:

$$r(t) = p(t) - \hat{p}_o(t). \tag{10.1}$$

The dimension of the residual vector r, n_s, depends on the number of inner pressure sensors installed in the DMA. In recent years, some optimal sensor placement algorithms have been developed to determine where the pressure head sensors should be installed inside the DMA with minimal economical costs (number of sensors), to guarantee a suitable performance regarding leak localization; see Pérez et al. (2011); Casillas et al. (2013), Sarrate et al. (2014) among others.

The number of potential leaks, $f \in \mathbb{R}^{n_n}$, is considered to be equal to the number of DMA nodes n_n, since from the modeling point of view, as proposed by Pérez et al. (2011) and Pérez et al. (2014), leaks are assumed to occur in these locations.

10.2.2 *Limitations of Sensitivity Analysis Approaches*

Most model-based leak localization approaches rely on the sensitivity-to-leak analysis (Pérez et al. 2011, 2014), where the theoretical pressure disturbances caused by all potential leaks are stored in the leak sensitivity matrix $\boldsymbol{\Omega} \in \mathbb{R}^{n_s \times n_n}$ (with as many rows as DMA inner pressure sensors, n_s, and as many columns as potential leaks in all nodes n_n). Leak isolation is then based on matching the residual vector (10.1) with the columns of the sensitivity matrix by using some metrics such as the correlation or the angle; see Casillas et al. (2012) for details. The leak sensitivity matrix can be mathematically formalized as follows:

$$\boldsymbol{\Omega} = \begin{pmatrix} \frac{\partial r_1}{\partial f_1} & \cdots & \frac{\partial r_1}{\partial f_{nn}} \\ \vdots & \ddots & \vdots \\ \frac{\partial r_{ns}}{\partial f_1} & \cdots & \frac{\partial r_{ns}}{\partial f_{nn}} \end{pmatrix}, \tag{10.2}$$

where each element $\boldsymbol{\Omega}_{i,j}$ measures the effect of the leak f_j in the residual r_i associated to the pressure at node i. In practice, it is extremely difficult to calculate $\boldsymbol{\Omega}$ analytically because a water distribution network is a large-scale multivariable nonlinear system such that its equations can only be solved numerically as discussed in the introduction. Thereby, the sensitivity matrix is generated by the simulation of the network model and the evaluation of the sensitivity $\Omega_{i,j}$ as

$$\Omega_{i,j} = \frac{\hat{p}_{i,f_j} - \hat{p}_{i,0}}{f_j}, \tag{10.3}$$

where $\hat{p}_{i,f_j}$ is the predicted pressure in the node i when a nominal leak f_j is forced in node j and $\hat{p}_{i,0}$ is the predicted pressure in the node i under a scenario free of leaks (Pérez et al. 2011). The sensitivity matrix is then obtained by repeating this process for all n_n potential faults.

An important drawback of the leak sensitivity approach is that the practical evaluation of (10.3) depends on the nominal leak f_j (Blesa et al. 2012, 2016). If the real leak size is different from the nominal one, the real sensitivity will be different from the one computed with (10.3). Moreover, the sensitivity is also affected by the nodal demand uncertainty (Cugueró-Escofet et al. 2015), since this demand is not measured but estimated by using historical records of water consumption and the aggregated DMA consumption pattern. These uncertainties will lead to the deterioration of the leak localization results obtained by using the sensitivity approach. The approach proposed in this chapter aims to overcome these difficulties.

10.3 Proposed Method

10.3.1 Basic Architecture and Operation

The method for online leak localization proposed in this chapter relies on the scheme depicted in Fig. 10.1, and it is based on computing pressure residuals and analyzing them with a classifier. The hydraulic model is built using the Epanet hydraulic simulator (Rossman 2000) by considering the DMA structure (pipes, nodes, and valves) and network parameters (pipe coefficients). After the corresponding calibration process by incorporating real data, it is assumed that the hydraulic model is able to precisely represent the WDN behavior. Note, however, that the model is fed with estimated water demands in the nodes ($\hat{d}_1, \ldots, \hat{d}_{n_n}$). In practice, nodal demands ($d_1, \ldots, d_{n_n}$) are not measured, except for some particular consumers, where automatic metering

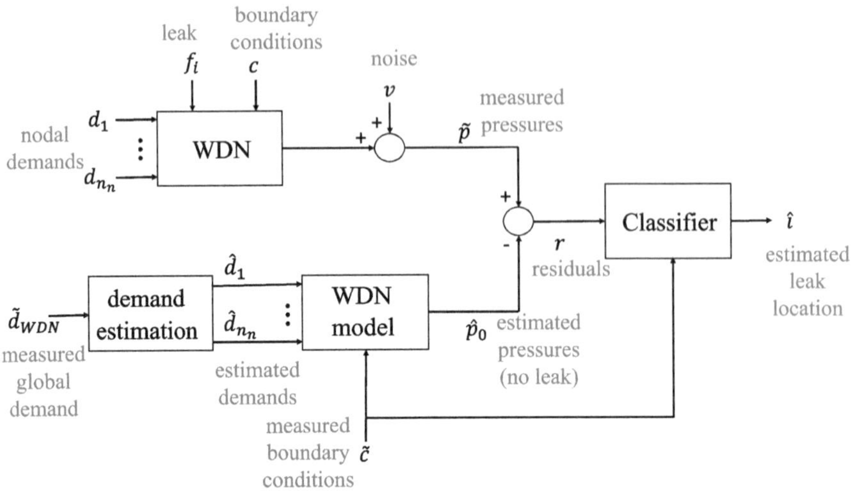

Fig. 10.1 Leak localization scheme

readers (AMRs) are available. Therefore, the nodal demands are typically obtained by the total measured DMA demand $\tilde{d}_{WDN}$ and they are distributed at nodal level by using historical consumption records. Hence, the residuals are not only sensitive to leaks but also to differences between the real demands and their estimated values. Additionally, pressure measurements are subject to the effect of sensor noise v, and this also affects the residuals. By taking all these effects into account, the classifier must be able to locate the real leak present in the WDN, which can be located in any node and with any (unknown) magnitude, while remaining robust to the demand uncertainty and the measurement noise. Finally, the operation of the network is constrained by some boundary conditions c (such as the position of internal valves, reservoir pressures, and flow consumption ($\tilde{d}_{WDN}$)) that are known (measured) and must be considered in the simulation and can also be used as inputs for the classifier.

10.3.2 Methodology Overview

The exploitation of the architecture presented in Fig. 10.1 relies on a methodology that distinguishes several offline and online procedures.

10.3.2.1 OffLine

The application of the architecture presented in Fig. 10.1 relies on an offline work that aims to attain a classifier able to distinguish the potential leaks under the described uncertainty conditions. In particular, the method proposed in this chapter considers an offline design based on the following stages:

- Modeling—A model for the WDN is obtained, calibrated, and implemented with Epanet. The model is basically built by taking into account the network structure and by applying flow balance conservation and pressure loss equations; see Pérez et al. (2011, 2014) for details.
- Data generation—The model implemented is extensively used to generate data in the residual space for each possible leak and for different operating and uncertainty conditions.
- Classifier training and evaluation—The classifier is first trained with a subset of the initial data set, then it is applied to the testing data in order to estimate its performance.

The data generation stage is critical since the availability of representative data is a necessary condition for obtaining a good classifier. Since the amount of data collected from the real monitored WDN is limited, a way to obtain a complete training data set is by using a hydraulic simulator. Hence, training and testing data are generated by applying the scheme depicted in Fig. 10.2, which is similar to the one presented in Fig. 10.1 but with the main difference of substituting the real WDN for a model that allows simulating the WDN not only in the absence but also in the presence of faults.

The presented scheme is exploited to achieve the following:

- Generate data for all possible leak locations, i.e., for all the different nodes in the WDN ($\bar{f}_i, \quad i = \{1, 2, ..., n_n\}$).
- Generate data for each possible leak location with different leak magnitudes within a given range ($\bar{f}_i \in [f_i^-, f_i^+]$).
- Generate sequences of demands and boundary conditions $\hat{c}_i$ that correspond to realistic, typical daily evolution in each node.
- Simulate differences between the real demands and the estimations computed by the demand estimation module ($(\bar{d}_1, ..., \bar{d}_{n_n}) \neq (\hat{d}_1, ..., \hat{d}_{n_n})$).

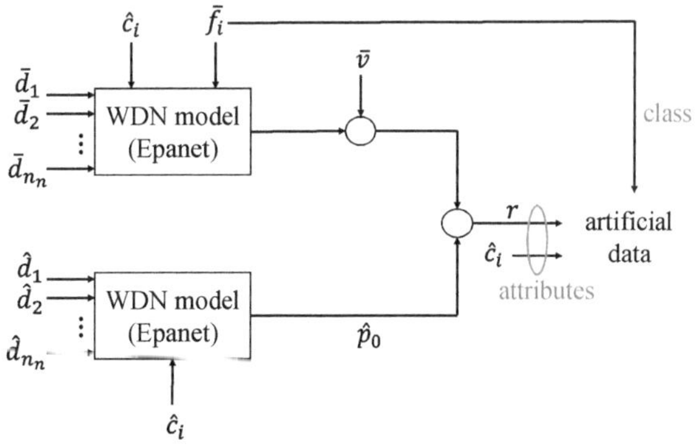

Fig. 10.2 Data generation scheme

Table 10.1 Confusion matrix Γ

	$\hat{f}_1$	$\cdots$	$\hat{f}_i$	$\cdots$	$\hat{f}_{n_n}$
f_1	$\Gamma_{1,1}$	$\cdots$	$\Gamma_{1,i}$	$\cdots$	Γ_{1,n_n}
$\vdots$	$\vdots$	$\vdots$	$\vdots$	$\vdots$	$\vdots$
f_i	$\Gamma_{i,1}$	$\cdots$	$\Gamma_{i,i}$	$\cdots$	Γ_{i,n_n}
$\vdots$	$\vdots$	$\vdots$	$\vdots$	$\vdots$	$\vdots$
f_{n_n}	$\Gamma_{n_n,1}$	$\cdots$	$\Gamma_{n_n,i}$	$\cdots$	Γ_{n_n,n_n}

- Take into account the measurement noise in pressure sensors by generating synthetic Gaussian noise ($\bar{\upsilon}$).

The artificial data obtained from simulations is divided into training and testing sets. The training stage is based on a learning procedure where the input is the (labeled) training data set, and the result is a classifier that must be able to accurately classify new data instances into the correct class. The generalization ability of the obtained classifier is checked in the validation stage, in which the performance indexes are computed for the testing data set.

The details of the training stage are particular to the type of classifier used. The results presented have been obtained by using two different well-known classifiers: the k-nearest neighbor (k-NN) classifier (Alpaydin 2010), which is nonparametric, and the Bayesian classifier, which is parametric. The details about the training of both classifiers will be provided in the next subsections.

The evaluation of classifiers normally relies on the use of the *confusion matrix* Γ, which summarizes the results obtained when the testing data set is applied to the classifier. The confusion matrix is a square matrix with as many rows and columns as nodes of the network (potential leak locations), when it is applied to the leak localization problem with the associated terminology. Each coefficient $\Gamma_{i,j}$ indicates how many times a leak in node i is recognized as a leak in node j. Table 10.1 illustrates the concept of the confusion matrix applied to leak localization.

In case of a perfect classification, Γ is diagonal, with $\Gamma_{i,i} = m$, for all $i = 1, \ldots, n_n$, and m is the size of the testing data set. In practice, nonzero coefficients will appear outside the main diagonal. For a leak in node i, the coefficient $\Gamma_{i,i}$ indicates the number of times that the leak is correctly identified as $\hat{f}_i$, while $\sum_{j=1}^{n_n} \Gamma_{i,j} - \Gamma_{i,i}$ indicates the number of times that it is wrongly classified. The overall accuracy (Ac) of the classifier is defined as

$$Ac = \frac{\sum_{i=1}^{n_n} \Gamma_{i,i}}{\sum_{i=1}^{n_n} \sum_{j=1}^{n_n} \Gamma_{i,j}}. \tag{10.4}$$

10.3.2.2 Online

Once the classifier has been trained and validated, it can be used online to localize leaks. According to Sect. 10.3.1, the classifier can be directly used to localize leaks based on the instantaneous values of the computed residuals.

However, this strategy may provide limited results, if there is a high level of uncertainty. The use of a temporal reasoning that takes into account not only the instantaneous values of the residuals but all the values within a time horizon is suggested in Casillas et al. (2014). This idea is implemented in different forms depending on the type of classifier that is used; details are provided in the next subsections.

10.3.3 k-NN Classifier Implementation

10.3.3.1 The *k*-NN Classifier

One of the well-accepted and established methods for classification is the *k*-NN algorithm (Alpaydin 2010), which is available in most numerical packages (e.g., Matlab, R, etc.). Its basic version works as follows. When a new data realization must be classified, the distance (typically, the Euclidean distance is used, but many other options are available) to all the instances of the training data set is computed. The *k*-nearest neighbors are then selected and a voting procedure is applied, where each neighbor votes for its own class, and the class with more votes is chosen. The process is illustrated in Fig. 10.3, where a value $k = 3$ is used and the new data instance is associated with class C_3 since two of the three minimal distances correspond to training instances of that class. The value for k is typically bigger than 1 to improve the robustness against outliers, and it must be smaller than the minimum number of instances of a single class from the training data set. The *k*-NN classifier is said to be a lazy classifier since the training procedure is limited to the storage of the training set, and all the computations are deferred until the classification process is performed.

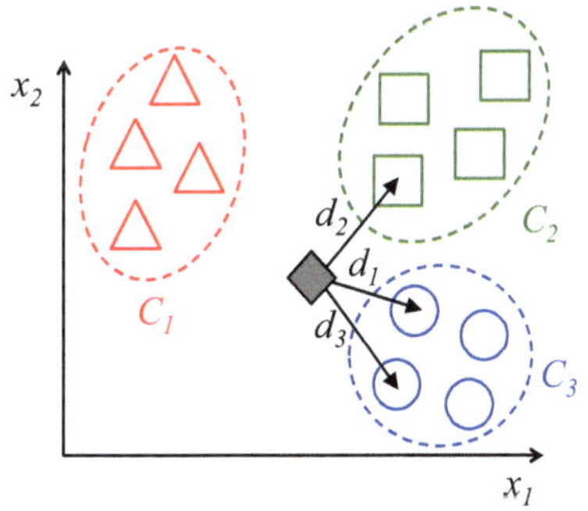

Fig. 10.3 The *k*-NN algorithm

10.3.3.2 Time Reasoning

If the uncertainty in the demands, the leak magnitude or the noise level are high, then the direct application of the classifier can provide poor leak localization results. This also happens when other ways of evaluating the pressure residuals are used (as the ones described in Sect. 10.2). To smooth the effect of demand uncertainty, leak magnitude and noise, typically the analysis of the residuals evolution is performed in a time horizon, i.e., the values for the residuals in the last N time instants are considered (Casillas et al. 2014).

A simple temporal reasoning can be based on incorporating the leak localization results provided by the classifier inside the time horizon and by applying a voting scheme, concluding that the candidate leak is located in the node that has been selected more frequently by the classifier.

A second and more sophisticated option could be to use the information contained in the confusion matrix. Hence, at each time instant t, when the classifier is providing a leak at node j as an explanation for the values of the residuals at the current time instant t, the whole column j of the confusion matrix is stored. This column provides an estimation of the probabilities $p(f_i|\hat{f}_j)$: the probabilities of a leak at node i when the classifier indicates that the leak is at node j, according to the information available for current time instant t. Then, the sum of column vectors stored along the time horizon N is computed. In the obtained vector, the position of the coefficient with the highest value indicates the most probable leak according to the information provided by the data in the time horizon $[t - N + 1, t]$.

10.3.4 Bayesian Classifier Implementation

10.3.4.1 Bayesian Classification

Given the residuals, the objective is to apply a Bayesian leak discrimination procedure in order to identify which leak may occur based on the observed behavior. Such a diagnosis procedure based on Bayesian reasoning is explained below.

At every time sample t, the probability of a leak occurrence is estimated as a result of the application of the Bayes Rule

$$P(f_i \mid \mathrm{r}(t)) = \frac{P(\mathrm{r}(t) \mid f_i)P(f_i)}{P(\mathrm{r}(t))}, \quad i = 1, ..., n_n, \tag{10.5}$$

where $P(f_i \mid \mathrm{r}(t))$ is the posterior probability that the leak f_i had caused the observed residual vector $\mathrm{r}(t) = (r_1(t) \cdots r_j(t))^T$, $P(\mathrm{r}(t) \mid f_i)$ is the likelihood of the residual $\mathrm{r}(t)$ assuming that the active leak is f_i, $P(f_i)$ is the prior probability for the leak f_i, and $P(\mathrm{r}(t))$ is a normalizing factor given by the Total Probability Law

$$P(\mathrm{r}(t)) = \sum_{i=1}^{n_n} P(\mathrm{r}(t) \mid f_i) P(f_i). \tag{10.6}$$

Regarding the prior probabilities, unless we have any additional information, an unprejudiced starting point is to consider all them equally probable as $P(f_i) = \frac{1}{n_n}$, $i = 1, ..., n_n$. To estimate the likelihood value $P(\mathrm{r}(t) \mid f_i)$, we need to perform a previous calibration task in order to obtain the joint probability density function for each leak in the residual space, $P(\mathrm{r}(t) \mid f_i), i = 1, ..., n_n$. The calibration stage is detailed in the next section. Note that, in contrast to standard naïve Bayesian classifiers, we do not need to assume independence between the residuals.

The application of (10.5) produces a set of values $P(f_i \mid \mathrm{r}(t))$, $\sum_{i=1}^{n_n} P(f_i \mid \mathrm{r}(t)) = 1$ that can be used to decide which leak is acting over the system. Note that, at each time sample t, we have information about the probability associated with each leak location. Thus, there can be many competing leaks, each one with a different probability value. The leak with the highest posterior probability can be selected as the most likely one, or all the leaks with a posterior probability above a prespecified threshold can be selected as leak candidates.

10.3.4.2 Recursivity

The results can be improved if (10.5) is recursively applied; if the posterior probability $P(f_i \mid \mathrm{r}(t))$ is used as the prior probability for the next time sample. This way, as long as new measurements data are available, the probabilities are updated, and many of the competing leaks can be discarded.

The only drawback is that if any of the leaks takes the posterior probability value of 1 at any time instant t, then all the remaining leaks take the 0 probability value, therefore preventing them from having a future value different from zero because of the recursive application of (10.5). This drawback can be easily overcomed by forcing all probabilities to have a maximum value of 0.99. When a leak f_i presents the probability $P(f_i \mid \mathrm{r}(t)) > 0.99$, we force it to be $P(f_i \mid \mathrm{r}(t)) = 0.99$, and we can force the remaining probabilities to be $P(f_n \mid \mathrm{r}(t)) = \frac{1-0.99}{n_n-1}$, $n = 1, ..., n_n, n \neq i$.

10.3.4.3 Bayesian Time Reasoning

Additionally, the results can be improved if a time horizon N is introduced. In this case, the posterior probability can be computed on the basis of the N previous time samples, that is, by computing $P(f_i \mid \mathrm{r}(t))$, we recursively can apply the following equation

$$P(f_i \mid \mathrm{r}(t-N+n)) = \frac{P(\mathrm{r}(t-N+n) \mid f_i)P(f_i \mid \mathrm{r}(t-N+n-1))}{P(\mathrm{r}(t-N+n))}, \quad i = 1, \ldots, n_n, \; n = 1, \ldots, N, \tag{10.7}$$

where an unprejudiced starting point may be $P(f_i \mid \mathrm{r}(t-N)) = \frac{1}{n_n}$, $i = 1, \ldots, n_n$.

10.3.4.4 Calibration of the Probability Density Functions

Unlike the k-NN classifier, the Bayes classifier requires a more elaborated training where a joint probability density function (PDF) for each leak class in the residual space, $P(\mathrm{r} \mid f_i)$, $i = 1, \ldots, n_n$, has to be estimated.

The first step is to decide the probability family. The law of large numbers states that most situations lead to a Gaussian distribution if the number of samples is high enough. Several tests can be applied to the residual values to assess if they are Gaussian distributed or not. For instance, we can apply the well-known Kolmogorov–Smirnov (Daniel 1990) or the Anderson–Darling (Stephens 1974) tests, among others.

Figure 10.4 shows the two leak distributions calibrated by means of the Gaussian probability density function. Leak 1 is better adjusted because it takes into account the cross-correlation between residuals r_1 and r_2. On the other hand, leak 2 is adjusted by assuming statistic independence between residuals r_1 and r_2, and therefore the fitting is not so accurate. Note also that other probability distribution families different from Gaussian could be used, including multimodal and nonparametric distributions.

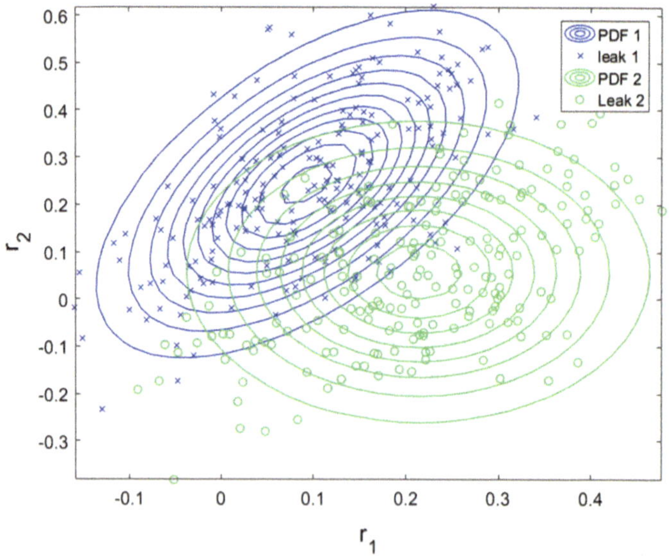

Fig. 10.4 Calibration for leaks 1 and 2

10.4 Case Studies

In this section, two DMA case studies of increasing size and complexity (Hanoi and Nova Icària) are introduced to assess the performance of the proposed methodology.

For these DMAs, leaks are considered at any of the demand nodes. The known variables are the input pressures and flows of the networks (reservoir boundary conditions) and some pressures at the inner nodes of the DMAs where sensors would be located; see Casillas et al. (2013), Sarrate et al. 2014 and Blesa et al. (2016) for details about optimal sensor location. It is considered that the demand pattern is known for all demand nodes but with some uncertainty as proposed by Cugueró-Escofet et al. (2015). The leak magnitude is assumed to be unknown but bounded by a known interval (minimum and maximum leak magnitudes). Finally, noise in pressure sensors is considered too.

For the two DMAs, leak localization results from different uncertainty scenarios are presented and discussed. Moreover, for the second (and largest) DMA, the results of localizing a real leak are also presented.

10.4.1 Hanoi Case Study

The proposed methodology has been first applied to the simplified model of the Hanoi (Vietnam) network, depicted in Fig. 10.5. This model consists of one reservoir, 34 pipes and 31 nodes. Measurements of two inner pressure sensors placed in nodes 14 and 30 are available as considered in other works (Soldevila et al. 2016).

In order to illustrate the performance of the proposed methodology, four different studies have been carried out under the following particular conditions:

- A leak size uncertainty study considering a leak range between 25 and 75 l/s (0.84 and 2.51% of the mean amount of total water daily consumption, which is 2991 l/s).

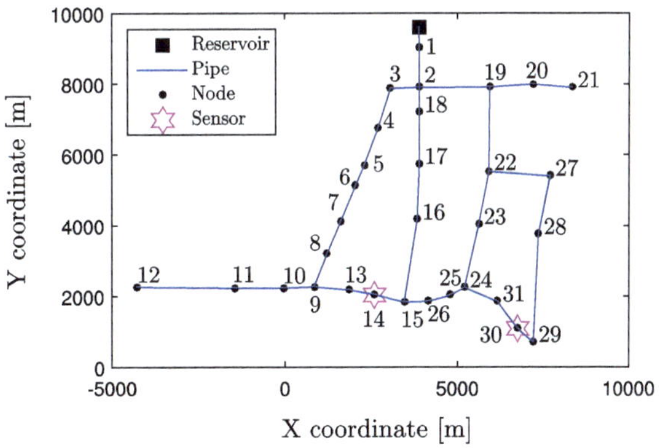

Fig. 10.5 Hanoi topological network

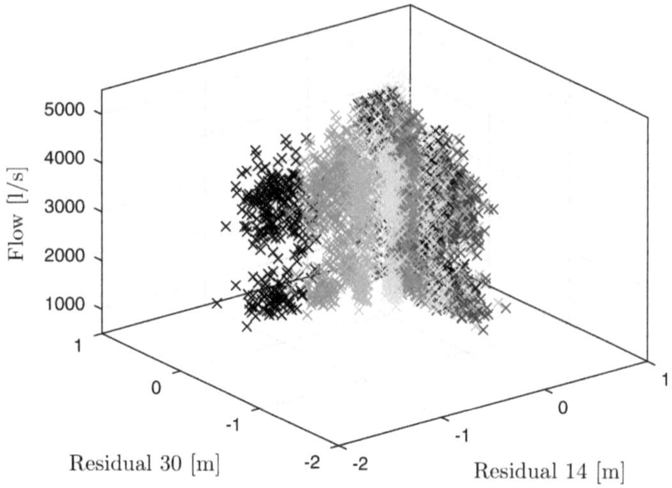

Fig. 10.6 Hanoi residual space with uncertainties, each color represents a leak at a different location

- A study considering noise in pressure measurements with an amplitude of $\pm 5\%$ of the mean value for all pressure residuals.
- A demand uncertainty study considering an uncertainty of $\pm 10\%$ of the nominal nodal demand values.
- A study considering that all the three uncertainties previously defined are simultaneously present in the DMA.

For each study, two complete data sets have been generated for each node (potential leak locations); one for training purposes and the other one for testing the leak localization performance. Each set used for testing, associated with a leak at a given node, is called a leak scenario. The variables conforming to the data are the input flow $\tilde{d}_{DMA}$ and the two residuals r_1 and r_2 associated with pressure measurements at nodes 14 and 30, respectively. The feature space used as input for the classifier is represented in Fig. 10.6. The sampling time used in the simulations is 10 min, but hourly average values of variables are used to improve the leak location performance. Different daily input flow patterns have been simulated as the one depicted in Fig. 10.7. According to the scheme presented in Fig. 10.1, the pressure residuals have been obtained by means of a WDN simulator (Epanet), where the uncertainties described above have been considered.

In order to determine whether the three classifier inputs (r_1, r_2 and $\tilde{d}_{DMA}$) follow a Gaussian distribution, a one-dimension Kolmogorov–Smirnov test on a training data set of 480 samples (for each of the 31 leak nodes) has been performed. As a result, the three inputs can be considered Gaussian distributed for a significance level of 3%.

The results obtained by the proposed method in the four different studies have been compared to the ones obtained by using the leak sensitivity analysis with the angle

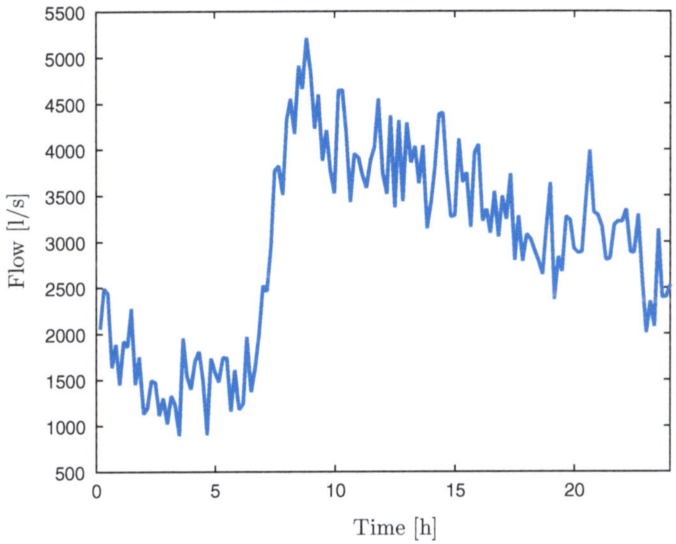

Fig. 10.7 Example of a daily Hanoi flow pattern

Table 10.2 Accuracy results in Hanoi network

Study	$N = 1$			$N = 24$		
	k-NN	Bayes	Angle	k-NN	Bayes	Angle
Leak uncertainty	60.21	83.60	76.61	77.41	83.87	77.41
Noise in measurements	69.62	83.19	73.79	83.87	83.87	70.96
Demand uncertainty	31.18	39.11	41.39	58.06	45.16	64.51
All together	32.12	48.25	36.96	74.19	83.87	54.83

metric proposed by Casillas et al. (2012) and summarized in Sect. 10.2. For the Angle method, only the two residuals are employed because the flow measurement has a great value and tends to reduce the effect of residuals in the diagnosis, thus resulting in worse results. The sensitivity matrix (10.2) has been computed by using (10.3) and by considering nominal leak conditions in every demand node ($\bar{v} = 0$, $\bar{d} = \hat{d}$ and $f_i = 50$ [l/s] $i = 1, ..., n_n$). The results obtained by using the Angle method and the two proposed methods (the k-NN classifier is tuned with k=1), in both cases considering only one sample ($N = 1$) and the equivalent number of samples of one day ($N = 24$) in the leak location diagnosis, are summarized in Table 10.2. The values presented in the table correspond to the overall accuracy Ac defined in (10.4).

It can be observed that the three methods exhibit good performances in the leak uncertainty case because of the linear directional variation of most of the residuals for this kind of uncertainty (Blesa et al. 2016). Note that in the case when only demand uncertainty is considered, the classifier-based methods perform worse than when all the uncertainties are considered together; this happens because the leak uncertainty

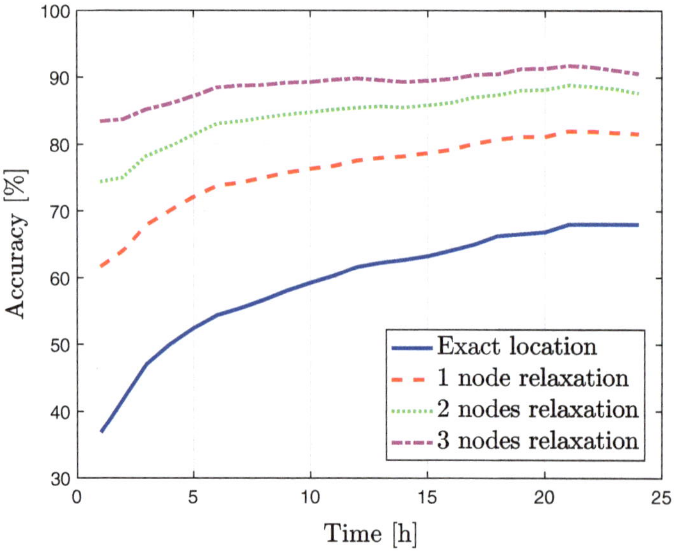

Fig. 10.8 Accuracy results over a time horizon for the k-NN classifier in the Hanoi network

spread the residual data providing a better separation. For the Bayesian classifier, the distribution tends to be more Gaussian.

When the time horizon and recursivity described in Sect. 10.3.4.1 are applied, we see that there is an improvement in the performance achieved in all uncertainty cases (except for the case of the noise uncertainty for the angle method, where the full performance is achieved since the first sample and then fluctuates within the time horizon around the same values).

The effect of the horizon length N in the performance (Accuracy) for the three studied methods is also analyzed by using the last study; to create the figures an extended data set (10 times larger) has been used. The results for the k-NN classifier are shown in Fig. 10.8, the results for the Bayesian classifier are shown in Fig. 10.9, and the results for the Angle method are shown in Fig. 10.10. The term "node relaxation" refers to the number of nodes in the topological distance between the node with the real leak and the node where the classifier predicts the leak for which the diagnosis is still considered correct.

As expected, the accuracy increases with the time horizon length N. It can be observed that it reaches a steady-state value when N is around 20 hours. This result justifies the use of a time horizon corresponding to one day, and it agrees with the results already presented by Casillas et al. (2012).

Finally, Fig. 10.11 shows a comparison of the three studied methods by using a different performance indicator, the average topological distance, which is the minimum distance in nodes between the node candidate and the node where the leak exists.

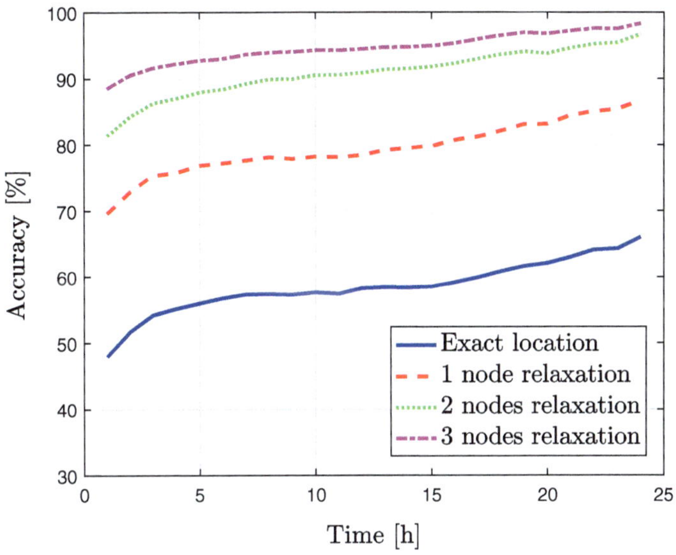

Fig. 10.9 Accuracy results over a time horizon for the Bayesian classifier in the Hanoi network

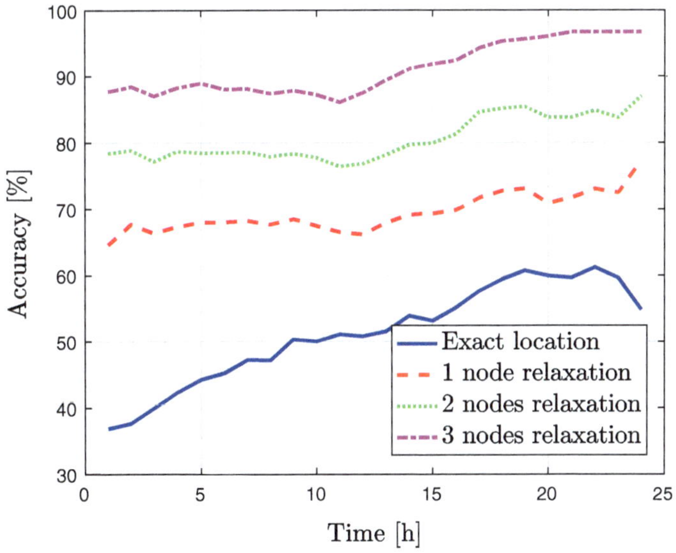

Fig. 10.10 Accuracy results over a time horizon for the Angle method in the Hanoi network

The results show the good performance of the classifiers, especially the Bayesian classifier, which works better than the k-NN classifier when the data have a clear distribution and a better reasoning over time. If not, the k-NN performs better as it can be seen in the Table 10.2 for the demand uncertainty case. In addition, in

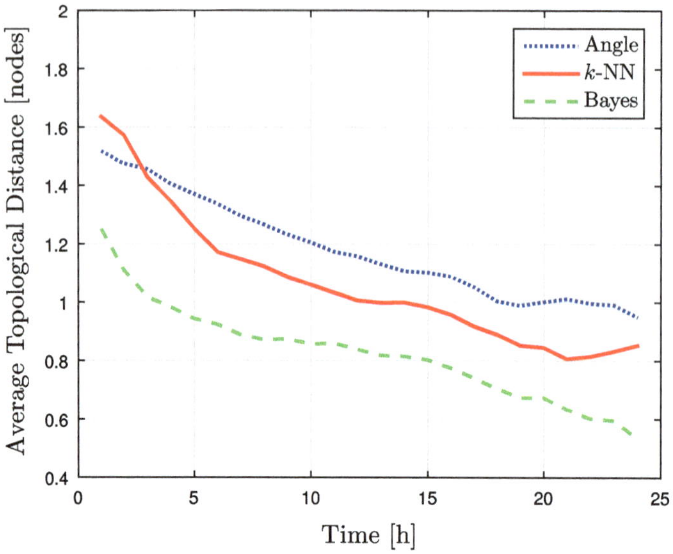

Fig. 10.11 Average topological distance results in a time horizon in the Hanoi network

Fig. 10.11, clearly the Bayesian classifier tends to point to a closer class when it fails as the k-NN classifier, but it can increase its performance at that point by choosing a bigger k value. In summary, the Bayesian classifier should be used when the classes present a Gaussian distribution, otherwise the k-NN classifier should be used.

10.4.2 Nova Icària Case Study

The Nova Icària network, shown in Fig. 10.12, is one of the DMA networks of the Barcelona WDN. This network consists of 1520 nodes, 1646 pipes, two reservoirs and two pressure valves, each one is located after the reservoirs with the aim of maintaining a certain pressure level. Inside the network, the pressures measured by five sensors installed at nodes 3, 4, 5, 6, and 7 are known, together with the flow entering the DMA and the set points for the pressure valves.

In this network a real case is studied. For this real case, experimental data captured under normal network operation and in the presence of a real leak is used. The leak was created when the water company operating the network opened a fire hydrant. The experiment took place on December 20, 2012 at 00:30 h and lasted around 30 hours with a leak size about 5.6 l/s, where the total demand of water in the network is in the range between 23.5 and 78 l/s approximately. Additionally, data captured in a normal operation scenario of five days before the leak scenario were also obtained. For more details, see Pérez et al. (2014). The sampling time of all data sensors is 10 minutes. In order to decrease the effect of uncertainties, the average value of every six

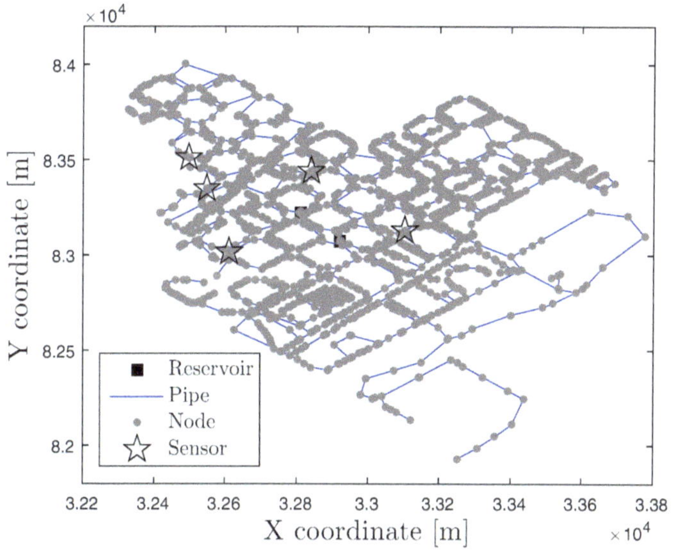

Fig. 10.12 Nova Icària topological network

samples has been considered every hour: 30 and 120 hourly samples are available for the leak and normal operation scenarios. An accurate Epanet model of the network and node demand estimations were provided as well.

First, the system has been simulated by considering the operating conditions of the fault-free scenario (input flow, boundary conditions and demand distributions). The differences between the 120 hourly samples of the five inner pressure sensors and the pressures estimated by the hydraulic model have been used to estimate the real uncertainty of the network (demand uncertainty, modeling errors and noise in the measurements). On the other hand, nominal hourly leak residuals $r_i^0(t), i = 1, \ldots, n_n, t = 1, \ldots, 24$ have been computed as the difference of the estimated pressures in the five inner sensors in a leak scenario and the ones estimated under the normal operation.

A k-NN classifier (with $k = 3$) has been trained for leak localization and validated. The inputs of the classifier are as follows: the five pressure residuals, the flow that enters the DMA, and the two set points of the valves. The data used in the training and testing stages are the 24 samples of nominal hourly residuals with real uncertainty added (120 samples): 96 samples for training and 48 for testing. The same training data sets generated help to calibrate the PDFs for the Bayesian classifier.

Figure 10.13 shows the result of the two proposed methods after applying 24 hourly samples: the k-NN classifier indicates that the leak is at node 3, while the real

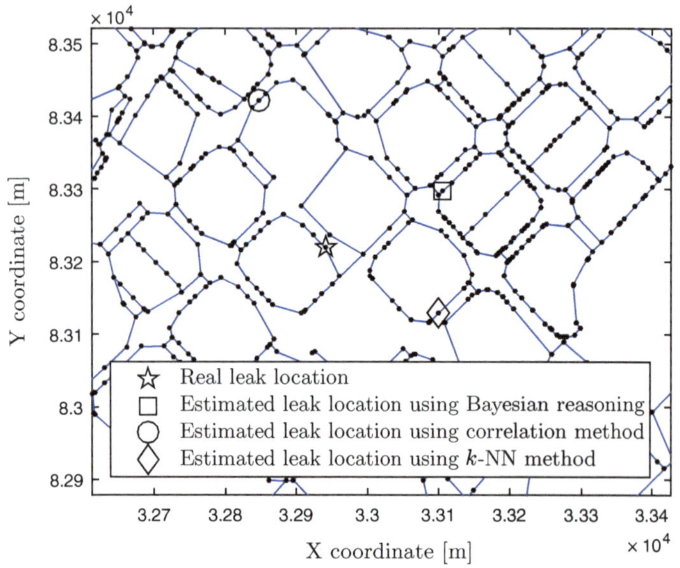

Fig. 10.13 Comparison of different leak location methods in the Nova Icària network

leak is at node 996. This means that the topological distance is 13 nodes, and the geographical linear distance is around 184 m. For the Bayesian classifier, the node candidate is 403, which has a topological distance of 10 nodes and a geographical linear distance of 183 m. As a comparison, the application of the correlation method (Pérez et al. 2014) provides as a node candidate node 1036, which is at a distance of 17 nodes and 222 m from the real leak location. This result is also shown in Fig. 10.13.

10.5 Conclusion

This chapter has proposed a new method for leak localization in WDNs that combines the use of pressure models with classifiers. A model of the considered WDN is used in a first stage to compute pressure residuals that are indicative of leaks. In a second stage, a classifier is applied to the obtained residuals with the aim of determining the leak location. This online scheme relies on a previous offline work in which the model is obtained and the classifier is trained with data generated in extensive simulations of the network under different leak conditions. These simulations consider leaks with different magnitudes in all the nodes of the network, differences between the estimated and the real water consumer demands and noise in measurements. The proposed method has been compared to previous leak localization methods described in the literature through its application to two DMA case studies of different size and complexity with satisfactory results.

Acknowledgements This work has been funded by the Ministerio de Economía, Industria y Competitividad (MEICOMP) of the Spanish Government and FEDER through the projects ECOCIS (ref. DPI2013-48243-C2-1-R) and HARCRICS (ref. DPI2014-58104-R) and through the grant IJCI-2014-20801, by the European Commission through contract EFFINET (ref. FP7-ICT2011-8-31 8556) and by the Catalan Agency for Management of University and Research Grants (AGAUR), the European Social Fund (ESF) and the Secretary of University and Research of the Department of Companies and Knowledge of the Government of Catalonia through the grant FI-DGR 2015 (ref. 2015 FI_B 00591).

References

Alpaydin, E. (2010). *Introduction to machine learning*. Cambridge: MIT Press.

Blanke, M., Kinnaert, M., Lunze, J., & Staroswiecki, M. (2006). *Diagnosis and fault-tolerant control*. Berlin/Heidelberg: Springer.

Blesa, J., Nejjari, F., & Sarrate, R. (2016). Robust sensor placement for leak location: Analysis and design. *Journal of Hydroinformatics*, *18*(1), 136–148.

Blesa, J., Puig, V., & Saludes, J. (2012). Robust identification and fault diagnosis based on uncertain multiple input multiple output linear parameter varying parity equations and zonotopes. *Journal of Process Control*, *22*(10), 1890–1912.

Casillas, M. V., Garza-Castañón, L. E., & Puig, V. (2012). Extended-horizon analysis of pressure sensitivities for leak detection in water distribution networks. In *8th IFAC symposium on fault detection, supervision and safety of technical processes* (pp. 570–575). Amsterdam: Elsevier.

Casillas, M. V., Garza-Castañón, L. E., & Puig, V. (2014). Model-based leak detection and location in water distribution networks considering an extended-horizon analysis of pressure sensitivities. *Journal of Hydroinformatics*, *16*(3), 649–670.

Casillas, M. V., Puig, V., Garza-Castañón, L. E., & Rosich, A. (2013). Optimal sensor placement for leak location in water distribution networks using genetic algorithms. *Sensors*, *13*(11), 14984–15005.

Colombo, A. F., Lee, P., & Karney, B. W. (2009). A selective literature review of transient-based leak detection methods. *Journal of hydro-environment research* (pp. 212–227).

Covas, D., & Ramos, H. (2001). Hydraulic transients used for leak detection in water distribution systems. In *4th International Conference on Water Pipeline Systems, BHR Group* (pp. 227–241).

Cugueró-Escofet, M. À., García, D., Quevedo, J., Puig, V., Espin, S., & Roquet, J. (2016). A methodology and a software tool for sensor data validation/reconstruction: Application to the Catalonia regional water network. *Control Engineering Practice*, *49*, 159–172.

Cugueró-Escofet, P., Blesa, J., Pérez, R., Cugueró-Escofet, M. À., & Sanz, G. (2015). Assessment of a leak localization algorithm in water networks under demand uncertainty. *IFAC Proceedings Volumes (IFAC-PapersOnline)*, *48*(21), 226–231.

Daniel, W. W. (1990). *Applied nonparametric statistics*. Boston: PWS-Kent.

Ferrante, M., & Brunone, B. (2003a). Pipe system diagnosis and leak detection by unsteady-state tests. 1. Harmonic analysis. *Advances in Water Resources*, *26*(1), 95–105.

Ferrante, M., & Brunone, B. (2003b). Pipe system diagnosis and leak detection by unsteady-state tests. 2. Wavelet analysis. *Advances in Water Resources*, *26*(1), 107–116.

Fuchs, H. V., & Riehle, R. (1991). Ten years of experience with leak detection by acoustic signal analysis. *Applied Acoustics*, *33*, 1–19.

Kepler, A., Covas, D., & Reis, L. (2011). Leak detection by inverse transient analysis in an experimental *PVC* pipe system. *Journal of Hydroinformatics*, *13*(2), 153–166.

Khulief, Y., Khalifa, A., Mansour, R., & Habib, M. (2012). Acoustic detection of leaks in water pipelines using measurements inside pipe. *Journal of Pipeline Systems Engineering and Practice*, *3*(2), 47–54.

Lambert, A. (2001). What do we know about pressure: Leakage relationships in distribution systems? In *IWA Conference System Approach to leakage control and water distribution system management. Brno, Czech Rebublic.*

Lambert, M. F., Simpson, A. R., Vítkovský, J. P., Wang, X. J., & Lee, P. J. (2003). A review of leading-edge leak detection techniques for water distribution systems. In *20th AWA Convention, Perth, Australia.*

Mashford, J., De Silva, D., Marney, D., & Burn, S. (2009). An approach to leak detection in pipe networks using analysis of monitored pressure values by support vector machine. In *3rd International Conference on Network and System Security* (pp. 534–539).

Muggleton, J. M., Brennan, M. J., & Pinnington, R. J. (2002). Wavenumber prediction of waves in buried pipes for water leak detection. *Journal of Sound and Vibration*, *249*, 939–954.

Pérez, R., Puig, V., Pascual, J., Quevedo, J., Landeros, E., & Peralta, A. (2011). Methodology for leakage isolation using pressure sensitivity analysis in water distribution networks. *Control Engineering Practice*, *19*(10), 1157–1167.

Pérez, R., Sanz, G., Puig, V., Quevedo, J., Nejjari, F., Meseguer, J., et al. (2014). Leak localization in water networks. *IEEE Control Systems Magazine*, *34*(4), 24–36.

Pudar, R. S., & Liggett, J. A. (1992). Leaks in pipe networks. *Journal of Hydraulic Engineering*, *118*(7), 1031–1046.

Puust, R., Kapelan, Z., Savić, D. A., & Koppel, T. (2010). A review of methods for leakage management in pipe networks. *Urban Water Journal*, *7*(1), 25–45.

Ragot, J., & Maquin, D. (2006). Fault measurement detection in an urban water supply network. *Journal of Process Control*, *16*, 887.

Rossman, L. A. (2000). *EPANET 2 Users Manual*. United States Environmental Protection Agency.

Sarrate, R., Blesa, J., Nejjari, F., & Quevedo, J. (2014). Sensor placement for leak detection and location in water distribution networks. *Water Science and Technology: Water Supply*, *14*(5), 795–803.

Savić, D. A., Kapelan, Z., & Jonkergouw, P. (2009). Quo vadis water distribution model calibration? *Urban Water Journal*, *6*(1), 3–22.

Soldevila, A., Blesa, J., Tornil-Sin, S., Duviella, E., Fernandez-Canti, R. M., & Puig, V. (2016). Leak localization in water distribution networks using a mixed model-based/data-driven approach. *Control Engineering Practice*, *55*, 162–173.

Stephens, M. A. (1974). EDF statistics for goodness of fit and some comparisons. *Journal of the American statistical Association*, *69*(347), 730–737.

Thornton, J., & Lambert, A. (2005). Progress in practical prediction of pressure: leakage, pressure: Burst frequency and pressure: Consumption relationships. In *Conference Proceedings of Leakage (2005), Halifax, Canada.*

Todini, E., & Pilati, S. (1988). A gradient algorithm for the analysis of pipe networks. *Computer applications in water supply: Vol. 1—systems analysis and simulation* (pp. 1–20). Taunton, UK: Research Studies Press Ltd.

Wachla, D., Przystalka, P., & Moczulski, W. (2015). A method of leakage location in water distribution networks using artificial neuro-fuzzy system. *IFAC-PapersOnLine*, *48*(21), 1216–1223.

Wu, Z. Y., & Sage, P. (2006). Water loss detection via genetic algorithm optimization-based model calibration. In *Systems analysis symposium* (pp. 1–11). ASCE.

Yang, J., Wen, Y., & Li, P. (2008). Leak location using blind system identification in water distribution pipeline. *Journal of Sound and Vibration*, *310*, 134–148.

Chapter 11
Sensor Placement for Classifier-Based Leak Localization in Water Distribution Networks

Adrià Soldevila, Joaquim Blesa, Sebastian Tornil-Sin, Rosa M. Fernandez-Canti and Vicenç Puig

Abstract This chapter presents a sensor placement method for the classifier-based approaches for leak localization in water distribution networks introduced in the previous chapter. The proposed approach formulates the sensor placement problem as a binary optimization problem. Because of the complexity of the problem, it is solved by applying genetic algorithms. In order to reduce the number of sensor configurations to test, a binary matrix that identifies pairs of sensors providing similar information is added as a constraint. The sensors are placed in an optimal way, which maximizes the accuracy of the leak localization. The proposed approach is first illustrated by means of the application to an academic example based on the reduced version of the Hanoi water distribution network. A more realistic case study is then proposed based on the Limassol district metered area.

11.1 Introduction

As already discussed in the previous chapter, leak detection and localization in water distribution networks (WDNs) is a subject of major concern for water companies. In the case of complex urban WDNs, this is not an easy task to handle. In order to manage the leak problem, and other problems such as pressure control, modern

A. Soldevila · S. Tornil-Sin · R.M. Fernandez-Canti · V. Puig (✉)
Research Center for Supervision, Safety and Automatic Control (CS2AC), Rambla Sant Nebridi, S/n, 08022 Terrassa, Spain
e-mail: vpuig@iri.upc.edu

A. Soldevila
e-mail: adria.soldevila@upc.edu

S. Tornil-Sin
e-mail: stornil@iri.upc.edu

R.M. Fernandez-Canti
e-mail: rosa.mari.fernandez@upc.edu

J. Blesa · S. Tornil-Sin · V. Puig
Institut de Robòtica i Informàtica Industrial (CSIC-UPC), Carrer Llorens Artigas, 4-6, 08028 Barcelona, Spain
e-mail: jblesa@iri.upc.edu

C. Verde and L. Torres (eds.), *Modeling and Monitoring of Pipelines and Networks*, Applied Condition Monitoring 7,
DOI 10.1007/978-3-319-55944-5_11

urban WDNs are usually divided into district metered areas (DMAs), where the *flow* and the *pressure* at the input are measured (Lambert et al. 2003; Puust et al. 2010). Leakages increase the flow and decrease the pressure measurements at the DMA entrance. Leak detection and localization, however, are not trivial tasks because of unpredictable variations in consumer demands and measurement noise, as well as long-term trends and seasonal effects. Leak detection can be implemented by means of the analysis of the minimum night flow in the DMAs, that can also provide an estimation of the leakage level (Puust et al. 2010). Leak localization usually requires the analysis of more than one measured variable, however, and it is a more complex problem. Regarding the type of sensors, although the use of flow measurements is feasible in large water supply networks, this is not the case in WDN, where there is a dense mesh of pipes with only flow measurements at the entrance of each DMA. In this case, water companies consider as a feasible approach the possibility of installing some pressure sensors inside the DMA, because they are less expensive and easier to install and maintain.

In the previous chapter, a classifier-based leak localization architecture and an associated methodology applicable to WDNs is proposed. In the first stage of the proposed architecture, residuals are obtained by comparing available pressure measurements with the estimations provided by a WDN hydraulic model. In a second stage, a classifier is applied to the residuals with the aim of determining the leak location. The classifier is trained with data generated by a simulation of the WDN under different leak scenarios and uncertainty conditions. Several classification approaches are considered and compared. As discussed in the previous chapter, in the last years, several techniques have been proposed for leak localization purposes, such as transient analysis, parameter estimation techniques, leak sensitivity analysis, and artificial intelligence methods. Among them, artificial intelligence methods relying on classifiers seem to be a suitable option to deal with the problem of the uncertainty in WDNs.

The problem of optimal sensor placement in WDN was first studied for contaminant detection (Berry et al. 2003; Shastri and Diwekar 2005). In recent years, some optimal pressure sensor placement algorithms have been developed for determining which pressure sensors must be installed inside the DMA such that, with minimum economical costs (the number of sensors), a suitable performance regarding leak localization is guaranteed. The main problem of optimal pressure sensor placement is that it leads to a combinatory optimization problem being infeasible to solve by evaluating all possible sensor locations. In order to solve this combinatory problem, the use of genetic algorithms (GAs) has been proposed by Pérez et al. (2009) considering a binary leak sensitivity matrix. In Casillas et al. (2013), GAs are used to solve an integer optimization problem based on projections between residuals and the nonbinarized leak sensitivity matrix. In Cugueró-Escofet et al. (2015a), the approach of Casillas et al. (2013) was extended to consider a relaxed isolation index that takes into account an acceptable isolation distance. More recently, the method proposed by Casillas et al. (2013) has been extended by Steffelbauer and Fuchs-Hanusch (2016) by considering uncertainty.

Alternatively, the use of clustering analysis to group sensors with similar behavior, reducing the number of combinations to be evaluated, is proposed by Sarrate et al. (2014b), combined with an efficient branch and bound search. In Blesa et al. (2016), the model uncertainties are considered in the selection. In Nejjari et al. (2015), the sensor placement has the aim of reducing the isolation error distance. More recently, in Perelman et al. (2016) the optimal sensor placement problem for the leak localization in WDNs is formulated as a minimum test cover problem.

In this chapter, a sensor placement method for the classifier-based approaches for leak localization in WDNs introduced in previous the chapter is presented. Given a number of pressure sensors to be installed in the demand nodes of a DMA, the proposed approach provides the locations of the sensors that maximize the accuracy of a leak localization method that combines the use of pressure models with classifiers (see Soldevila et al. (2016) and the previous chapter). The proposed method requires data generated from extensive network simulations. These simulations consider leaks with different magnitudes in all the nodes of the network, differences between the estimated and real consumer water demands, and noise in pressure sensors for all the operating points. Therefore, the presence of model uncertainty is considered in the sensor placement method. Every sensor configuration determines the data that will be used to train the classifier used for the leak localization task. In order to tackle the combinatorial number of sensor configurations to be considered, the use of GAs in combination with a sensor distance matrix constraint is proposed for obtaining the optimal placement.

11.2 Background

11.2.1 Leak Localization Using Pressure Residuals and Classifiers

In the previous chapter, an online leak localization method based on computing pressure residuals r and analyzing them by a classifier is proposed; see Fig. 11.1. Residuals are computed as differences between pressure measurements $\tilde{p}$ provided by pressure sensors installed inside the DMA and pressure estimations $\hat{p}_0$ provided by a hydraulic model simulated under leak-free conditions. It is assumed that the network has n_n nodes and that only a limited number of sensors can be installed according to budget constraints such that $n_s << n_n$. The WDN model is built by using a hydraulic simulator such as Epanet, and it is assumed to be able to accurately represent the WDN behavior after the corresponding calibration process by using real data. Note, however, that the model is fed with estimated water demands in the nodes $(\hat{d}_1, \ldots, \hat{d}_{n_n})$ since in practice real nodal demands $(d_1, \ldots, d_{n_n})$ are not measured except for some particular consumers where automatic metering readers (AMRs) are available. The estimated water demands are typically obtained by the total measured DMA demand $\tilde{d}_{WDN}$ and distributed at the nodal level according to

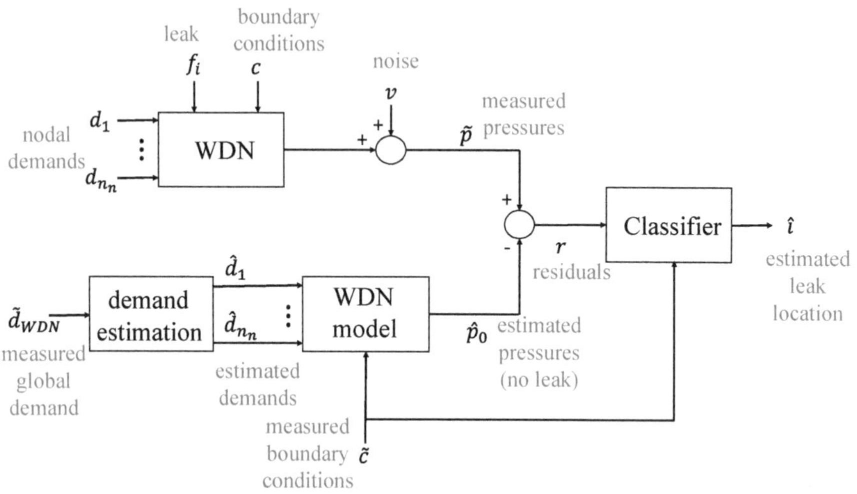

Fig. 11.1 Leak localization scheme

historical consumption records. Hence, the residuals are not only sensitive to leaks but also to the differences between the real demands and their estimated values. Additionally, pressure measurements are subject to the effect of sensor noise $\boldsymbol{v}$, and this also affects the residuals. Taking all these effects into account, the classifier must be able to locate the real leak present in the DMA, which can be in any node and with any (unknown) magnitude, while being robust to the demand uncertainty and the measurement noise. Finally, note that the operation of the network is constrained by some boundary conditions (for instance, the position of internal valves and reservoir pressures) that are known (measured). These conditions are taken into account in the simulation and can also be used as inputs for the classifier.

11.2.2 Data Generation

The application of the architecture described above (Fig. 11.1) relies on an offline work with the main goal of training and validating a classifier that can distinguish the potential leaks under the described uncertainty conditions. In this process, the data generation stage is critical. Since the data that can be obtained from the real monitored WDN can be really limited, and inexistent for the particular sensor placement case the way to attain a complete training data set is by using the hydraulic simulator. Hence, training (and validation and testing) data are generated by applying the scheme depicted in Fig. 11.2, similar to the one presented in Fig. 11.1 but with the main difference of substituting the real WDN for a model that allows simulating the WDN not only in the absence but also in the presence of faults.

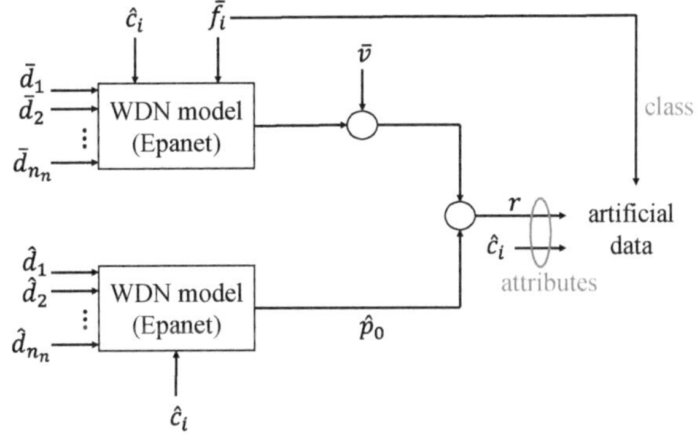

Fig. 11.2 Data generation scheme

The presented scheme is used to undertake the following:

- Generate data for all possible leak locations at every potential sensor placement i.e., for all the different nodes in the WDN ($\bar{f}_i, \quad i = 1, 2, \ldots, n_n$).
- Generate data for each possible leak location and magnitude inside a given range ($\bar{f}_i \in [f_i^-, f_i^+]$).
- Generate sequences of demands ($\bar{d}_1, \ldots, \bar{d}_{n_n}$) and boundary conditions $\hat{c}_i$ that correspond to realistic, typical daily evolution in each node.
- Simulate differences between the real demands and the estimations computed by the demand estimation module $\left((\bar{d}_1, \ldots, \bar{d}_{n_n}) \neq (\hat{d}_1, \ldots, \hat{d}_{n_n})\right)$.
- Take into account the measurement noise in pressure sensors, by generating synthetic Gaussian noise ($\bar{\nu}$).

We emphasize that the hydraulic model computes the internal pressures at all the network nodes and that the presented data generation scheme allows the generation of a complete data set that can be analyzed to determine which pressure measurements are more useful for leak localization purposes.

11.2.3 Classifier Evaluation

As in the previous chapter, to evaluate the trained classifier for a given sensor configuration, the *confusion matrix* Γ can be computed, which summarizes the results obtained when the classifier is applied to a validation data set. We should recall that when applied to the leak localization problem, the confusion matrix is a square matrix with as many rows and columns as nodes in the network (potential leak locations), where each coefficient $\Gamma_{i,j}$ indicates how many times a leak in node i is recognized as a leak in node j, see Table 11.1.

Table 11.1 Confusion matrix Γ

	$\hat{f}_1$	$\cdots$	$\hat{f}_i$	$\cdots$	$\hat{f}_{n_n}$
f_1	$\Gamma_{1,1}$	$\cdots$	$\Gamma_{1,i}$	$\cdots$	Γ_{1,n_n}
$\vdots$	$\vdots$	$\vdots$	$\vdots$	$\vdots$	$\vdots$
f_i	$\Gamma_{i,1}$	$\cdots$	$\Gamma_{i,i}$	$\cdots$	Γ_{i,n_n}
$\vdots$	$\vdots$	$\vdots$	$\vdots$	$\vdots$	$\vdots$
f_{n_n}	$\Gamma_{n_n,1}$	$\cdots$	$\Gamma_{n_n,i}$	$\cdots$	Γ_{n_n,n_n}

In case of a perfect classification, the confusion matrix should be diagonal, with $\Gamma_{i,i} = m$, for all $i = 1, \ldots, n_n$ being m the size of the validation (or testing) data set. In practice, nonzero coefficients will appear outside the main diagonal. For a leak in node i, the coefficient $\Gamma_{i,i}$ indicates the number of times that the leak f_i is correctly identified, while $\sum_{j=1}^{n_n} \Gamma_{i,j} - \Gamma_{i,i}$ indicates the number of times that it is wrongly classified. The overall accuracy (Ac) of the classifier is defined as

$$Ac = \frac{\sum_{i=1}^{n_n} \Gamma_{i,i}}{\sum_{i=1}^{n_n} \sum_{j=1}^{n_n} \Gamma_{i,j}}. \tag{11.1}$$

11.3 Sensor Placement Problem

11.3.1 Problem Formulation

As already discussed, the objective of this chapter is to develop an approach to place a given number of sensors (n_s) in a DMA of a WDN to obtain a sensor configuration with a maximized leak isolability performance when using the leak localization method scheme presented in the previous section. This problem can be recast into the feature selection (also known as variable or attribute selection) problem (Tang et al. 2014). The solution of this problem aims at selecting a subset of relevant variables (sensor measurements) for use in the classifier construction to maximize its performance. A feature selection algorithm combines a search technique for proposing new variable subsets, along with an evaluation measure which scores the different feature subsets to obtain the best subset of features. The simplest algorithm is to test each possible subset of features finding the one which minimizes the error rate. This is an exhaustive search of the space, however, this is computationally intractable except for small feature sets.

To select a configuration with n_s sensors, the following binary vector is defined

$$q = \left[q_1, \cdots, q_{n_n} \right], \tag{11.2}$$

where $q_i = 1$ if the pressure in the node i is measured and $q_i = 0$ otherwise. The vector q indicates which sensors are installed.

In order to evaluate the quality of a sensor configuration regarding its ability to locate a leak at node $i \in \{1, \ldots, n_n\}$, assuming the case of a single leak, a performance index based on the classifier accuracy (11.1) is proposed.

This performance index depends on the considered sensor configuration parameterized in terms of the binary variable q to determine the best selection

$$Ac(q) = \frac{\sum_{i=1}^{n_n} \Gamma_{i,i}(q)}{\sum_{i=1}^{n_n} \sum_{j=1}^{n_n} \Gamma_{i,j}(q)}. \tag{11.3}$$

Note that for a given sensor configuration q, $100Ac(q)$ is the percentage of correctly located leaks.

Based on the vector q and the performance index $Ac(q)$, the sensor placement problem can be translated into an optimization problem formulated as follows

$$\max_q Ac(q) \qquad \text{s.t.} \qquad \sum_{i=1}^{n_n} q_i = n_s, \tag{11.4}$$

where q is defined in (11.2) and $n_s \in \{1, \ldots, n_n\}$ is the number of sensors that we want to place.

11.3.2 Sensor Placement Using Genetic Algorithms

The optimal sensor placement problem, formulated as the classifier feature selection problem described in the previous section, is solved by using genetic algorithms.

The overall procedure can be seen in the Fig. 11.3, where the "*Nominal Sensitivity Matrix*" is a data set containing only residuals without uncertainties.

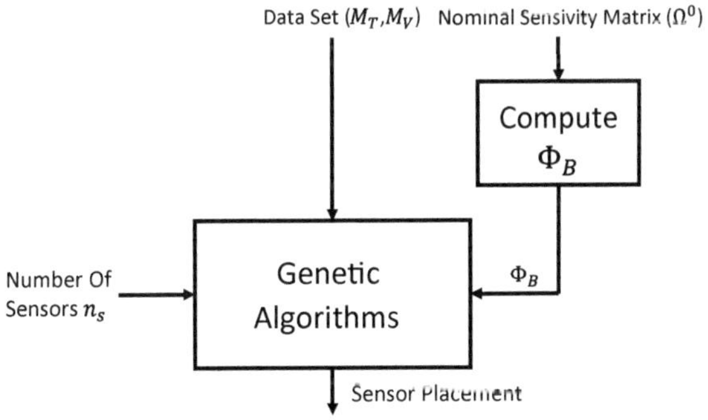

Fig. 11.3 Scheme of optimization process

11.3.2.1 Sensor Distance Matrix

In order to reduce the amount of sensor configurations to be tested in the GA heuristic search, sensor configurations (defined by q) that have at least a pair of sensors with similar behavior in the residual space can be discarded. To measure the different behavior of a pair of sensors in the residual space, the leak sensitivity matrix defined in the previous chapter can be approximately generated by simulation for a given operating point defined by a nominal network inflow (unique value of water consumption), nominal demand distribution (fixed nodal demand consumption, $\bar{d}$ in this case) and nominal leak size (f^0) (Blesa et al. 2014), approximating the sensitivity components $\Omega_{i,j}$ by

$$\Omega^0_{i,j} = \frac{\hat{p}^0_{i,f_j} - \hat{p}^0_{i,0}}{f^0}, \tag{11.5}$$

where superscript ^{0} denotes the nominal conditions mentioned before.

A criterion that can be used for determining the similarity between sensors is based on comparing the rows of the nominal sensitivity matrix as proposed by Sarrate et al. (2014a). If we consider a nominal sensitivity matrix using the following notation

$$\boldsymbol{\Omega}^0 = \begin{pmatrix} s^0_1 \\ \vdots \\ s^0_{n_n} \end{pmatrix}, \tag{11.6}$$

where s^0_i $i = 1, \ldots, n_n$ are row vectors

$$s^0_i = \left(\Omega^0_{i,1}, \ldots, \Omega^0_{i,n_n}\right), \tag{11.7}$$

with components computed by using (11.5), a sensor distance matrix Φ can be defined as

$$\Phi_{i,j} = \|s^0_i - s^0_j\|_1 \qquad \forall i = 1, \ldots, n_n \text{ and } j = 1, \ldots, n_n. \tag{11.8}$$

Φ is a symmetric square matrix of dimension n_n and diagonal 0. A threshold σ can be determined in order to decide whether two sensors display different behavior in the residual space or not. A binary matrix $\Phi^{(B)}$ that collects the information of which pairs of sensor combinations are suitable for a sensor configuration or not according to their dissimilarity can then be computed as

$$\Phi^{(B)}_{i,j} = \begin{cases} 0, \text{ if } \left|\Phi_{i,j}\right| < \sigma \\ 1, \text{ if } \left|\Phi_{i,j}\right| \geq \sigma \end{cases}. \tag{11.9}$$

11.3.2.2 Sensor Placement Algorithm

The optimization problem (11.4) solved by the genetic algorithms has as objective function to optimize the accuracy defined in (11.1). The accuracy will be assessed after the classifier training process has ended by using a validation data set as described in Sect. 11.2. A training matrix M_T and a validation matrix M_V with data from all the candidate sensor locations will be provided to the sensor placement algorithm. For every sensor placement solution, the accuracy obtained by using the training and validation data sets corresponding to the selected sensors will be evaluated.

The training matrix M_T has $n_n + l + 1$ columns where the first n_n columns are the node pressure measurements, the next l columns are the l added attributes (e.g., the measurement of the total water inflow) used by the classifier, and the last column corresponds to the label where each data scenario belongs. The number of rows corresponds to the total number of data scenarios used to train the classifier. The validation matrix M_V has the same column format $(n_n + l + 1)$, and the number of rows corresponds to the total number of data scenarios used for validation purposes.

The pseudocode of the algorithm is shown in Algorithm 1. First, we initialize the variables of the GA (line 1), including the bit string type population, the tolerance, the population size p_s and the elite count in order to save part of the previously analyzed results. We then declare the search constraints (line 2), which are n_s, as the constraint of the set of possible solutions for each variable and the number of sensors. In the optimization process (lines 4 to 24), an initial matrix with random sensor positions is delivered by the GA (line 6). Before proceeding to the objective function optimization, it is checked (with the function GetUsed() in line 9) if the sensor configuration has already been considered earlier (as proposed by Oh et al. (2004)). The stored value is retrieved with the function GetAc() (line 20). If not yet considered, the sensor configuration is considered, if the pairs of sensors are allowed, to be tested. If the new sensor configuration is not yet tested, and if all the sensor pairs are suitable for a sensor configuration according to the binary sensor distance matrix (11.9) and the function CheckCombinations() (which returns 1 if the configuration is allowed (line 10)), the sensor placement configuration is tested by evaluating the objective function (line 13), and the combination is stored as used with the function SetUsed() (line 14). The Ac value obtained is then also stored with the function SetAc() (line 15). If there is at least one forbidden pair of sensors in the sensor configuration, the configuration is discarded, and a zero value is assigned to the objective function, and also stored with the combination (line 17). The binary vector q allows the selection of the adequate columns of the matrices M_T and M_V in order to train (line 11), validate (line 12) and compute Ac (line 13) for the classifier according to the selected nodes to be measured. Once the Ac value has been obtained for all members of the matrix I, the maximum value is obtained (line 23). If some stopping criteria is fulfilled (maximum number of generations or improvement less than the fitness function tolerance), then the optimization is finished, and the sensor placement selected is the one that provides the best Ac value, if not a new generation is created using the best Ac value.

Algorithm 1 Sensor placement based on Genetic Algorithms

Require: A training matrix M_T and a validation matrix M_V. The number of features to select n_s, the number of nodes n_n, the population size p_s and the binarized matrix $\Phi^{(B)}$.
Ensure: A near-optimal sensors configuration q with error index Ac_{max}.

$init \leftarrow$ InitVarGA()
$constraint \leftarrow$ SetConstraints(n_s)
Inputs: $init, constraint, M_T, M_V, p_s, n_n, \Phi^{(B)}$.
while An optimization criterion is not reached **do**
  GA based search:
  Generate I matrix of size $(p_s \times n_n)$ where each row is a member of a generation.
  for $k = 1, \cdots, p_s$ **do**
    $q(k) \leftarrow I(k)$
    if GetUsed(q(k)) $= 0$ **then**
      if CheckCombinations($\Phi^{(B)}, q(k)$) $= 1$ **then**
        $C(q(k)) \leftarrow$ Train($M_T(q(k))$)
        Γ(q(k)) $\leftarrow$ Validate($C(q(k)), M_V(q(k))$)
        $Ac(q(k)) \leftarrow \frac{\sum_{i=1}^{n_n} \Gamma_{ii}(q(k))}{\sum_{i=1}^{n_n}\sum_{j=1}^{n_n} \Gamma_{ij}(q(k))}$
        SetUsed($q(k)$)
        SetAc($Ac(q(k), q(k)$)
      else
        $Ac(q(k)) = 0$
      end if
    else
      $Ac(q(k)) =$ GetAc($q(k)$)
    end if
  end for
  Find $\{q, Ac_{max}\}$ such that $Ac_{max} = \max_q(Ac(q(1), \ldots, Ac(q(p_s)))$.
end while

11.4 Case Studies and Results

The proposed sensor placement approach is tested in two different networks. On the one hand, a small network (Hanoi) is used since it allows the comparison between the proposed approach and the results obtained by using the exhaustive search method. On the other hand, a medium-sized network (Limassol) shows the performance in a more realistic scenario.

All the results have been obtained by using a PC with an INTEL(R) CORE(TM) i7-4720HQ CPU @ 2.60 [GHz], 8 [GB] of memory RAM, a Windows 10 Home 64-bits operating system and by using the MATLAB 2015a software (MAT 2015). The sensor placement approach follows Algorithm 1 and uses MATLAB GA Toolbox, which for the examined case studies has been used considering the following parameters:

- Fitness function tolerance: 10^{-6}.
- Population size: $p_s = 5$.
- Elite count: $0.05p_s$, but considering that at least one survives (which is the case given with the p_s selected).
- Maximum number of generations: 50.

11.4.1 Hanoi Case Study

The Hanoi (Vietnam) network, presented in Fig. 11.4, is a simplified network of the real one and consists of one reservoir, 31 consumer nodes and 36 pipes. The water consumption has a daily pattern similar to the one depicted in Fig. 11.5; all the water consumption patterns have been generated from an unique pattern distribution obtained from the average values of five days adding an uncertainty of ±12.5%. Given the size of the network, the placement of only two pressure sensors is considered, as presented by Casillas et al. (2013). These sensors and the flow sensor at the inlet are considered to operate with a sampling time of 10 min.

To generate the data sets, three different uncertainty sources (as proposed by Cugueró-Escofet et al. (2015b)) are considered in the following way:

- The demand uncertainty source has a magnitude of ±10% of the nominal node consumption value.

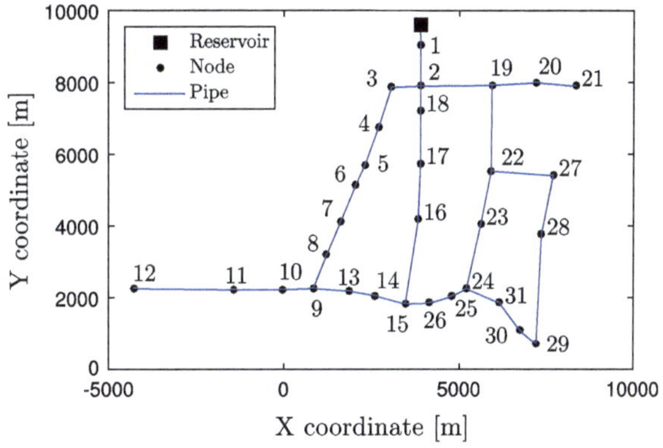

Fig. 11.4 Hanoi topological network

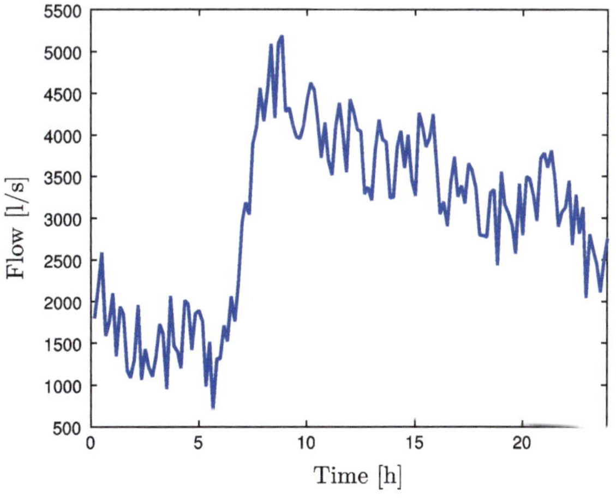

Fig. 11.5 Example of a typical daily flow consumption in the Hanoi network

- The leak size varies from 25 to 75 l/s.
- The measurement noise magnitude is considered as ±5% of the average value of all pressure residuals.

By using all of these uncertainty levels, 10 complete data sets (in order to have a large enough number of different scenarios to avoid outliers or strange cases) have been created by simulating the pressure measurements at each possible position (class) where data are generated every 10 min. The hourly average value has then been computed with the aim of reducing the uncertainty and removing outliers. Thus, each complete data set is composed of a training data set with five days of data (120 samples for each class) and a validation data set with one day of data (24 samples for each class). Finally, a unique testing data set with 10 days of data (240 samples per class) is generated. For the nominal sensitivity matrix (11.5), one instance is generated for each class and sensor (complete nominal sensitivity matrix) with a value of total consumption of water of 2991.1 l/s and leak size of 50 l/s.

Classifiers use as attributes the flow measurement at the inlet and the two pressure residuals from the nodes where the sensor configuration is assessed. The proposed sensor placement method using GA and with/without the $\Phi^{(B)}$ matrix, where a σ value of the average value of all the Φ except the diagonal is used, is compared to the exhaustive search. The results for the k-nearest neighbors (k-NN) classifier for a k value equal to one, since a proper k value must be selected when the sensor placement is fixed, are summarized in Table 11.2. The results for the case of the Bayesian classifier with variable dependent Gaussian probability density functions (PDFs) are presented in Table 11.3. The genetic algorithms are designed to store only the best member of each generation, and each generation with size equal to the population size is fixed to have five members. To compute the Ac value in both tables, the same testing data set is used for each sensor placement obtained (with their respective training data set). The time units in the tables are seconds, and the Ac values are in %. The best configurations (highest accuracy performance over the testing data set) obtained are highlighted in bold for each method.

From these results, clearly both methods present an important improvement in terms of computational time when GAs are used, and the GA stand-alone method and GA plus $\Phi^{(B)}$ are able to avoid the local minima and find the global optima in some cases, i.e., the best sensor placements are attained. Moreover, notice that on average the introduction of the $\Phi^{(B)}$ matrix not only significantly reduces the computational time compared to the purely GA method but also increases the accuracy. Finally, compared to the k-NN classifier, the Bayesian classifier is more time-consuming but its accuracy is better. This is probably because the allowed combinations are better: the criteria used to select the permitted pairs of sensor configurations work better for this classifier than for the k-NN classifier.

To decide the best sensor configuration, the one with the highest accuracy value is chosen. Therefore, for the technique of the genetic algorithms plus the use of $\Phi^{(B)}$ matrix, in case of the k-NN classifier, the best sensor placement obtained is at nodes 14 and 28. On the other hand, for the Bayesian classifier, the best sensor placement is

Table 11.2 Sensor placement results in the Hanoi network for the k-NN classifier

Data set	Exhaustive search			Genetic algorithm			Genetic algorithm + $\Phi^{(B)}$			
	Sensors	Time	Ac	Sensors	Time	Ac	Sensors	Time filter	Time GA	Ac
1	14, 27	474	39.11	9, 15	71	31.19	14, 27	0.06	63	39.11
2	14, 29	471	38.14	14, 29	70	38.14	14, 29	0.06	38	38.68
3	14, 28	470	38.68	14, 28	47	38.68	14, 31	0.06	19	34.34
4	**14, 28**	**499**	**41.06**	**14, 28**	**44**	**41.06**	**14, 28**	**0.06**	**33**	**41.06**
5	14, 28	472	39.03	1, 30	14	16.80	14, 28	0.06	46	39.03
6	14, 27	473	38.02	10, 15	45	32.58	14, 27	0.06	36	36.07
7	15, 28	473	38.52	15, 28	55	38.52	15, 28	0.06	43	38.52
8	15, 28	477	38.02	26, 28	29	35.55	15, 28	0.06	23	38.02
9	14, 27	469	38.89	5, 14	50	31.16	14, 29	0.06	36	37.56
10	14, 28	474	38.91	15, 29	51	36.88	4, 15	0.06	15	29.04
Average	–	475	38.83	–	47	34.05	–	0.06	35	37.14

Table 11.3 Sensor placement results in the Hanoi network for the Bayesian classifier

Data set	Exhaustive search			Genetic algorithm			Genetic algorithm + $\Phi^{(B)}$			
	Sensors	Time	Ac	Sensors	Time	Ac	Sensors	Time filter	Time GA	Ac
1	14, 28	537	51.57	9, 15	71	42.00	14, 28	0.06	47	51.57
2	14, 28	535	51.65	14, 29	69	51.65	14, 28	0.06	57	51.65
3	14, 28	544	52.01	14, 28	47	52.01	14, 28	0.06	55	52.01
4	**14, 28**	**545**	**52.55**	**14, 28**	**44**	**52.55**	**14, 28**	**0.06**	**51**	**52.55**
5	14, 28	578	51.41	1, 30	13	46.47	4, 13	0.06	29	35.53
6	26, 27	583	46.72	10, 15	44	43.99	7, 28	0.06	32	40.13
7	14, 28	537	52.12	14, 28	85	52.12	14, 29	0.06	72	50.73
8	13, 28	536	47.33	13, 28	55	47.33	13, 28	0.06	62	47.33
9	14, 28	599	52.37	5, 14	50	43.56	14, 28	0.06	53	52.37
10	15, 28	535	46.92	6, 26	40	37.29	15, 30	0.06	50	46.57
Average	–	553	50.46	–	58	46.90	–	0.06	51	48.04

also at nodes 14 and 28. In both cases, the accuracy is assessed by using a time horizon scheme (see previous chapter) in Fig. 11.6 for the k-NN classifier (with $k = 1$) and in Fig. 11.7 for the Bayesian classifier, both were obtained by using the first data set as the training data and the testing data set. The term "node relaxation" refers to the number of nodes in topological distance between the node with the real leak and the node, where the classifier predicts the leak for which the diagnosis is still considered correct.

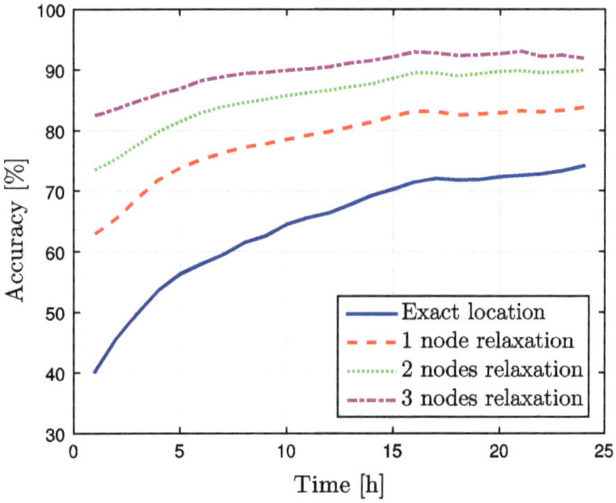

Fig. 11.6 Accuracy curves for the k-NN classifier in the Hanoi network with sensor placement at nodes 14 and 28

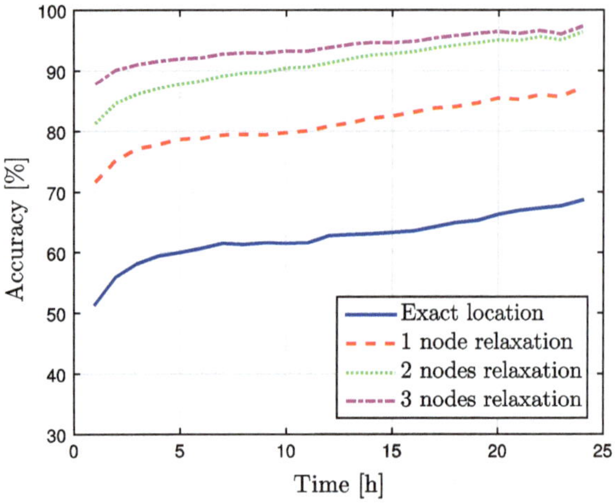

Fig. 11.7 Accuracy curves for the Bayesian classifier in the Hanoi network with sensor placement at nodes 14 and 28

The average topological distance, which is the average value of the minimum distance in nodes between the node predicted by the classifier and the real node with the leak, is depicted in Fig. 11.8 for both classifiers.

The results in this network show that the best performance is achieved with the Bayesian classifier being in agreement with the results presented in the previous chapter. The final sensor placement result can be seen in Fig. 11.9.

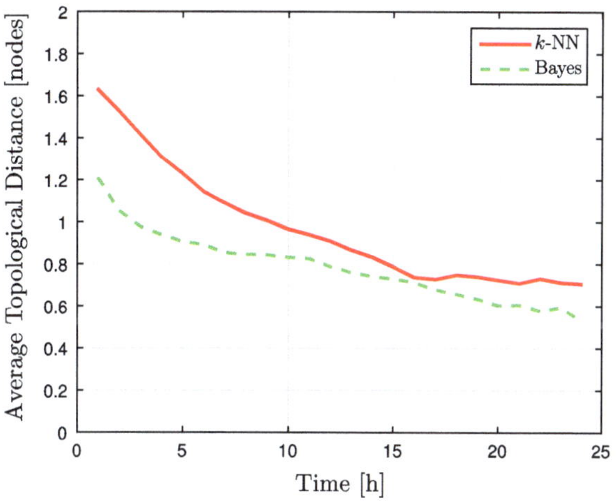

Fig. 11.8 Average topological distance for both classifiers in the Hanoi network

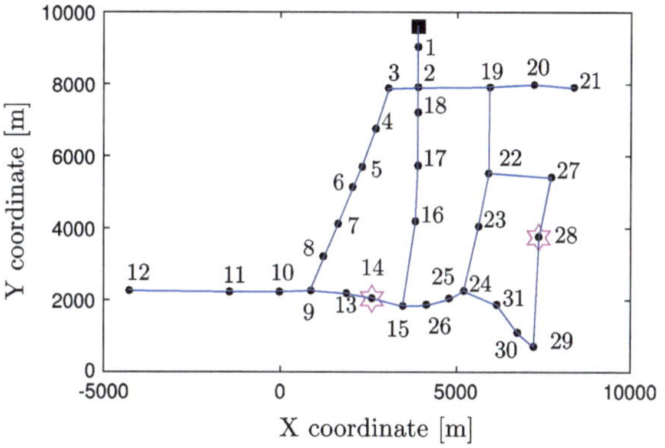

Fig. 11.9 Sensor placement for the k-NN and Bayesian classifiers in the Hanoi network

11.4.2 Limassol Case Study

The Limassol (Cyprus) DMA network (presented in Fig. 11.10) is medium-sized with one reservoir, 197 consumer nodes and 236 pipes. The consumption of water has a pattern (depicted in Fig. 11.11) generated as in the previous case (with different scale). For this network, it is decided to place three pressure sensors.

The data sets are generated considering similar uncertainties (in this case the leak varies from 2 to 6 l/s) as the Hanoi case, same sampling time (and computing the hourly average value) and the same number of examples for each class. The classifiers are built as in the Hanoi case, but using four attributes: the flow measurement at the inlet and the three pressure residuals. For the sensitivity matrix, the considered values are 492.2 l/s for the total water consumption and 4 l/s for the leak size.

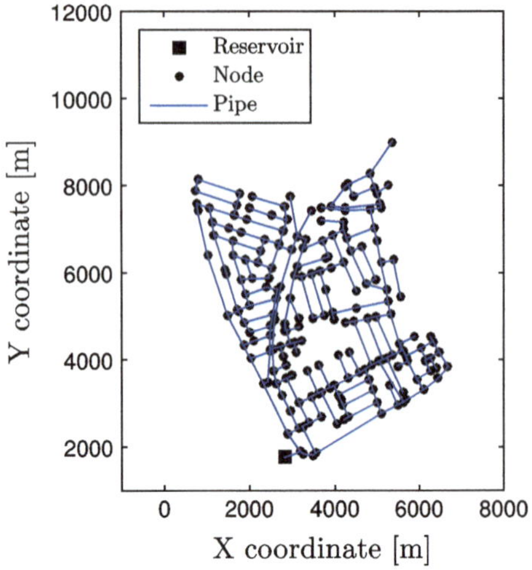

Fig. 11.10 Limassol network

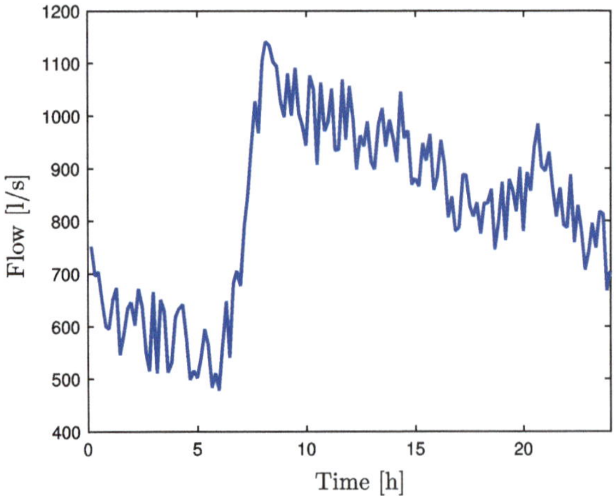

Fig. 11.11 Example of a typical daily flow consumption in the Limassol network

The results are summarized in Table 11.4 for the case of the k-NN classifier ($k = 1$ as in the Hanoi case) and in Table 11.5 for the Bayesian classifier. In this case, the σ value is equal to the mean value of the Φ matrix except for the diagonal. The best configurations are highlighted in bold for each method.

In this network, as in the Hanoi case, the use of the $\Phi^{(B)}$ matrix reduces the computation time in most cases and in the average value. In this network, the computational time reduction is similar to Hanoi's, but the degradation of the accuracy value is worse. This is probably caused by the low population size compared with the total number of possible combinations. The best result obtained for the k-NN classifier is to place the sensors at nodes 5, 11, and 124, and for the Bayesian classifier

Table 11.4 Sensor placement results in the Limassol network for the k-NN classifier

Data set	Genetic algorithm			Genetic algorithm + $\Phi^{(B)}$			
	Sensors	Time	Ac	Sensors	Time filter	Time GA	Ac
1	**15, 46, 113**	**8628**	**11.03**	1, 7, 195	6.4	7348	9.04
2	1, 7, 11	8602	10.89	8, 102, 182	7.5	16154	10.19
3	8, 183, 195	14906	9.84	52, 128, 133	6.8	8020	9.66
4	124, 183, 185	13957	8.09	7, 195, 197	6.2	6580	9.66
5	3, 7, 8	5508	9.66	1, 7, 195	7.1	1595	8.80
6	6, 8, 11	13652	10.72	104, 183, 195	6.8	5909	8.36
7	129, 185, 190	15308	8.03	1, 2, 197	6.3	479	5.38
8	1, 3, 7	2010	9.07	13, 40, 167	6.3	8294	10.37
9	87, 124, 128	12434	9.55	104, 124, 167	8.0	10551	10.37
10	3, 166, 181	4995	7.11	**5, 11, 124**	**6.2**	**13304**	**10.39**
Average	–	9690	9.39	–	6.7	7793	9.22

Table 11.5 Sensor placement results in the Limassol network for the Bayesian classifier

Data set	Genetic algorithm			Genetic algorithm + $\Phi^{(B)}$			
	Sensors	Time	Ac	Sensors	Time filter	Time GA	Ac
1	39, 77, 133	21348	17.35	11, 46, 133	6.3	21255	18.90
2	7, 19, 23	18701	19.28	17, 166, 181	6.3	17845	19.60
3	45, 110, 185	16696	16.40	40, 75, 156	6.8	14742	15.74
4	7, 11, 110	47590	19.34	**17, 46, 181**	**6.7**	**25515**	**19.63**
5	11, 91, 186	28032	17.43	91, 188, 190	6.5	20117	15.67
6	39, 48, 485	28106	19.13	39, 93, 189	6.7	15775	17.23
7	94, 133, 166	7730	19.22	100, 124, 183	6.9	15765	18.90
8	124, 189, 192	29217	16.61	14, 167, 185	6.4	13200	17.93
9	13, 22, 100	21904	18.86	93, 188, 190	6.4	11453	15.97
10	**40, 66, 104**	**15221**	**20.10**	13, 22, 190	6.5	6416	17.63
Average		23454	18.37	–	6.5	16208	17.72

to place the sensors at nodes 17, 46, and 181. The accuracy curves (using the first training data set and the testing data set) for both sensor placements are depicted in Figs. 11.12 and 11.13, for the k-NN classifier case and the Bayesian classifier case respectively.

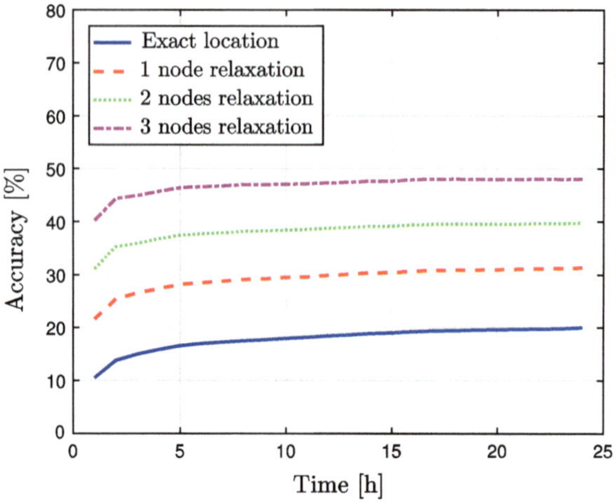

Fig. 11.12 Accuracy curves for the k-NN classifier in the Limassol network with sensor placement at nodes 5, 11 and 124

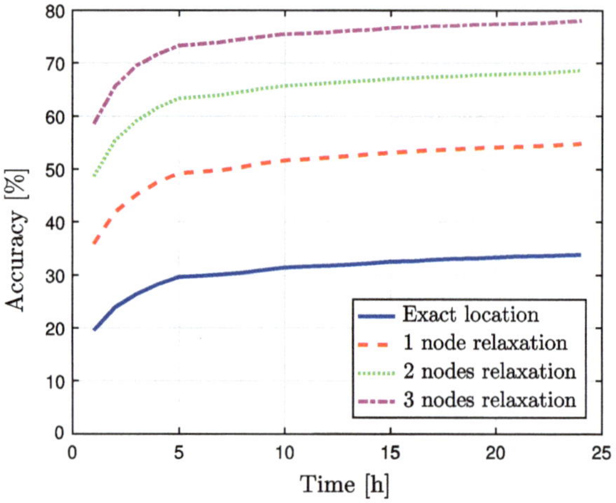

Fig. 11.13 Accuracy curves for the Bayesian classifier in the Limassol network with sensor placement at nodes 17, 46 and 181

The average topological distance for k-NN and Bayesian classifiers for the sensor placements obtained can be seen in Fig. 11.14.

Finally, the resulting sensor placement for the last proposed method (GA + $\Phi^{(B)}$) for both classifiers is depicted in Fig. 11.15.

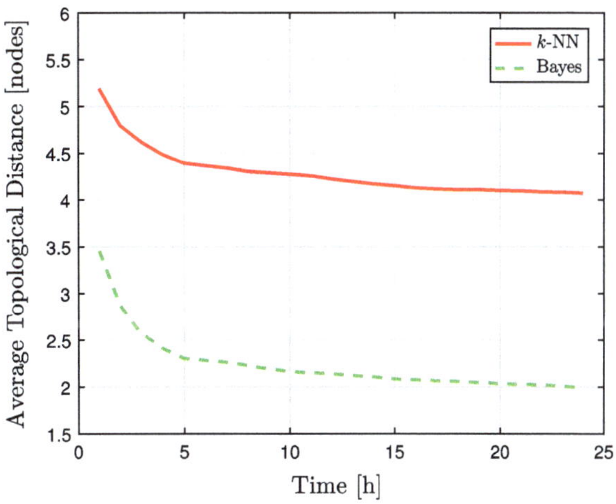

Fig. 11.14 Average topological distance for both classifiers in the Limassol network

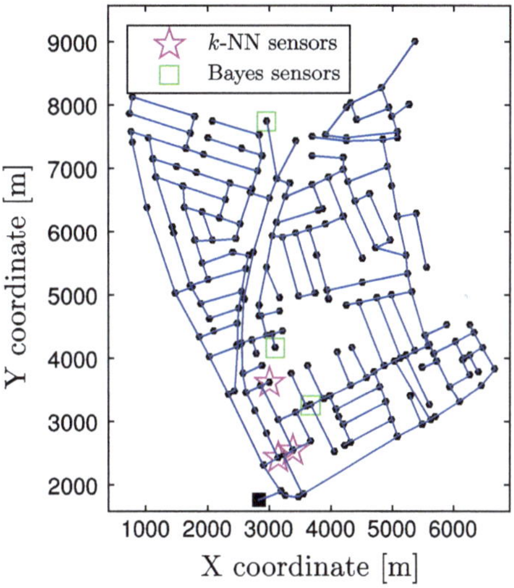

Fig. 11.15 Sensor placement for the k-NN and Bayesian classifiers in the Limassol network

11.5 Conclusions

In this paper, an optimal sensor placement method for placing a given number n_s of pressure sensors in WDNs to be used for classifier-based leak localization has been presented. The obtained sensor configuration is optimal in the sense that it maximizes the leak isolability when using a classifier-based leak localization method. In order to tackle the complexity of the optimal sensor placement, the use of genetic algorithms and a method to reduce the number of sensor configurations by discarding has been proposed. The performance of the proposed method has been illustrated by means of the application to the Hanoi and Limassol networks. The simulation results show that the genetic algorithm provides on average a similar performance index as that of the exhaustive search algorithm, whereas the computation load decreases significantly. Thus, this method becomes suitable for networks of growing complexity. Additionally, the use of classifiers allows the direct introduction of different sources of uncertainty and leads to good isolability results.

Acknowledgements This work has been partially funded by the Ministerio de Economía, Industria y Competitividad (MEICOMP) of the Spanish Government through the project CICYT ECOCIS (ref. DPI2013-48243-C2-1-R), by MEICOMP and FEDER through the project CICYT HARCRICS (ref. DPI2014-58104-R), by MEICOMP through the grant IJCI-2014-20801, by EFFINET grant FP7-ICT-2012-318556 of the European Commission, by the DGR of Generalitat de Catalunya (SAC group Ref. 2014/SGR/374) and by the Catalan Agency for Management of University and Research Grants (AGAUR), the European Social Fund (ESF) and the Secretary of University and Research of the Department of Companies and Knowledge of the Government of Catalonia through the grant FI-DGR 2015 (ref. 2015 FI_B 00591).

References

MATLAB. (2015). *MATLAB R2015a version*. Natick, MA: The Mathworks, Inc.

Berry, J., Fleischer, L., Hart, W., & Phillips, C. (2003). Sensor placement in municipal water networks. *Journal of Water Resources Planning and Management*, *131*(3), 3383–3392.

Blesa, J., Nejjari, F., & Sarrate, R. (2014). Robustness analysis of sensor placement for leak detection and location under uncertain operating conditions. *Procedia Engineering*, *89*, 1553–1560.

Blesa, J., Nejjari, F., & Sarrate, R. (2016). Robust sensor placement for leak location: Analysis and design. *Journal of Hydroinformatics*, *18*(1), 136–148.

Casillas, M. V., Puig, V., Garza-Castañón, L. E., & Rosich, A. (2013). Optimal sensor placement for leak location in water distribution networks using genetic algorithms. *Sensors*, *13*(11), 14984–15005.

Cugueró-Escofet, M. À., Puig, V., Quevedo, J., & Blesa, J. (2015a). Optimal pressure sensor placement for leak localisation using a relaxed isolation index: Application to the Barcelona water network. *IFAC-PapersOnLine*, *48*(21), 1108–1113.

Cugueró-Escofet, P., Blesa, J., Pérez, R., Cugueró-Escofet, M. À., & Sanz, G. (2015b). Assessment of a leak localization algorithm in water networks under demand uncertainty. *IFAC-PapersOnLine*, *48*(21), 226–231.

Lambert, M. F., Simpson, A. R., Vítkovský, J. P., Wang, X. J., & Lee, P. J. (2003). A review of leading-edge leak detection techniques for water distribution systems. In *20th AWA Convention, Perth, Australia*.

Nejjari, F., Sarrate, R., & Blesa, J. (2015). Optimal pressure sensor placement in water distribution networks minimizing leak location uncertainty. *Procedia Engineering*, *119*(1), 953–962.

Oh, I. S., Lee, J. S., & Moon, B. R. (2004). Hybrid genetic algorithms for feature selection. *IEEE Transactions on Pattern Analysis and Machine Intelligence*, *26*(11), 1424–1437.

Perelman, L. S., Abbas, W., Koutsoukos, X., & Amin, S. (2016). Sensor placement for fault location identification in water networks: A minimum test cover approach. *Automatica*, *72*, 166–176.

Pérez, R., Puig, V., Pascual, J., Peralta, A., Landeros, E., & Jordanas, L. (2009). Pressure sensor distribution for leak detection in Barcelona water distribution network. *Water Science and Technology: Water Supply*, *9*(6), 715–721.

Puust, R., Kapelan, Z., Savić, D. A., & Koppel, T. (2010). A review of methods for leakage management in pipe networks. *Urban Water Journal*, *7*(1), 25–45.

Sarrate, R., Blesa, J., & Nejjari, F. (2014a). Clustering techniques applied to sensor placement for leak detection and location in water distribution networks. In *2014 22nd Mediterranean Conference of Control and Automation (MED)* (pp. 109–114).

Sarrate, R., Blesa, J., Nejjari, F., & Quevedo, J. (2014b). Sensor placement for leak detection and location in water distribution networks. *Water Science and Technology: Water Supply*, *14*(5), 795–803.
Shastri, Y., & Diwekar, U. (2005). Optimal sensor placement for water distribution network security. In *AIChE Annual Meeting, Conference Proceedings* (pp. 6713–6736).
Soldevila, A., Blesa, J., Tornil-Sin, S., Duviella, E., Fernandez-Canti, R. M., & Puig, V. (2016). Leak localization in water distribution networks using a mixed model-based/data-driven approach. *Control Engineering Practice*, *55*, 162–173.
Steffelbauer, D. B., & Fuchs-Hanusch, D. (2016). Efficient sensor placement for leak localization considering uncertainties. *Water Resources Management*, 1–17.
Tang, J., Alelyani, S., & Liu, H. (2014). Feature selection for classification: A review. In C. Aggarwal (Ed.), *Data classification* (pp. 37–64). Chapman and Hall/CRC.

Chapter 12
Water Leak Diagnosis in Pressurized Pipelines: A Real Case Study

Jorge A. Delgado-Aguiñaga and Ofelia Begovich

Abstract This chapter presents a successful leak diagnosis for a real pipeline. The diagnosis was performed on the water pipeline West 5 located in Guadalajara City, Mexico, which is supervised by Inter-Municipal System of Potable Water and Sewage (SIAPA). Herein, the authors try to highlight those difficulties that arise when facing a real leak problem, especially if the fluid line is not monitored online or records of the flow rate and pressure are not available. By considering that only one leak is present in the system and by using the pipeline configuration information provided by the SIAPA staff, a discrete-time extended Kalman filter, used as a state observer, was designed in order to isolate the leak. The final decision about the leak location was based on the results of four different database analyses in which three showed a similar tendency.

12.1 Introduction

Pipeline networks are clearly a part of modern society, since liquids and gases are distributed through these in most major cities and even in many rural areas. In the last century, such systems have increased around the world in step with the growing population to satisfy its requirements. This brings unavoidable problems if the maintenance is not adequate. In fact, faults in these systems such as leaks have caused accidents with high economic costs, environmental damage, and human deaths. Leaks

J.A. Delgado-Aguiñaga (✉) · O. Begovich
CINVESTAV Guadalajara, Av. del Bosque 1145, colonia el Bajío,
45019 Zapopan, Jalisco, Mexico
e-mail: adelgado@gdl.cinvestav.mx

O. Begovich
e-mail: ofelia.begovich@cinvestav.mx

C. Verde and L. Torres (eds.), *Modeling and Monitoring of Pipelines and Networks*, Applied Condition Monitoring 7,
DOI 10.1007/978-3-319-55944-5_12

are mainly due to corrosion of the pipeline material, welding problems at the joints, the water hammer effect, illegal extractions, and other natural events. In response, the scientific community has tried to develop some strategies for identifying faults quickly enough to prevent disasters or at least to reduce them to the barest minimum. Recently, with improvements in technology, several methodologies have been developed to face the problem. Currently, there are many works in the literature dealing with the pipeline monitoring problem from different perspectives (Billman and Isermann 1987; Begovich et al. 2012; Verde 2001; Delgado-Aguiñaga et al. 2016b; Torres et al. 2011; Besançon et al. 2013; Delgado-Aguiñaga et al. 2016a; Cataldo et al. 2016; Ostapkowicz 2014; Zhang et al. 2015). In most of those approaches, it is assumed that pressure and flow rate measurements are available at the pipeline ends. Although this is not very demanding, in practice it could result in a serious problem for data acquisition if the system being monitored lacks the necessary instrumentation.

When considering the applicability of such methods in real systems such as water pipelines, additional problems always arise including those related to sensor placement, data acquisition, and knowledge of the pipeline parameters for example. Especially for underground pipelines, the parameters are difficult to estimate with accuracy.

Unfortunately, faults can be detected some time after they have appeared, possibly not until an accident occurs or other warnings are detected, such that moisture on the ground. Considering a worst case, the fault condition is reported long after its occurrence, and there are no registered data; one could then guess the presence of more than one fault.

In this chapter, a single leak location was successfully achieved in a water pipeline in Guadalajara City, Mexico. The full procedure is described here, beginning with the sensor placement issue, followed by the data acquisition stage once the leak has been reported, and concluding with the application of a leak detection and identification (LDI) methodology based on a discrete-time extended Kalman filter (EKF) as a state observer. The final decision about the leak location was made based on an offline analysis of a set of experiments, which showed a similar tendency in the leak position.

This presentation is organized as follows: Sect. 12.2 describes the sequence of events which motivated the study and the implementation of a feasible proposition for the leak location with its hypotheses; Sect. 12.3 presents a minimal order fluid model by including the leak effect and the observer design; Sect. 12.4 describes the data acquisition stage and database analysis. The second part of the chapter, starts with the Sect. 12.5 which presents an analysis of the results and the final decision. Finally, Sect. 12.6 shows some photographs of the excavation and the pipeline before and after the damage reparation and Sect. 12.7 offers some conclusions.

Table 12.1 Technical specifications of $HydrINS$ 2

Measurement range	Bidirectional flow measurement from 0.02 to 5 [m/s]
Units	Selectable: [mm], [m], [l], [m^3], [ft^3],...; [s], [m], [hour],...
Power	Internal lithium batteries providing up to 4 years of battery life
Software	WinFluid
Temperature range	$-20\,[°C]$ to $60\,[°C]$
Maximum pressure	$20\,[bars]$ $(290\,[psi])$. Integral to BSP quick-fit pressure connector
Data acquisition	$Vista+$ loggers for local recording and download

12.2 Problem Description and Hypotheses

The leak report began some months before the methodology presented here was applied. It was detected when the staff of SIAPA performed some routine inspections at different points of the entire pipeline network, and a discrepancy was observed in the flow rate measurements in a specific section. The conclusion was that there was at least one leak in the underground pipeline. After this first step, other close access points were studied to determine a smaller section in which the abnormal events were present. It was concluded that between the access points of Matatlán Avenue (MA) and López Cotilla Road (LPR) the damage was uncovered. At this stage, the measurement data were absent, and the event occurrence time was unknown.

Subsequently, three different attempts were carried out by the SIAPA staff to find the leak, but all were unsuccessful. In the first attempt, the SIAPA staff found that some water was springing to the surface, so a laboratory analysis was performed for chlorine content in order to confirm that it was potable water and not groundwater. This analysis revealed a small amount of chlorine, and with this result, the SIAPA staff decided to execute the procedure for finding the leak. The leak was not found, however. After that, the team believed the leak to be nearby, and another two additional attempts were executed both upstream and downstream of the first one. Again the result was unsuccessful, however. The mission was then interrupted because of the high economic cost.

On February 18th 2016, the Centro de Investigación y de Estudios Avanzados Guadalajara (CINVESTAV-Gdl) research group began to work with SIAPA. The first obstacle was the lack of sensors in the water pipeline. As a first step, the sensor collocation procedure was executed: two sensors were inserted in the underground pipeline. The upstream sensor was fixed at the corner of Periférico Avenue and MA, and the downstream sensor was fixed at Periférico Avenue and LCR. Both sensors include an electromagnetic flowmeter (HydrINS 2) manufactured by HYDREKA® and are equipped with a data logger that allows the storage of the variables see; Fig. 12.1. This kind of instrumentation can measure both the flow rate and the pressure head. Technical specifications are given in Table 12.1.

As a result, a section of pipeline between the sensors was analyzed, and the characteristics depicted in Table 12.2 were obtained.

Table 12.2 Parameters of the section of the water pipeline

Parameter	Symbol	Value	Units
Length between sensors	L	362.5	[m]
Inner diameter	D	0.9	[m]
Wall thickness	ϵ	0.089	[m]
Roughness	η	0.001	[m]
Piezometric head due to slope	Θ	0.8	[m]

The other pipeline parameters depend on the water temperature and/or on the flow rate. To compute these parameters, a constant water temperature $T = 20$ [$°C$] was used, which was provided by the SIAPA staff on the basis of statistical records. The procedure for computing those variables can be found in Delgado-Aguiñaga et al. (2016a). In particular, the friction factor f was approximated by using the Haaland expression (12.1), which is suitable for this type of pipe made of prestressed concrete, since both the roughness and the flow regime are inside of the considered limits (Genic et al. 2011). The roughness η was obtained from the datasheet given by the pipe manufacturer COMECOP:

$$f(Q,T) = \left\{-1.8log\left[\left(\frac{\varepsilon}{3.7}\right)^{1.11} + \frac{6.9}{Re(Q,T)}\right]\right\}^{-2} \qquad (12.1)$$

in which Re is the Reynolds number in the interval [$4000 - 10^8$] and $\varepsilon = \eta/D$ is the relative pipe roughness in the interval [$1 \times 10^{-6} - 0.05$]. Assuming a constant

Fig. 12.1 Installed HydrINS 2 at the upstream end

temperature of the water, hereinafter the friction factor depends only on the flow rate $f(Q)$.

12.2.1 Hypotheses

From the point of view of fault detection, a single leak in a pipeline can be located with a steady-state model if flows and the pressure head at the ends of a pipeline, together with the physical parameters given in Table 12.2, are known (González et al. 2016), even in a branched pipeline. This allows the computation of the parameters associated with the leak through diverse algorithms. If the data are noisy, a dynamic model is suggested instead of the static for the improvement of the leak parameters estimation. If there is more than one leak, the knowledge of steady-state conditions at the ends of pipeline is not sufficient for the leaks' location. The reason is the lack of isolability condition with this scenario. Persistent input signals could be applied (Besançon 2007) or the transient pressure wave in the fluid must be generated for the events' isolability (Rentería et al. 2016); both scenarios are not, however, feasible in many practical applications. To design the leak's estimator, we assumed the following conditions:

1. The pipeline's equivalent length is equal to the real length, because the couplings between each straight section causes a small drop in the pressure head and can be neglected (Wylie and Streeter 1993).
2. Only one leak is present.

12.3 Leak Diagnosis Scheme

12.3.1 Pipeline Modeling with Leak Effect

The dynamics describing the transient response in closed conduits transporting a pressurized fluid are given by a pair of partial differential equations (PDEs) known as water hammer equations (WHE) (Roberson et al. 1998). The momentum equation is then stated as

$$\frac{\partial Q(z,t)}{\partial t} + gA\frac{\partial H(z,t)}{\partial z} + \mu Q(z,t)\left|Q(z,t)\right| = 0 \qquad (12.2)$$

and the continuity equation as

$$\frac{\partial H(z,t)}{\partial t} + \frac{b^2}{gA}\frac{\partial Q(z,t)}{\partial z} = 0. \qquad (12.3)$$

Both equations are obtained by using momentum and mass conservation equations and by considering the following assumptions: the pipeline is straight, the fluid is slightly compressible with a constant density, the duct wall is slightly deformable, the convective velocity changes are negligible, and the pipeline's cross-sectional area is constant.

Here Q denotes the flow rate [m^3/s], H is the pressure head [m], z is the length coordinate [m], t is the time coordinate [s], g is the gravity acceleration [m/s^2], A is the cross-sectional area [m^2], b is the pressure wave speed in the fluid [m/s], with $\mu = f/2DA$, D is the inner diameter [m], and f is the friction factor.

In this work, to approximate the WHE, the pressure heads at the pipeline ends were considered as boundary conditions, which are measured by sensors and denoted as

$$\begin{aligned} H(z=0,t) &= H_{in}(t) \\ H(z=L,t) &= H_{out}(t). \end{aligned} \tag{12.4}$$

Internal conditions, such as leaks or obstructions, can cause a discontinuity in the solution of the governing equations. In particular, the presence of a single leak arbitrarily located at position $z_l \in (0, L)$ can be handled as a new boundary condition in (12.2) and (12.3), with outflow $Q_l = C_d A_l \sqrt{2g}\sqrt{H_l}$ in which C_d is the discharge coefficient and A_l is the leak cross-sectional area. Now by defining $\lambda \equiv C_d A_l \sqrt{2g}$, the leak outflow rate Q_l can be expressed as Crowe et al. (2000)

$$Q_l = \mathcal{H}_{t_l} \lambda_l \sqrt{H_l}, \tag{12.5}$$

where H_l is the pressure head at the leak point $[m]$, λ_l is the leak coefficient [m$^{5/2}$/s], and $\mathcal{H}_{t_l}$ is the Heaviside unit step function associated with the leak occurrence at time t_l.

The momentum and continuity equations (12.2) and (12.3) are quasilinear and hyperbolic partial differential equations; in addition, an analytical solution of these equations is not available such that a finite approximation method must be considered instead (Chaudhry 2014). Here the finite difference method was used. Now by using a discretized version of the governing equations and by introducing the presence of a leak precisely at the single interior discretization node, arbitrarily located, the following finite approximation was obtained (Chaudhry 2014):

$$\begin{aligned} \dot{Q}_1 &= \frac{-gA}{\Delta z_1}(H_2 - H_{in}) - \frac{f(Q_1)}{2DA} Q_1|Q_1| \\ \dot{H}_2 &= \frac{-b^2}{gA\Delta z_1}\left(Q_2 - Q_1 + \mathcal{H}_{t_l}\lambda_l\sqrt{H_2}\right) \\ \dot{Q}_2 &= \frac{-gA}{\Delta z_2}(H_{out} - H_2) - \frac{f(Q_2)}{2DA} Q_2|Q_2|, \end{aligned} \tag{12.6}$$

where the two sections reduce to $\Delta z_1 = z_l$ and $\Delta z_2 = L - z_l$ with z_l as the leak position.

12.3.2 Observer Design

To reconstruct the leak parameters (magnitude and position), they can be considered as additional state variables and be included in a new augmented state vector (Delgado-Aguiñaga et al. 2016b; Torres et al. 2011). Therefore, the leak diagnosis problem can be solved by designing a state observer of the system that resulted from the augmented state vector. To do that, let us consider the pressure head $u = [H_{in}\ H_{out}]^T$ as an input variable and the end flow rates $y = [Q_1\ Q_2]^T$ as an output variable. Finally, by considering

$$\dot{z}_l = 0; \quad \dot{\lambda}_l = 0 \tag{12.7}$$

the derivatives of the leak parameters z_l, λ_l as new states, the extended model from (12.6) is given by

$$\begin{bmatrix} \dot{Q}_1 \\ \dot{H}_2 \\ \dot{Q}_2 \\ \dot{z}_l \\ \dot{\lambda}_l \end{bmatrix} = \begin{bmatrix} \frac{-gA}{z_l}(H_2 - u_1) - \frac{f(Q_1)}{2DA} Q_1|Q_1| \\ \frac{-b^2}{gAz_l}\left(Q_2 - Q_1 + \mathcal{H}_{t_l}\lambda_l\sqrt{H_2}\right) \\ \frac{-gA}{L-z_l}(u_2 - H_2) - \frac{f(Q_2)}{2DA} Q_2|Q_2| \\ 0 \\ 0 \end{bmatrix} \tag{12.8}$$

Hence, the model (12.8) can be written in compact form as

$$\dot{x} = \xi(x, u), \tag{12.9}$$

with $x = [Q_1\ H_2\ Q_2\ z_l\ \lambda_l]^T$ and $\xi(.)$ a nonlinear function. Now to estimate the leak parameters z_l and λ_l, a discrete-time EKF is designed from the nonlinear model (12.9); practical examples can be found in Delgado-Aguiñaga et al. (2016b, 2015). Here, model (12.8) is discretized by using Heun's method. Thus, the approximated solution for the initial value problem

$$\dot{x}(t) = \xi(x(t), u(t)), \quad x(t_0) = x_0 \tag{12.10}$$

is given by Golub and Ortega (1992):

$$x^{i+1} = x^i + \frac{\Delta t}{2}[\xi(x^i, u^i) + \xi(x^i + \Delta t \xi(x^i, u^i), u^{i+1})] \tag{12.11}$$

where Δt is the time step and i is the index of discrete time. Finally, the ordinary differential equations in (12.8) are transformed by using (12.11) into following difference equations:

$$\begin{aligned}
Q_1^{i+1} &= Q_1^i + \frac{\Delta t}{2}(-\frac{gA}{z_l^i}(H_2^i - u_1^i) - \frac{1}{2DA}f(Q_1^i)Q_1^i|Q_1^i| \\
&\quad - \frac{gA}{z_l^i}(\tilde{H}_2^{i+1} - u_1^{i+1}) - \frac{1}{2DA}f(\tilde{Q}_1^{i+1})\tilde{Q}_1^{i+1}|\tilde{Q}_1^{i+1}|) \\
H_2^{i+1} &= H_2^i + \frac{\Delta t}{2}(-\frac{b^2}{gAz_l^i}(Q_2^i - Q_1^i + \mathcal{H}_{t_l}\lambda_l^i\sqrt{H_2^i}) \\
&\quad - \frac{b^2}{gAz_l^i}(\tilde{Q}_2^{i+1} - \tilde{Q}_1^{i+1} + \mathcal{H}_{t_l}\lambda_l^i\sqrt{\tilde{H}_2^{i+1}})) \\
Q_2^{i+1} &= Q_2^i + \frac{\Delta t}{2}(-\frac{gA}{L - z_l^i}(u_2^i - H_2^i) - \mu f(Q_2^i)Q_2^i|Q_2^i| \\
&\quad - \frac{gA}{L - z_l^i}(u_2^{i+1} - \tilde{H}_2^{i+1}) - \frac{1}{2DA}f(\tilde{Q}_2^{i+1})\tilde{Q}_2^{i+1}|\tilde{Q}_2^{i+1}|) \\
z_l^{i+1} &= z_l^i \\
\lambda_l^{i+1} &= \lambda_l^i,
\end{aligned} \tag{12.12}$$

where

$$\begin{aligned}
\tilde{Q}_1^{i+1} &= Q_1^i + \Delta t\left(-\frac{gA}{z_l^i}\left(H_2^i - u_1^i\right) - \frac{1}{2DA}f\left(Q_1^i\right)Q_1^i|Q_1^i|\right) \\
\tilde{H}_2^{i+1} &= H_2^i + \Delta t\left(-\frac{b^2}{gAz_l^i}\left(Q_2^i - Q_1^i + \mathcal{H}_{t_l}\lambda_l^i\sqrt{H_2^i}\right)\right) \\
\tilde{Q}_2^{i+1} &= Q_2^i + \Delta t\left(-\frac{gA}{L - z_l^i}\left(u_2^i - H_2^i\right) - \frac{1}{2DA}f(Q_2^i)Q_2^i|Q_2^i|\right).
\end{aligned} \tag{12.13}$$

The discretized version can be written in compact form:

$$x^{i+1} = \xi_d(x^i, u^{i+1}, u^i); \quad y^i = Hx^i, \tag{12.14}$$

where

$$x^i = [Q_1^i \; H_2^i \; Q_2^i \; z_l^i \; \lambda_l^i]^T; \quad u^i = [H_{in}^i \; H_{out}^i]^T$$

$$H = \begin{bmatrix} 1\,0\,0\,0\,0 \\ 0\,0\,1\,0\,0 \end{bmatrix}.$$

A discrete-time EKF as the state observer for the system (12.14) can be chosen such as in Simon (2006):

$$\hat{x}^i = \hat{x}^{i^-} + K^i(y^i - H\hat{x}^{i^-}),$$

where $\hat{x}^{i^-}$ is the a priori estimate of x^i

$$\hat{x}^{i^-} = \xi_d(\hat{x}^{i-1}, u^i, u^{i-1}),$$

K^i is the Kalman gain

$$K^i = P^{i^-} H^T (H P^{i^-} H^T + R)^{-1}$$

Moreover, the a priori and posteriori covariance matrix P^{i^-} and P^i are given respectively by

$$P^{i^-} = F^i P^{i-1} (F^i)^T + Q, \qquad P^i = (I - K^i H) P^{i^-}$$

and F^i corresponds to the Jacobian matrix

$$F^i = \frac{\partial \xi_d(x, u)}{\partial x}|_{x=\hat{x}^i}$$

In this recursive estimation, the matrices P^{0^-}, R and Q correspond to the covariance matrices of the initial conditions, output and process noises, respectively, with the property

$$P^{0^-} = (P^{0^-})^T > 0 \quad R = R^T > 0, \quad \text{and} \quad Q = Q^T > 0.$$

In particular, for the estimation of the leak parameter the following covariance matrices of measure and process were used:

$$Q = diag[1 \times 10^{-3},\ 1 \times 10^{-1},\ 1 \times 10^{-3},\ 600,\ 1 \times 10^{-4}], \tag{12.15}$$

and

$$R = diag[1 \times 10^{-3},\ 1 \times 10^{-3}]. \tag{12.16}$$

To obtain moving average values of the leak parameters, the estimation generated by the EKF was filtered with the equation

$$v_F(k) = \frac{1}{2N+1}\left(v(k+N) + v(k+N-1) + \cdots + v(k-N)\right), \tag{12.17}$$

where $v_F(k)$ is the smoothed value for the variable at time k, N is the number of neighboring data taken on either side of $v_F(k)$, and $2N + 1$ is the span dimension. The span must be odd, and it is adjusted for data points that cannot accommodate the specified number of neighbors on either side. A span equal to 101 was used for the position parameter z_l and a span of 21 for the leak magnitude λ_l.

Table 12.3 Data packets

	Total sampled interval I	Estimation interval I_E
Data packet 1	$22{:}00-25/feb\ to$ $11{:}38-26/feb$	$03{:}00-26/feb\ to$ $06{:}40-26/feb$
Data packet 2	$16{:}44-27/feb\ to$ $11{:}37-28/feb$	$23{:}24-27/feb\ to$ $01{:}54-28/feb$
Data packet 3	$15{:}58-29/feb\ to$ $10{:}54-01/mar$	$15{:}58-29/feb\ to$ $22{:}30-29/feb$
Data packet 4	$13{:}26-02/mar\ to$ $13{:}30-03/mar$	$19{:}15-02/mar\ to$ $21{:}45-02/mar$

12.4 Data Acquisition and Its Analysis

Once the sensors were placed at the specific points of the pipe under study, the data acquisition stage began. The staff at SIAPA performed this task by obtaining four data packets with a sampling rate of 0.016 [Hz]. The total time of the data for each packet was denoted as I_i, and the selected time interval for the estimation was denoted as I_{Ei}. Before applying the designed observer, all variables of each data packet were synchronized. Thus, I_i and I_{Ei} with $i = 1, \ldots, 4$ stand for the considered interval according to the $i - th$ data packet (Table 12.3).

In order to analyze these data packets, we selected a measurement interval with a steady-state condition. This was done because the extended system (12.8) is only locally observable. Thus, the convergence of any observer can only be guaranteed for specific space regions. This fact has been studied with real data of the pilot PP-R pipeline circuit built at CINVESTAV-Gdl. From the variables of the four packets shown in the set of Figs. 12.2, 12.3, 12.9, 12.10, 12.16, 12.17, 12.23 and 12.24 one identifies abrupt changes in some time intervals. This phenomenon was justified because flow control valves are placed near the studied section. A branch junction is

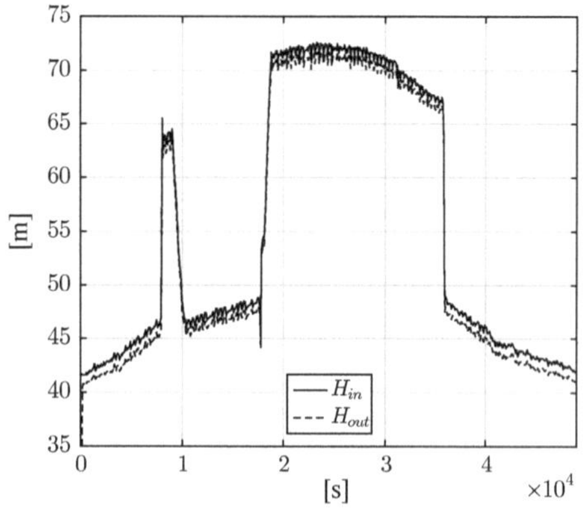

Fig. 12.2 Pressure heads during the interval I_1

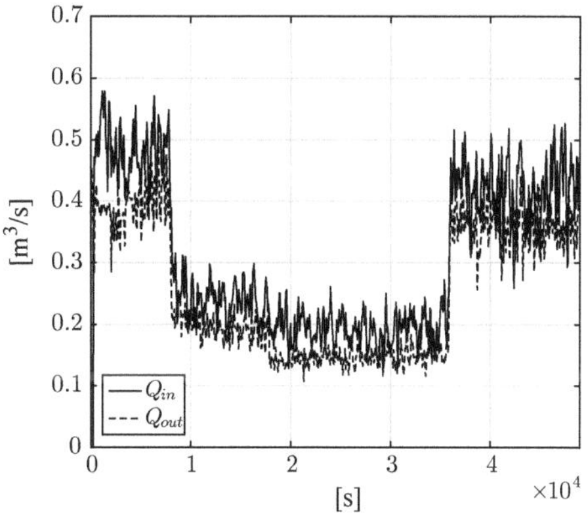

Fig. 12.3 Flow rates during the interval I_1

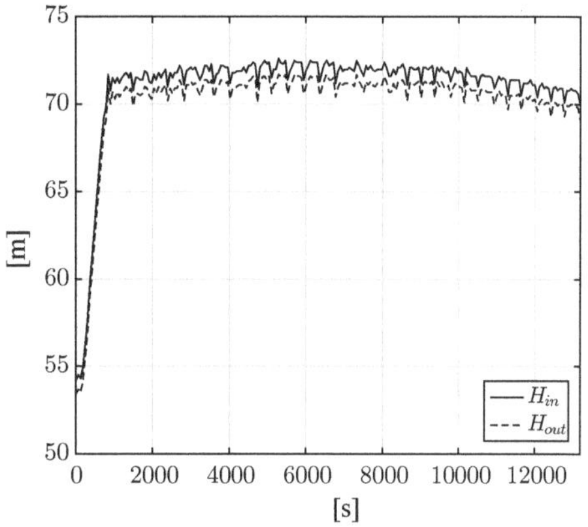

Fig. 12.4 Pressure heads during the interval I_{E1}

located approximately 8 [m] from the upstream sensor which is a disturbance for the estimator. The autoregressive filter (12.17) for each leak's estimation was applied during the intervals I_{E_i} respectively.

12.4.1 Analysis of Data Packet 1

In Figs. 12.2 and 12.3, the pressures heads (H_{in}, H_{out}) and flow rates (Q_{in}, Q_{out}) that were recorded during the time interval I_1 are plotted. The respective data in the estimation interval I_{E1} are shown in Figs. 12.4 and 12.5.

The data during the interval I_{E1} were injected into the EKF for the estimation of the leak parameters. The initial values of the state for the EKF are shown in Table 12.4.

The evolution of the estimation of the leak position $\hat{z}_l$ calculated with the EKF and its corresponding filter evolution are shown in Fig. 12.6.

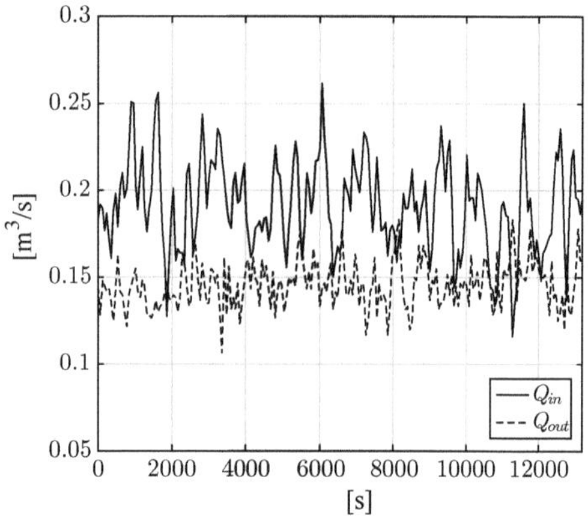

Fig. 12.5 Flow rates during the interval I_{E1}

Table 12.4 EKF initialization value corresponding to data packet 1

Variable	Value	Units
$\hat{Q}_1^0$	0.1835	$[\mathrm{m}^3/s]$
$\hat{H}_2^0$	41.6081	[m]
$\hat{Q}_2^0$	0.1450	$[\mathrm{m}^3/s]$
$\hat{z}_l^0$	308.125	[m]
$\hat{\lambda}_l^0$	0	$[\mathrm{m}^{5/2}/s]$

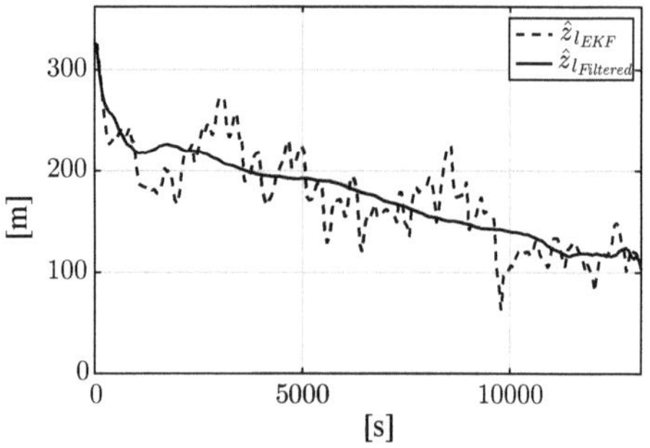

Fig. 12.6 Estimated leak position $\hat{z}_l$ in I_{E1}

The corresponding friction factors $f(Q_1)$ and $f(Q_2)$ computed by using the Haaland expression (12.1) are shown in Fig. 12.7; like the estimation of the position, Fig. 12.8 shows the behavior of the estimated and filtered orifice parameter associated with the leak magnitude.

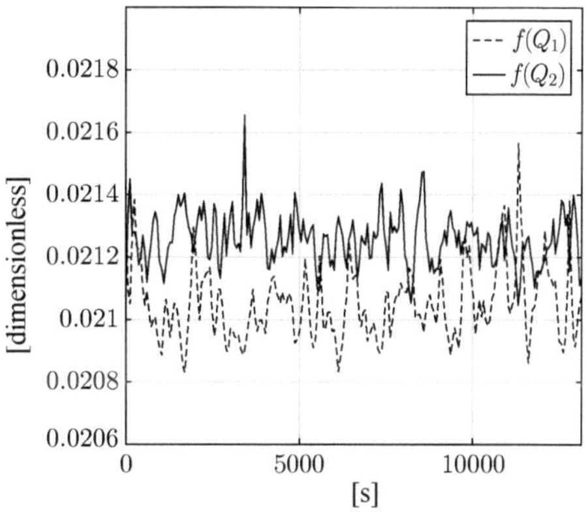

Fig. 12.7 Friction coefficients $f(Q_1)$ and $f(Q_2)$ in I_{E1}

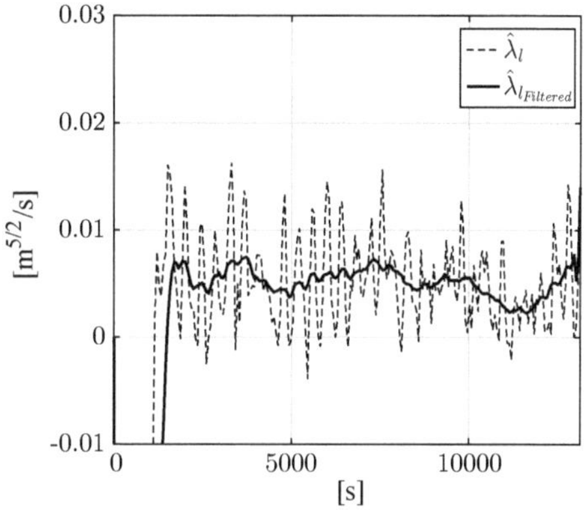

Fig. 12.8 Leak orifice parameter $\hat{\lambda}_l$ and the filtered signal $\hat{\lambda}_{l_F}$ in I_{E1}

Table 12.5 EKF initialization corresponding to data packet 2

Variable	Value	Units
$\hat{Q}_1^0$	0.1806	$[\mathrm{m}^3/s]$
$\hat{H}_2^0$	37.6571	[m]
$\hat{Q}_2^0$	0.1278	$[\mathrm{m}^3/s]$
$\hat{z}_l^0$	181.25	[m]
$\hat{\lambda}_l^0$	0	$[\mathrm{m}^{5/2}/s]$

12.4.2 *Analysis of Data Packet 2*

As for packet 1, Figs. 12.9 and 12.10 show the pressures heads (H_{in}, H_{out}) and the flow rates (Q_{in}, Q_{out}) that were recorded during the interval I_2. The behavior of the variables during the estimated interval I_{E2} are shown in Figs. 12.11 and 12.12.

The initial state values for the EKF are shown in Table 12.5.

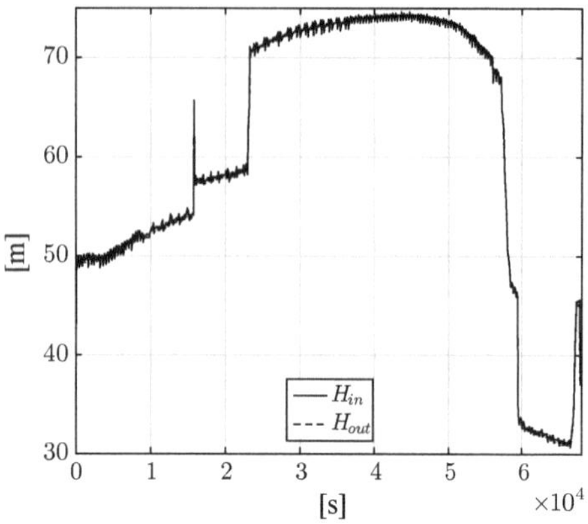

Fig. 12.9 Pressure head during the interval I_2

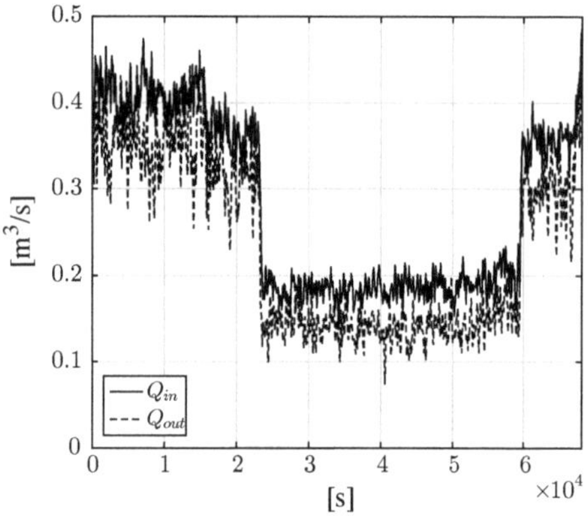

Fig. 12.10 Flow rate during the interval I_2

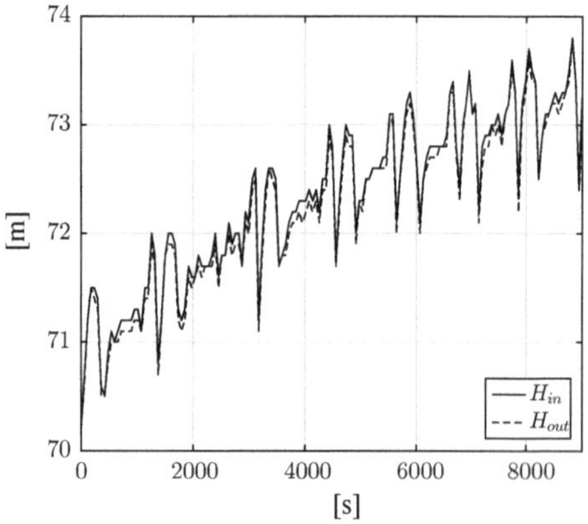

Fig. 12.11 Pressure heads during the interval I_{E2}

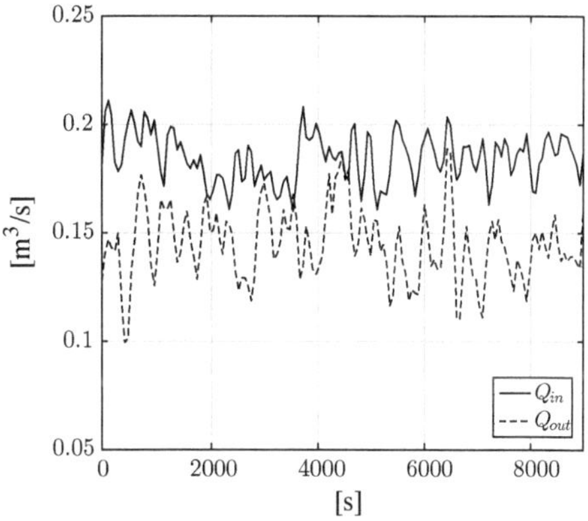

Fig. 12.12 Flow rates during the interval I_{E2}

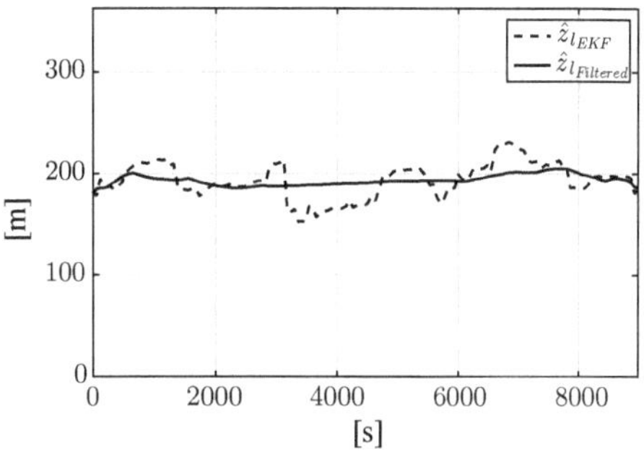

Fig. 12.13 Estimated leak position $\hat{z}_l$ in I_{E2}

The leak position estimation can be seen in Fig. 12.13, and from the time evolution one can say that the estimation remains around the initial condition. This is contrary to the value obtained with the package I_{E_1}. This could be explained because demands and valves handled by operators generate high-frequency transients in the water. In particular, during the time interval I_{E_2} one can identify transient behaviors and a small pressure drop between upstream and downstream points. Both effects produce a drastic regime change in the water with a respective flow rate reduction. This fluid behavior cannot be followed by the simple model (1.1) with the rough discretization used in the design of the EKF.

Similarly to packet 1, the corresponding friction factors are shown in Fig. 12.14; furthermore, the parameter estimation and its filtered version are depicted in Fig. 12.15, respectively, for the orifice associated with the leak.

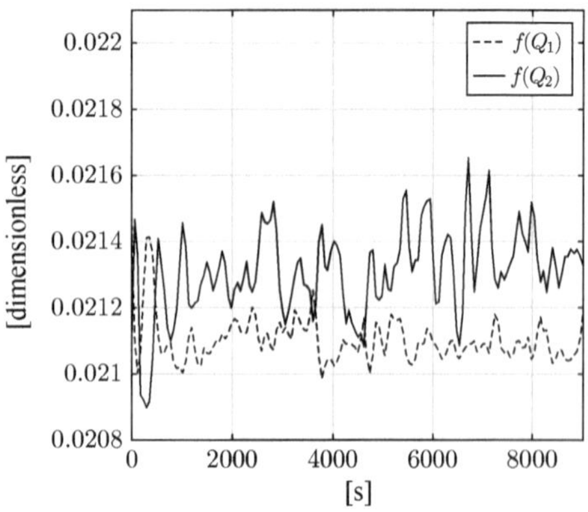

Fig. 12.14 Friction coefficients $f(Q_1)$ and $f(Q_2)$ in I_{E2}

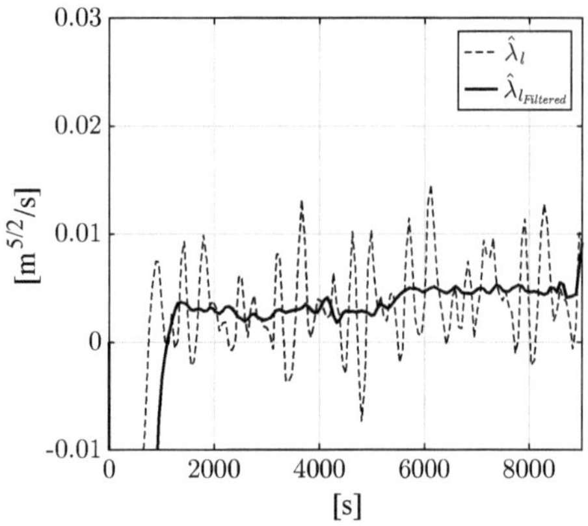

Fig. 12.15 Leak magnitude $\hat{\lambda}_l$ and the filtered version $\hat{\lambda}_{l_F}$ in I_{E2}

12.4.3 *Analysis of Data Packet 3*

Figures 12.16 and 12.17 show the pressure heads and flow rates (H_{in}, H_{out}) and (Q_{in}, Q_{out}) that were registered during the interval I_3. Figures 12.18 and 12.19 show the behavior of the variables in the interval I_{E3}.

The initial state of the EKF is shown in Table 12.6.

The leak position estimation can be seen in Fig. 12.20. As in the first data packet, the position estimation deviates around a value near 115 [m].

In the same way as before, the corresponding friction factors are shown in Fig. 12.21; likewise, the estimation and the filtered version of the parameter associated with the orifice are depicted in Fig. 12.22, respectively.

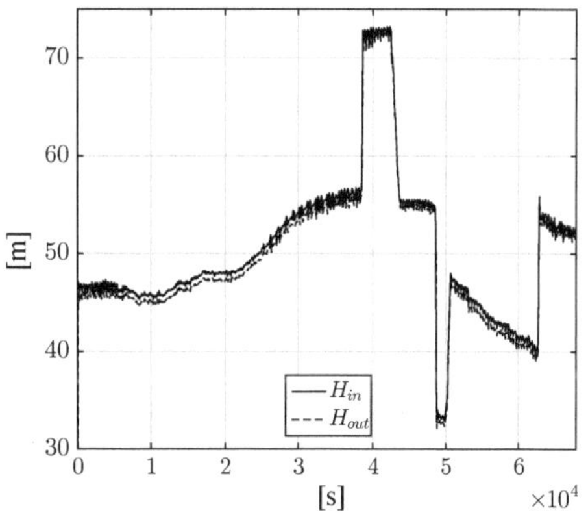

Fig. 12.16 Pressure heads during the interval I_3

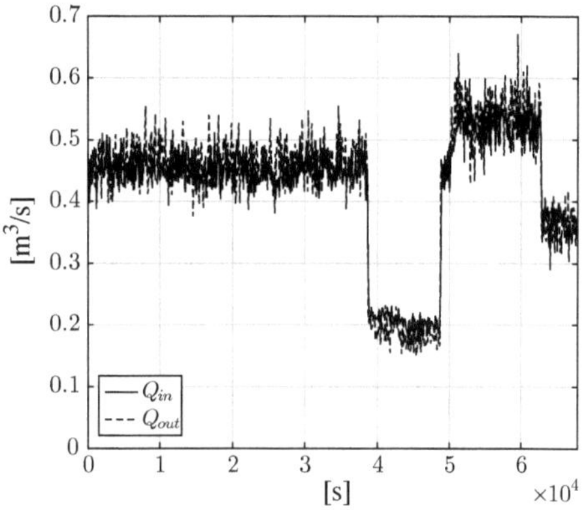

Fig. 12.17 Flow rates during the interval I_3

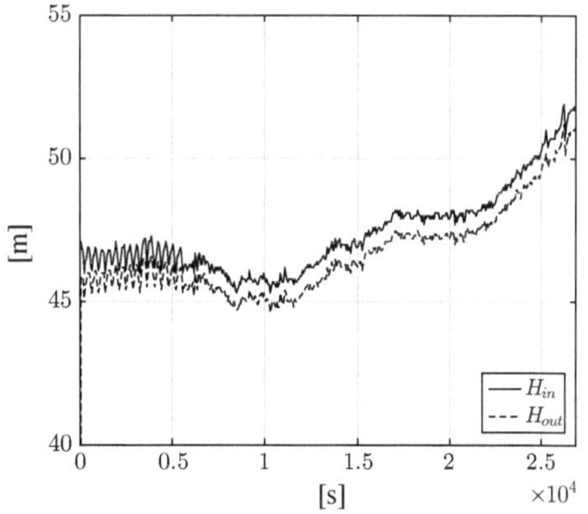

Fig. 12.18 Pressure heads during the interval I_{E3}

Table 12.6 EKF initialization corresponding to data packet 3

Variable	Value	Units
$\hat{Q}_1^0$	0.4761	$[\mathrm{m}^3/s]$
$\hat{H}_2^0$	46.5475	[m]
$\hat{Q}_2^0$	0.4072	$[\mathrm{m}^3/s]$
$\hat{z}_l^0$	344.375	[m]
$\hat{\lambda}_l^0$	0	$[\mathrm{m}^{5/2}/s]$

Table 12.7 EKF initialization corresponding to data packet 4

Variable	Value	Units
$\hat{Q}_1^0$	0.2824	$[\mathrm{m}^3/s]$
$\hat{H}_2^0$	40.7078	[m]
$\hat{Q}_2^0$	0.2617	$[\mathrm{m}^3/s]$
$\hat{z}_l^0$	181.25	[m]
$\hat{\lambda}_l^0$	0	$[\mathrm{m}^{5/2}/s]$

12.4.4 Analysis of Data Packet 4

The pressure head and flow rate variables (H_{in}, H_{out}) and flow rates (Q_{in}, Q_{out}) of the last packet can be seen in Figs. 12.23 and 12.24. Figures 12.25 and 12.26 show the variables in the interval I_{E4}.

The initial state of the EKF is shown in Table 12.7.

The leak position estimation can be seen in Fig. 12.27. The leak position estimation deviates also around a similar value as with the first and third data packets. In the same way as before, the corresponding friction factors are shown in Fig. 12.28; likewise, the estimation and the filtered version of the parameter associated with the leak orifice are depicted in Fig. 12.29, respectively. From these redundant results, i.e., those with the same tendency, one can deduce that the leak position could be near to the estimations considering packets 1, 3 and 4.

12.5 Final Decision

After the data packets' analyses, the leak positions were compared, and some conclusions were formulated. Three of the four analyses have the same tendency; the variance of such results is however significant, and providing a reliable position estimation is not possible. A statistical analysis was then performed on these similar results in order to give a leak position in terms of the mean and to discard whatever had a different tendency. To analyze the similar results, a time interval was chosen in each case by considering that the leak position estimation has reached a steady value.

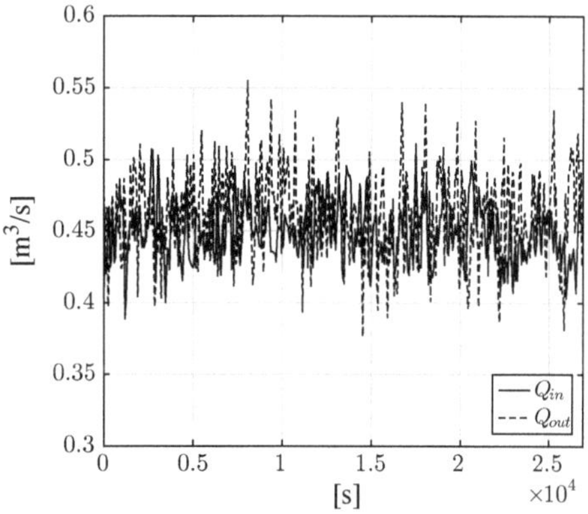

Fig. 12.19 Flow rates during the interval I_{E3}

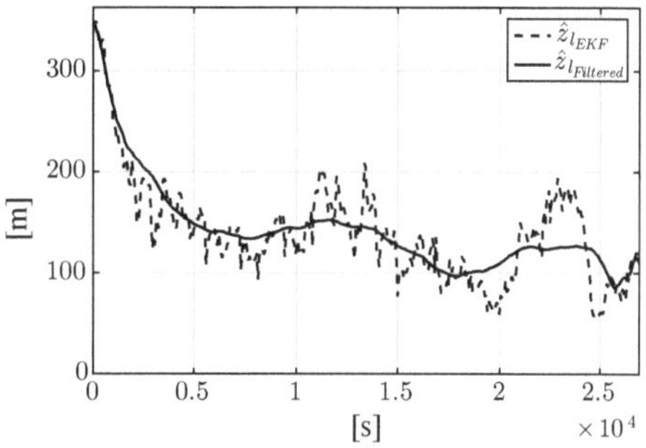

Fig. 12.20 Estimated leak position $\hat{z}_l$ in I_{E3}

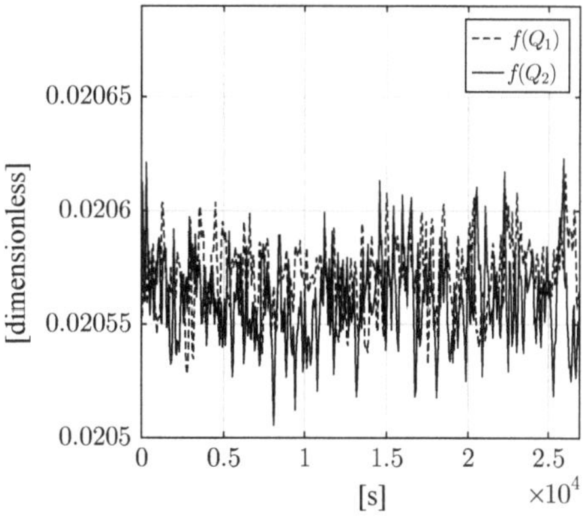

Fig. 12.21 Friction coefficients $f(Q_1)$ and $f(Q_2)$ in I_{E3}

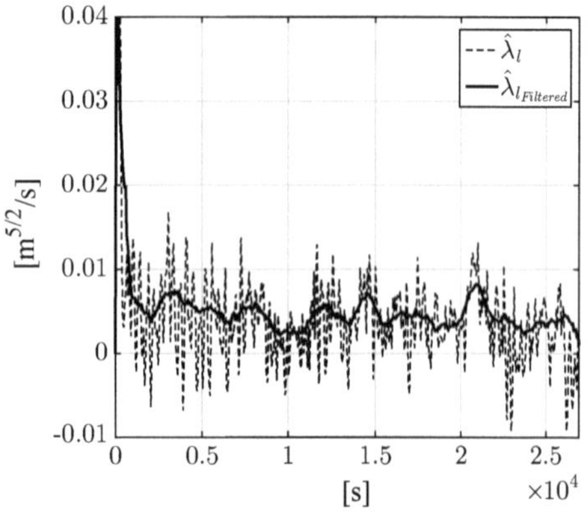

Fig. 12.22 Leak magnitude $\hat{\lambda}_l$ and the filtered version $\hat{\lambda}_{l_F}$ in I_{E3}

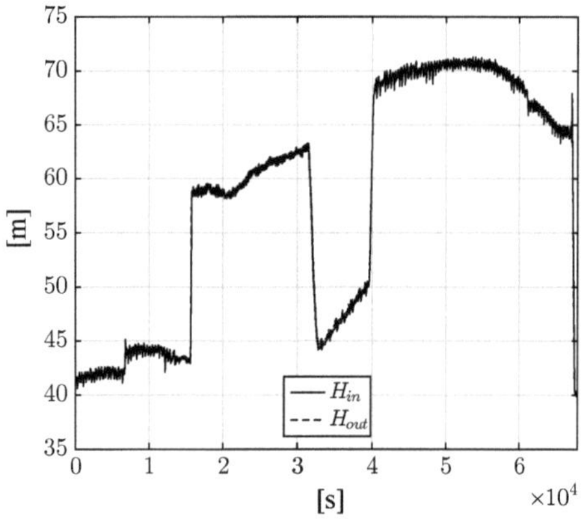

Fig. 12.23 Pressure heads during the interval I_4

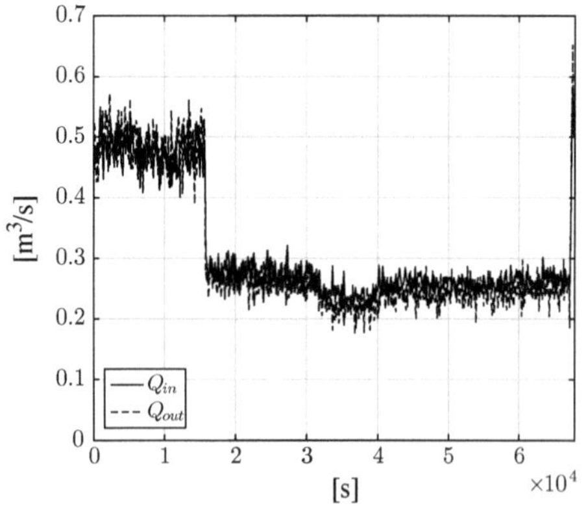

Fig. 12.24 Flow rates during the interval I_4

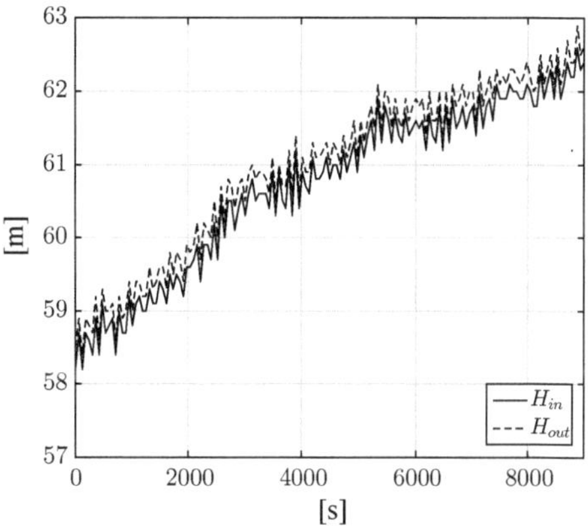

Fig. 12.25 Pressure heads during the interval I_{E4}

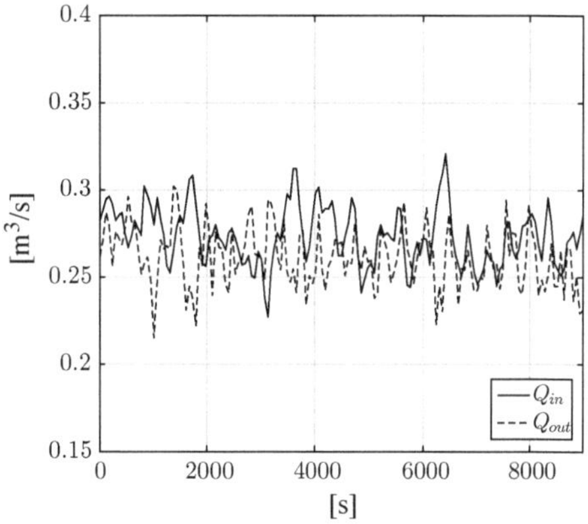

Fig. 12.26 Flow rates during the interval I_{E4}

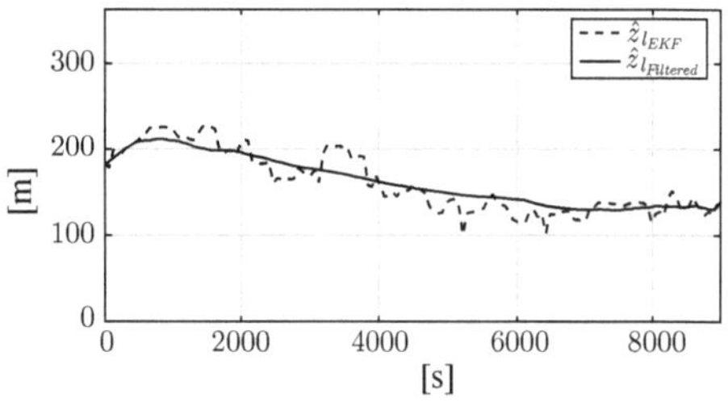

Fig. 12.27 Estimated leak position $\hat{z}_l$ in I_{E4}

This interval was denoted as I_C. Table 12.8 summarizes these statistical analyses: the average of the leak position in the time interval I_C is given in each case.

The average of the leak position estimation shows a similar tendency, with the exception of data packet 2. With these similar results, an average is calculated, which results in a global leak position of $\hat{z}_{l_g} \approx 121.3$ [m]. Thus, the research group provided an estimation of the leak position to the SIAPA staff, and the excavation procedure

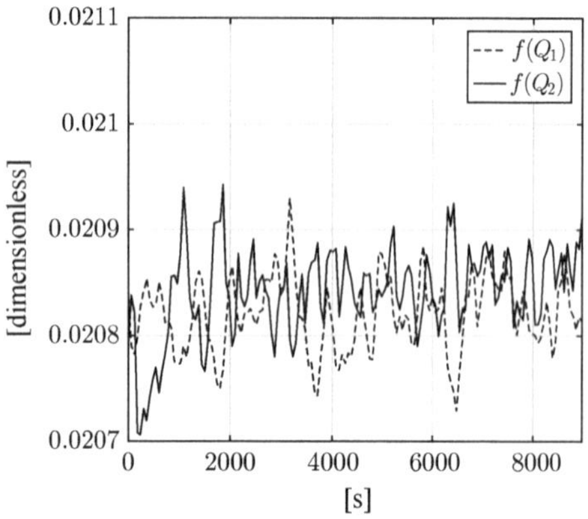

Fig. 12.28 Friction coefficients $f(Q_1)$ and $f(Q_2)$ in I_{E4}

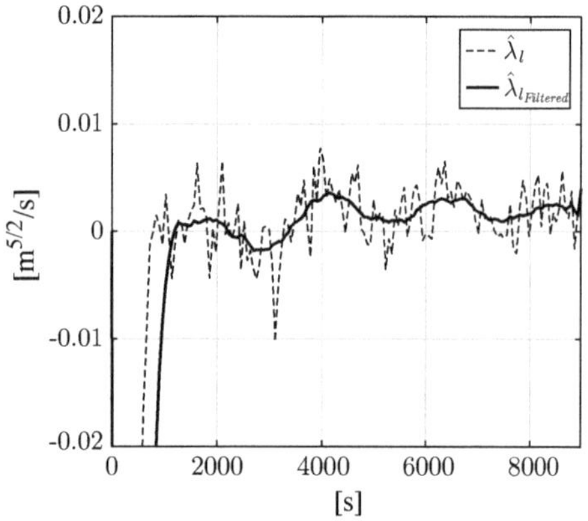

Fig. 12.29 Leak magnitude $\hat{\lambda}_l$ and the filtered version $\hat{\lambda}_{l_F}$ in I_{E4}

Table 12.8 Data analysis

Data packet	I_C	mean $[m]$
data packet 1	$05:50/26/feb\ to\ 06:40/26/feb$	$\hat{z}_l \approx 117.7$
data packet 2	$23:54/27/feb\ to\ 01:54/28/feb$	$\hat{z}_l \approx 192$
data packet 3	$20:30/29/feb\ to\ 22:30/29/feb$	$\hat{z}_l \approx 115.7$
data packet 4	$20:45/02/mar\ to\ 21:45/02/mar$	$\hat{z}_l \approx 130.7$

Fig. 12.30 SIAPA staff looking for the leak

began. The excavation region was approximately between 115 and 125 [m] upstream, and the real leak position was $z \approx 116.5$ [m].

12.6 Leak Location

Figure 12.30 illustrates the moment before the leak was found, and the leak jet can be clearly visualized in Fig. 12.31.

The water jet was located just between two pipe stretches, and it was caused by an injury in the ring-shaped rubber gasket that hermetically seals two consecutive pipe sections. This damage could be caused by the abrupt changes in the operating point, since the effect of water hammer causes pressure waves inside the pipeline which affects the weaker part, being precisely the gaskets. An earthquake could also be the origin of the damage.

Fig. 12.31 The leak was located at $z \approx 116.5\,[\mathrm{m}]$

Fig. 12.32 Leak position estimation, data packet 1

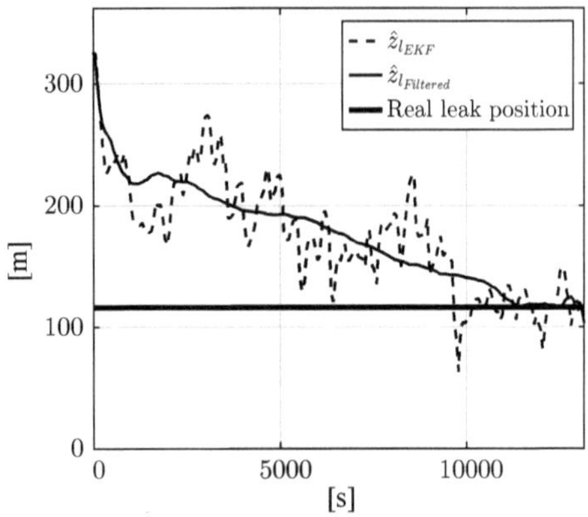

On the other hand, the results obtained from the data packet analysis show a similar tendency in the real leak location in three of the experiments; see Figs. 12.32, 12.34 and 12.35. Conversely, the estimated result shown in Fig. 12.33 deviates from the real position and has been correctly dismissed in the final decision.

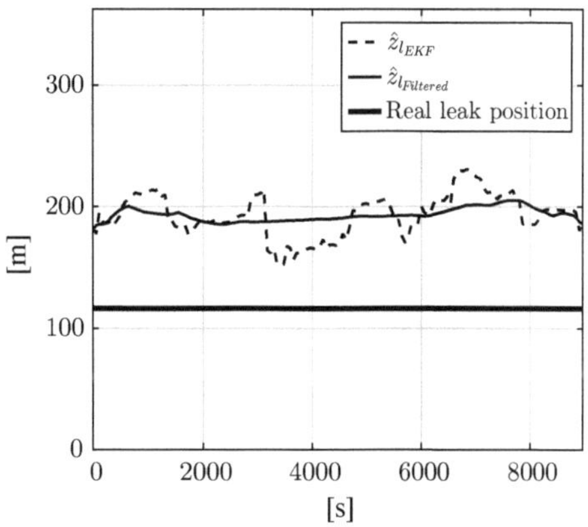

Fig. 12.33 Leak position estimation, data packet 2

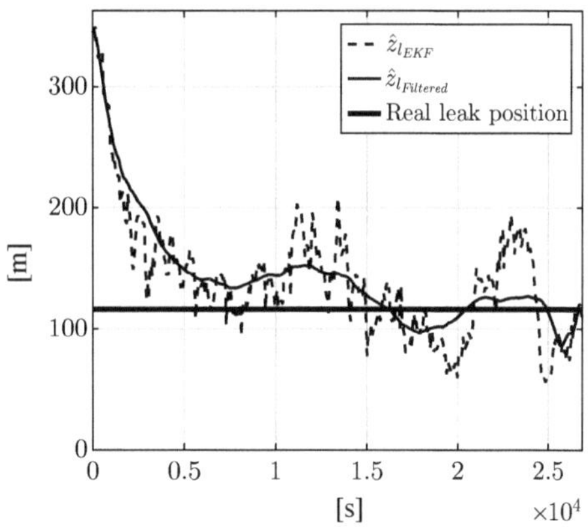

Fig. 12.34 Leak position estimation, data packet 3

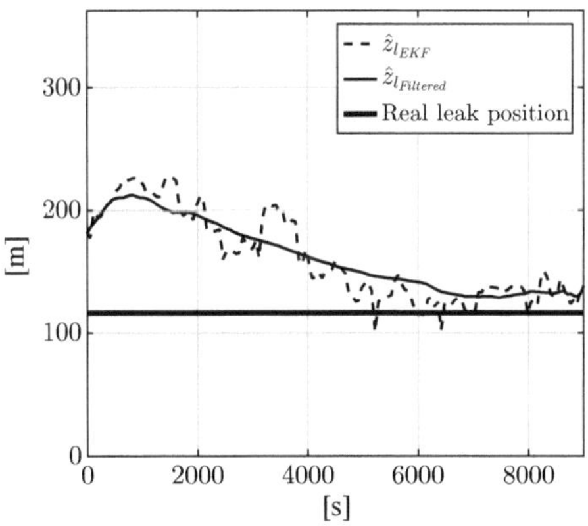

Fig. 12.35 Leak position estimation, data packet 4

Fig. 12.36 The leak disappears by using a provisional patch

Fig. 12.37 The leak disappears permanently by using a concrete patch

Figure 12.36 depicts the provisional repair with a lead plate. This allows for preparing the definitive patch, which is done with concrete, as shown in Fig. 12.37.

12.7 Conclusions

On the base of this experience, the CINVESTAV-Gdl group concluded: in three of the experiments, the leak estimations had the same tendency, while the other was dismissed. By considering those coincident results, the decision to search for the leak was made on the basis of statistical analysis, and a successful leak location was achieved. For an offline real leak diagnosis of the dimension of the Guadalajara line, the bigger the database is, the more reliable and robust the leak estimation is.

Acknowledgements We are grateful to all those who participated in this research. We thank Prof. Gildas Besançon always afforded us his assistance and valuable advice. We also thank the editors Dr. Lizeth Torres and Dr. Cristina Verde for considering this research as part of the book dedicated to the monitoring of pipeline networks. Finally, we are grateful to the director of SIAPA, C.E. Aristeo Mejía, and in general to all the staff at SIAPA who trusted us to locate the leak position. **This work was supported by CONACYT through project No. 177656.**

References

Begovich, O., Pizano, A., & Besançon, G. (2012). On-line implementation of a leak isolation algorithm in a plastic pipeline prototype. *Latin American Applied Research*, *57*(6), 131–140. https://hal.archives-ouvertes.fr/hal-00756134.

Besançon, G. (2007). *Nonlinear observers and applications*. Heidelberg: Springer.

Besançon, G., Scola, I. R., Guillén, M., Dulhoste, J. F., Santos, R., & Georges, D. (2013). Observer-based detection and location of partial blockages in pipelines. In *2013 Conference on Control and Fault-Tolerant Systems (SysTol)* (pp. 694–699). doi:10.1109/SysTol.2013.6693934.

Billman, L., & Isermann, R. (1987). Leak detection methods for pipelines. *Automatica*, *23*, 381–385.

Cataldo, A., Benedetto, E. D., Cannazza, G., Masciullo, A., Giaquinto, N., D'Aucelli, G. M., et al. (2016). Recent advances in the TDR-based leak detection system for pipeline inspection. *Measurement*. doi:10.1016/j.measurement.2016.09.017. http://www.sciencedirect.com/science/article/pii/S0263224116305206.

Chaudhry, M. H. (2014). *Applied hydraulic transients*. New York: Springer.

Crowe, C. T., Elger, D. F., & Roberson, J. A. (2000). *Engineering fluid mechanics*. New Jersey: Wiley.

Delgado-Aguiñaga, J., Besançon, G., & Begovich, O. (2015). Leak isolation based on extended Kalman filter in a plastic pipeline under temperature variations with real data validation. In *23th Mediterranean Conference on Control and Automation (MED), 2015* (pp. 316–321). doi:10.1109/MED.2015.7158769.

Delgado-Aguiñaga, J., Begovich, O., & Besançon, G. (2016a). Exact-differentiation-based leak detection and isolation in a plastic pipeline under temperature variations. *Journal of Process Control*, *42*, 114–124. doi:10.1016/j.jprocont.2016.04.005. http://www.sciencedirect.com/science/article/pii/S0959152416300300.

Delgado-Aguiñaga, J., Besançon, G., Begovich, O., & Carvajal, J. (2016b). Multi-leak diagnosis in pipelines based on extended Kalman filter. *Control Engineering Practice*, *49*, 139–148. doi:10.1016/j.conengprac.2015.10.008. http://www.sciencedirect.com/science/article/pii/S0967066115300356.

Genic, S., Arandjelovic, I., Kolendic, P., Jaric, M., Budimir, N., & Genic, V. (2011). A review of explicit approximations of Colebrook's Equation. *FME Transactions*, *39*, 67–71.

Golub, G. H., & Ortega, J. M. (1992). *Scientific computing and differential equation: An introduction to numerical methods*. Cambridge: Academic Press.
González, O., Verde, C., & Torres, L. (2016). Leak estimation method for complex pipelines with branch junctions. *Journal of Pressure Vessel Technology, ASME, 139*.
Ostapkowicz, P. (2014). Leakage detection from liquid transmission pipelines using improved pressure wave technique. *Eksploatacja i Niezawodność Maintenance and Reliability, 16*, 9–16.
Rentería, F., Verde, C., & Torres, L. (2016). Locating leaks by using the step response of a pipeline. In V. Puig (Ed.), *3rd Conference on Control and Fault-Tolerant Systems Systol 2016-IEEE*. Spain: Barcelona
Roberson, J. A., Cassidy, J. J., & Chaudhry, M. H. (1998). *Hydraulic engineering*. Wayne Anderson.
Simon, D. (2006). *Optimal state estimation*. New Jersey: Wiley.
Torres, L., Besançon, G., Georges, D., Navarro, A., & Begovich, O. (2011). Examples of pipeline monitoring with nonlinear observers and real-data validation. In *8th International Multi-Conference on Systems, Signals and Devices*.
Verde, C. (2001). Multi-leak detection and isolation in fluid pipelines. *Control Engineering Practice, 9*, 673–682.
Wylie, E. B., & Streeter, V. L. (1993). *Fluid transients in systems*. New Jersey: Prentice-Hall.
Zhang, T., Tan, Y., Zhang, X., & Zhao, J. (2015). A novel hybrid technique for leak detection and location in straight pipelines. *Journal of Loss Prevention in the Process Industries, 35*, 157–168. doi:10.1016/j.jlp.2015.04.012. http://www.sciencedirect.com/science/article/pii/S0950423015001163.

Index

C. Verde and L. Torres (eds.), *Modeling and Monitoring of Pipelines and Networks*, Applied Condition Monitoring 7,
DOI 10.1007/978-3-319-55944-5

MIX
Papier aus verantwortungsvollen Quellen
Paper from responsible sources
FSC® C105338

If you have any concerns about our products,
you can contact us on
ProductSafety@springernature.com

In case Publisher is established outside the EU,
the EU authorized representative is:
Springer Nature Customer Service Center GmbH
Europaplatz 3, 69115 Heidelberg, Germany

Printed by Libri Plureos GmbH
in Hamburg, Germany